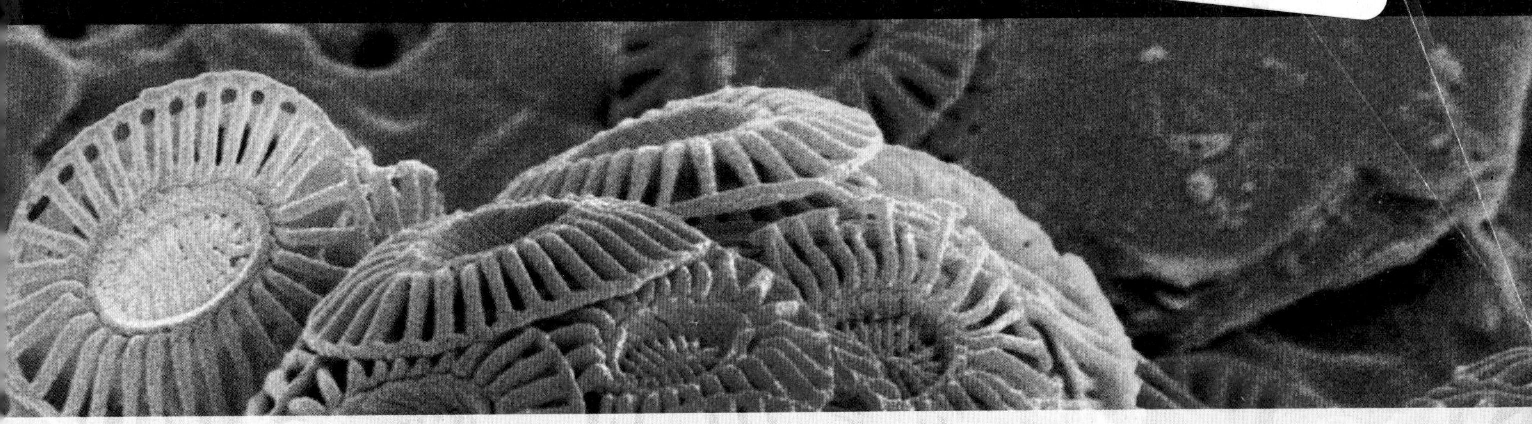

Project Directors

Angela Hall, Emma Palmer, Robin Millar, Mary Whitehouse

Editors

Angela Hall, Mary Whitehouse, Emma Palmer

Authors

Cris Edgell, John Lazonby, Robin Millar, Mike Shipton, Mike Kalvis, Ted Lister, Cliff Porter, Carol Tear

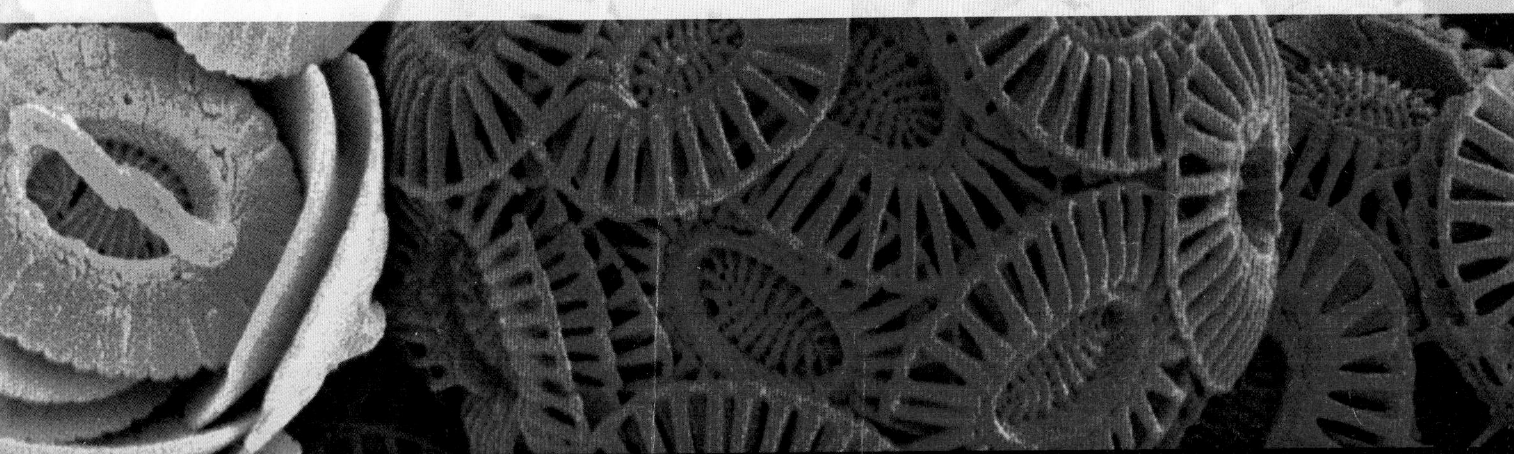

THE UNIVERSITY of York

OXFORD UNIVERSITY PRESS

Official Publisher Partnership

Contents

How to use this book

Welcome to Twenty First Century Science. This book has been specially written by a partnership between OCR, The University of York Science Education Group, The Nuffield Foundation Curriculum Programme, and Oxford University Press.

On these two pages you can see the types of page you will find in this book, and the features on them. Everything in the book is designed to provide you with the support you need to help you prepare for your examinations and achieve your best.

Module Openers

Why study?: This explains why what you are about to learn is useful to scientists.

Find out about: Every module starts with a short list of the things you'll be covering.

Ideas about Science: Here you can read about the key ideas about science covered in this module.

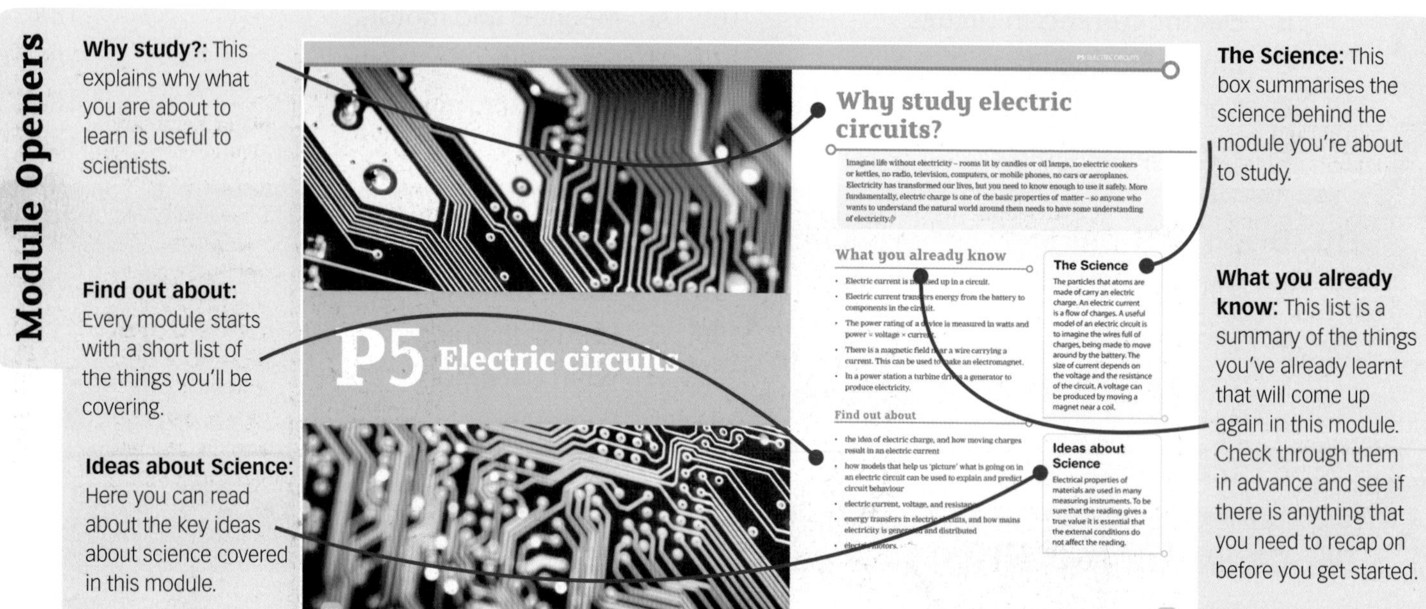

The Science: This box summarises the science behind the module you're about to study.

What you already know: This list is a summary of the things you've already learnt that will come up again in this module. Check through them in advance and see if there is anything that you need to recap on before you get started.

Main Pages

Find out about: For every part of the book you can see a list of the key points explored in that section.

Key words: The words in these boxes are the terms you need to understand for your exams. You can look for these words in the text in bold or check the glossary to see what they mean.

Worked examples: These help you understand how to use an equation or to work through a calculation. You can check back whenever you use the calculation in your work to make sure you understand it.

Questions: Use these questions to see if you've understood the topic.

Science Explanations

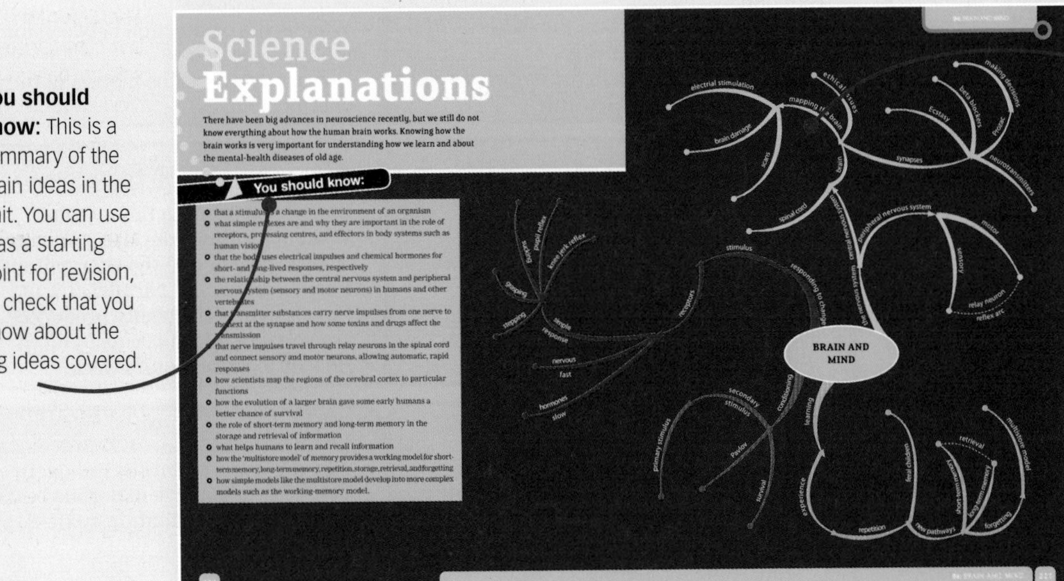

You should know: This is a summary of the main ideas in the unit. You can use it as a starting point for revision, to check that you know about the big ideas covered.

Visual summary: Another way to start revision is to use a visual summary, linking ideas together in groups so that you can see how one topic relates to another. You can use this page as a starting point for your own summary.

Ideas about Science and Review Questions

Ideas about Science

Review Questions

Ideas about Science: For every module this page summarises the ideas about science that you need to understand.

Review Questions: You can begin to prepare for your exams by using these questions to test how well you know the topics in this module.

Structure of assessment

Matching your course

What's in each module?

As you go through the book you should use the module opener pages to understand what you will be learning and why it is important. The table below gives an overview of the main topics each module includes.

B4	C4	P4
• How do chemical reactions take place in living things? • How do plants make food? • How do living organisms obtain energy?	• What are the patterns in the properties of elements? • How do chemists explain the patterns in the properties of elements? • How do chemists explain the properties of Group 1 and Group 7 elements?	• How can we describe motion? • What are forces? • What is the connection between forces and motion? • How can we describe motion in terms of energy changes?
B5	**C5**	**P5**
• How do organisms develop? • How does an organism produce new cells? • How do genes control growth and development within the cell?	• What types of chemicals make up the atmosphere? • What reactions happen in the hydrosphere? • What types of chemicals make up the lithosphere? • How can we extract useful metals from minerals?	• Electric current – a flow of what? • What determines the size of current in an electric circuit and the energy it transfers? • How do series and parallel circuits work? • How is mains electricity produced? How are voltages and currents induced? • How do electric motors work?
B6	**C6**	**P6**
• How do organisms respond to changes in their environment? • How is information passed through the nervous system? • What can we learn through conditioning? • How do humans develop complex behaviour?	• Chemicals and why we need them. • Planning, carrying out, and controlling a chemical synthesis.	• Why are some materials radioactive? • How can radioactive materials be used and handled safely, including wastes?

How do the modules fit together?

The modules in this book have been written to match the specification for GCSE Additional Science. In the diagram to the right you can see that the modules can also be used to study parts of GCSE Biology, GCSE Chemistry, and GCSE Physics courses.

	GCSE Biology	GCSE Chemistry	GCSE Physics
GCSE Science	B1	C1	P1
GCSE Science	B2	C2	P2
GCSE Science	B3	C3	P3
GCSE Additional Science	B4	C4	P4
GCSE Additional Science	B5	C5	P5
GCSE Additional Science	B6	C6	P6
	B7	C7	P7

GCSE Additional Science assessment

The content in the modules of this book matches the modules of the specification.

Twenty First Century Science offers two routes to the GCSE Additional Science qualification, which includes different exam papers depending on the route you take.

The diagrams below show you which modules are included in each exam paper. They also show you how much of your final mark you will be working towards in each paper.

Route	Unit	Modules Tested			Percentage	Type	Time	Marks Available
Route 1	A162	B4	B5	B6	25%	Written Exam	1 h	60
	A172	C4	C5	C6	25%	Written Exam	1 h	60
	A182	P4	P5	P6	25%	Written Exam	1 h	60
	A154	Controlled Assessment			25%		4.5–6 h	64
Route 2	A151	B4	C4	P4	25%	Written Exam	1 h	60
	A152	B5	C5	P5	25%	Written Exam	1 h	60
	A153	B6	C6	P6	25%	Written Exam	1 h	60
	A154	Controlled Assessment			25%		4.5–6 h	64

Command words

The list below explains some of the common words you will see used in exam questions.

Calculate
Work out a number. You can use your calculator to help you. You may need to use an equation. The question will say if your working must be shown. (Hint: don't confuse with 'Estimate' or 'Predict'.)

Compare
Write about the similarities and differences between two things.

Describe
Write a detailed answer that covers what happens, when it happens, and where it happens. Talk about facts and characteristics. (Hint: don't confuse with 'Explain'.)

Discuss
Write about the issues related to a topic. You may need to talk about the opposing sides of a debate, and you may need to show the difference between ideas, opinions, and facts.

Estimate
Suggest an approximate (rough) value, without performing a full calculation or an accurate measurement. Don't just guess – use your knowledge of science to suggest a realistic value. (Hint: don't confuse with 'Calculate' and 'Predict'.)

Explain
Write a detailed answer that covers how and why a thing happens. Talk about mechanisms and reasons. (Hint: don't confuse with 'Describe'.)

Evaluate
You will be given some facts, data, or other kind of information. Write about the data or facts and provide your own conclusion or opinion on them.

Justify
Give some evidence or write down an explanation to tell the examiner why you gave an answer.

Outline
Give only the key facts of the topic. You may need to set out the steps of a procedure or process – make sure you write down the steps in the correct order.

Predict
Look at some data and suggest a realistic value or outcome. You may use a calculation to help. Don't guess – look at trends in the data and use your knowledge of science. (Hint: don't confuse with 'Calculate' or 'Estimate'.)

Show
Write down the details, steps, or calculations needed to prove an answer that you have given.

Suggest
Think about what you've learnt and apply it to a new situation or context. Use what you have learnt to suggest sensible answers to the question.

Write down
Give a short answer, without a supporting argument.

Top Tips

Always read exam questions carefully, even if you recognise the word used. Look at the information in the question and the number of answer lines to see how much detail the examiner is looking for.

You can use bullet points or a diagram if it helps your answer.

If a number needs units you should include them, unless the units are already given on the answer line.

Making sense of graphs

Scientists use graphs and charts to present data clearly and to look for patterns in the data. You will need to plot graphs or draw charts to present data in the practical investigation and then describe and explain what the data is showing. Examination questions may also ask you to describe and explain what a graph is telling you.

Describing the relationship between variables

The pattern of points plotted on a graph can show whether two **factors** are related.

To describe the relationship in detail you should:

- read the axes and check the units used
- identify distinct phases of the graph – where the gradient is different
- use data when describing the changes.

Gradient of the graph

The **gradient** of the graph describes the way one variable changes relative to the other. Often the x axis is the time axis. The gradient then describes the rate of change.

Look at this graph, which shows the product of a chemical reaction being produced over time. The gradient of the graph gives the rate of the chemical reaction.

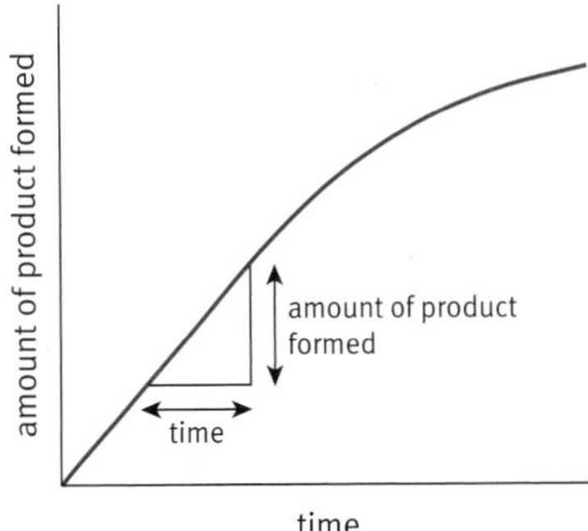

Graph showing how rate is calculated for a chemical reaction.

Look at this graph, which shows the distant travelled by a car over a period of time. The gradient of the graph gives the speed of the car.

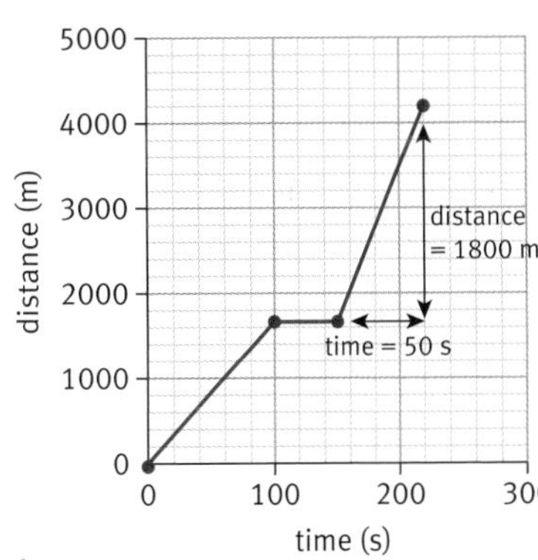

$$\text{speed} = \frac{\text{distance travelled}}{\text{time taken}}$$

$$\text{speed} = \frac{1800\text{ m}}{50\text{s}} = 36\text{ m/s}$$

Calculating the gradient of a distance–time graph to find the speed.

Calculating reacting masses and percentage yields

Problem *How much aluminium powder do you need to react with 8.0 g of iron oxide?*

$2Al(s) + Fe_2O_3(s) \longrightarrow Al_2O_3(s) + 2Fe(s)$

Work out the relative formula masses (RFM) by using **multiplication** and **addition.**

Reactants:

RFM of Al = 27;

RFM of $Fe_2O_3 = (2 \times 56) + (3 \times 16) = 160$

Products:

RFM of $Al_2O_3 = (2 \times 27) + (3 \times 16) = 102$

RFM of Fe = 56

To find the relative masses in this reaction **multiply** the RFM by the numbers in front of each formula in the equation. Then convert the relative masses to reacting masses by including units. These can be g, kg or tonnes depending on the data in the question. The units must be the same for each of the values.

$2Al(s)$	+	$Fe_2O_3(s)$	$\longrightarrow$	$Al_2O_3(s)$	+	$2Fe(s)$
$(2 \times 27) = 54$ g		160 g		102 g		$(2 \times 56) = 112$ g

To find the quantities required, scale the reacting masses to the known quantities by using **simple ratios.** Always include the **correct units** when substituting values.

$$\frac{\text{mass of aluminium required}}{\text{reacting mass of aluminium}} = \frac{\text{mass of iron oxide used}}{\text{reacting mass of iron oxide}}$$

$$\frac{\text{mass of aluminium required}}{54\text{ g}} = \frac{8\text{ g}}{160\text{ g}}$$

Rearrange the equation, to find the mass of aluminium.

$$\text{Mass of aluminium required} = \frac{8\text{ g}}{160\text{ g}} \times 54\text{ g} = \mathbf{2.7g}$$

Problem *What is the percentage yield of iron for the same reaction, if 4.9 g of iron is actually produced?*

To work out the theoretical yield of iron, use ratios as before

$$\frac{\text{mass of iron yielded}}{\text{reacting mass of iron}} = \frac{\text{mass of iron oxide used}}{\text{reacting mass of iron oxide}}$$

$$\frac{\text{mass of iron yielded}}{112\text{ g}} = \frac{8\text{ g}}{160\text{ g}}$$

and then rearrange the equation.

$$\text{Theoretical yield of iron} = \frac{8\text{ g}}{160\text{ g}} \times 112\text{ g} = 5.6\text{ g}$$

Use this equation to calculate the percentage yield

$$\textbf{percentage yield} = \frac{\textbf{actual yield (g)}}{\textbf{theoretical yield (g)}} \times \mathbf{100\ \%}$$

Actual yield of iron given in the question = 4.9 g

Substitute the quantities.

$$\text{percentage yield} = \frac{4.9\text{ g}}{5.6\text{ g}} \times 100 = \mathbf{87.5\ \%}$$

Frequency data

Frequency graphs or charts show the number of times a data value occurs. For example, if four students have a pulse rate of 86, then the data value 86 has a frequency of four.

A large data set with lots of different values can be arranged into class intervals (or groups). Collecting data in class intervals can be done by tallying. It works well to have data arranged in five or six class intervals.

Class interval	Tally	Frequency
60–65	|	1
65–70	||||	4
70–75	卌 卌 ||	12
75–80	卌 |||	8
80–85	卌	5
85–90	|	1
	Total	**31**

A data set of pulse rates from a class of 31 pupils tallied in a frequency table.

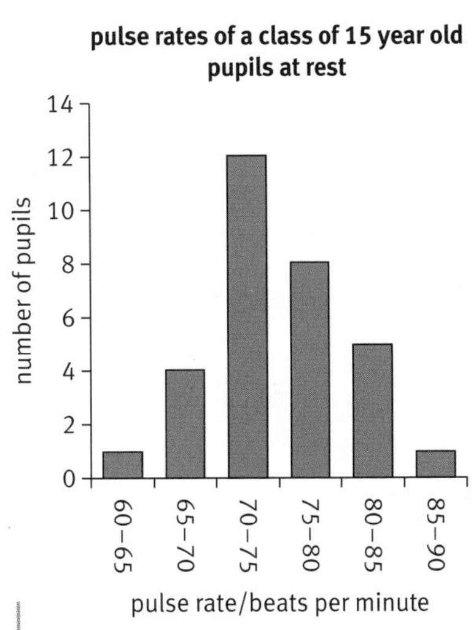

Frequency data can be shown in a bar chart.

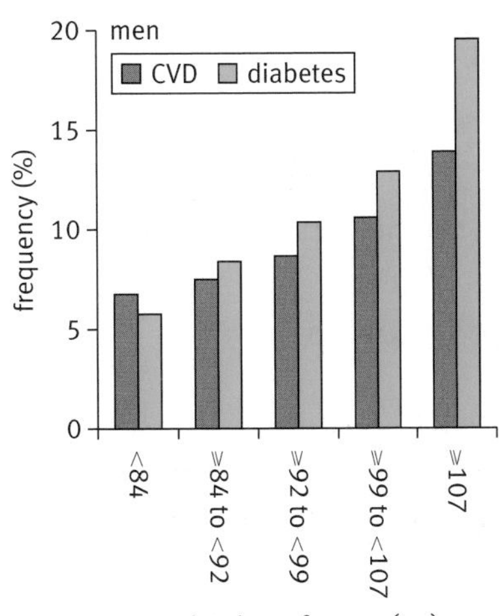

Sometimes frequency graphs have % as the units on the y axis. CVD: cardiovascular disease.

Q Write down a question that can be answered from each of these graphs.

Range and mean

Statistics are used to describe data. Useful statistics to describe the pulse rate data are the range and mean of the data.

The range of this data set can be expressed as 'between x beats per minute (lowest value) and y beats per minute (highest value)'

The mean is the total of all the values divided by the number of data points.

Q Write down two statements about the pulse rate data.

Q If you are comparing the pulse rates of two different classes in your school, why would it be useful to have both statistics (mean and range)?

Controlled assessment

In GCSE Additional Science the controlled assessment counts for 25% of your total grade. Marks are given for a practical investigation.

Your school or college may give you the mark schemes for this.

This will help you understand how to get the most credit for your work.

Practical investigation (25%)

Investigations are carried out by scientists to try and find the answers to scientific questions. The skills you learn from this work will help prepare you to study any science course after GCSE.

To succeed with any investigation you will need to:

- choose a question to explore
- select equipment and use it appropriately and safely
- design ways of making accurate and reliable observations
- relate your investigation to work by other people investigating the same ideas.

Your investigation report will be based on the data you collect from your own experiments. You will also use information from other people's research. This is called secondary data.

You will write a full report of your investigation. Marks will be awarded for the quality of your report. You should:

- make sure your report is laid out clearly in a sensible order
- use diagrams, tables, charts, and graphs to present information
- take care with your spelling, grammar, and punctuation, and use scientific terms where they are appropriate.

Marks will be awarded under five different headings.

Strategy

- Develop a hypothesis to investigate.
- Choose a procedure and equipment that will give you reliable data.
- Carry out a risk assessment to minimise the risks of your investigation.
- Describe your hypothesis and plan using correct scientific language.

Collecting data

- Carry out preliminary work to decide the range.
- Collect data across a wide enough range.
- Collect enough data and check its reliability.
- Control factors that might affect the results.

Analysis

- Present your data to make clear any patterns in the results.
- Use graphs or charts to indicate the spread of your data.
- Use appropriate calculations such as averages and gradients of graphs.

Evaluation

- Describe and explain how you could improve your method.
- Discuss how repeatable your evidence is, accounting for any outliers.

Review

- Comment, with reasons, on your confidence in the secondary data you have collected.
- Compare the results of your investigation to the secondary data.
- Suggest ways to increase the confidence in your conclusions.

Tip

The best advice is 'plan ahead'. Give your work the time it needs and work steadily and evenly over the time you are given. Your deadlines will come all too quickly, especially if you have coursework to do in other subjects.

Secondary data

Once you have collected the data from your investigation you should look for some secondary data relevant to your hypothesis. This will help you decide how well your data agrees with the findings of other scientists. Your teacher will give you secondary data provided by OCR, but you should look for further sources to help you evaluate the quality of all your data. Other sources of information could include:

- experimental results from other groups in your class or school
- text books
- the Internet.

When will you do this work?

Your school or college will decide when you do your practical investigation. If you do more than one investigation, they will choose the one with the best marks.

Your investigation will be done in class time over a series of lessons.

You may also do some research out of class.

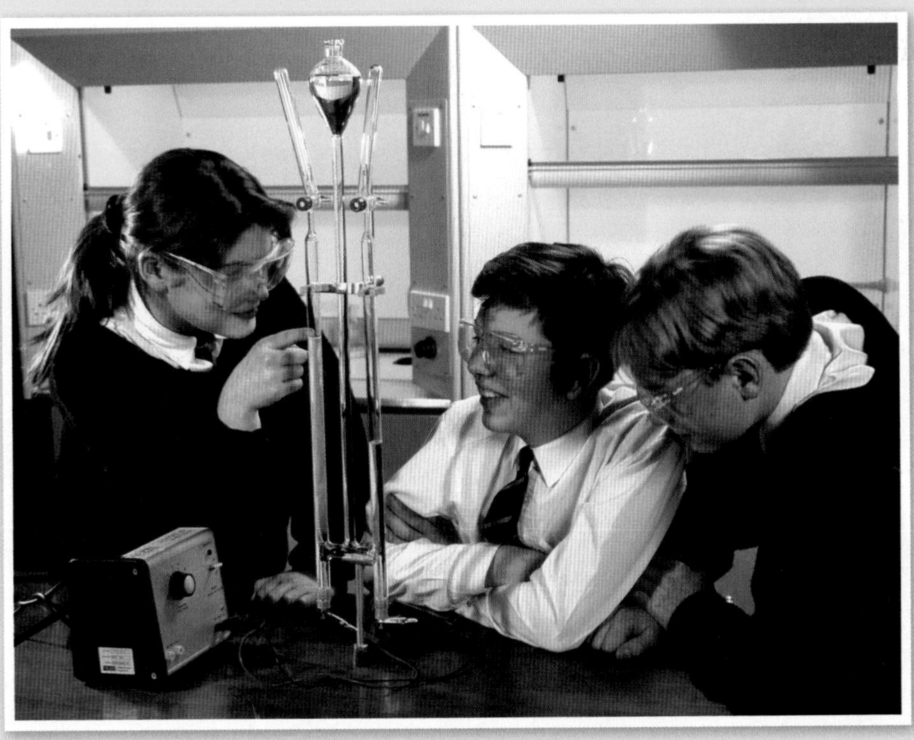

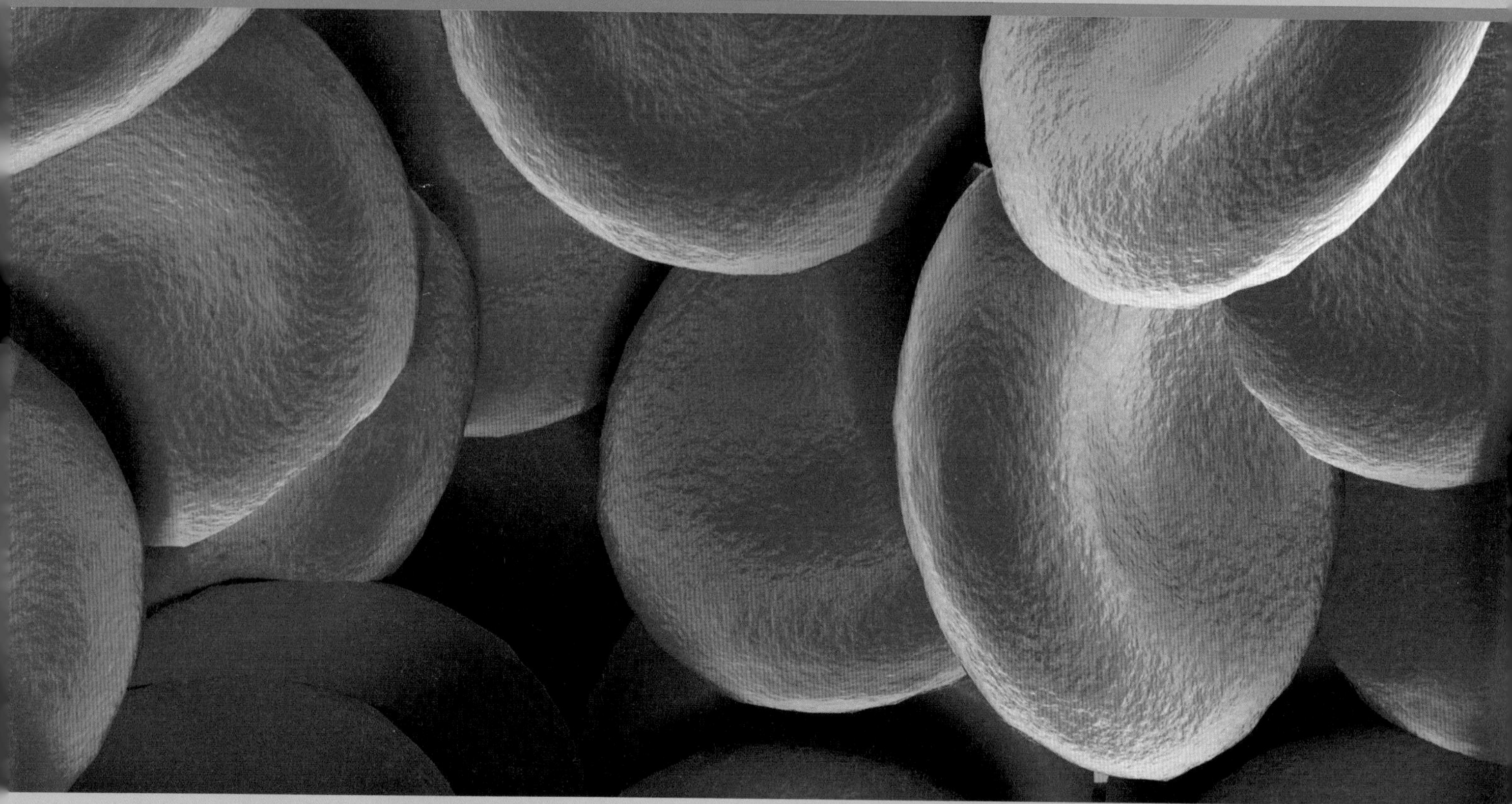

B4 The processes of life

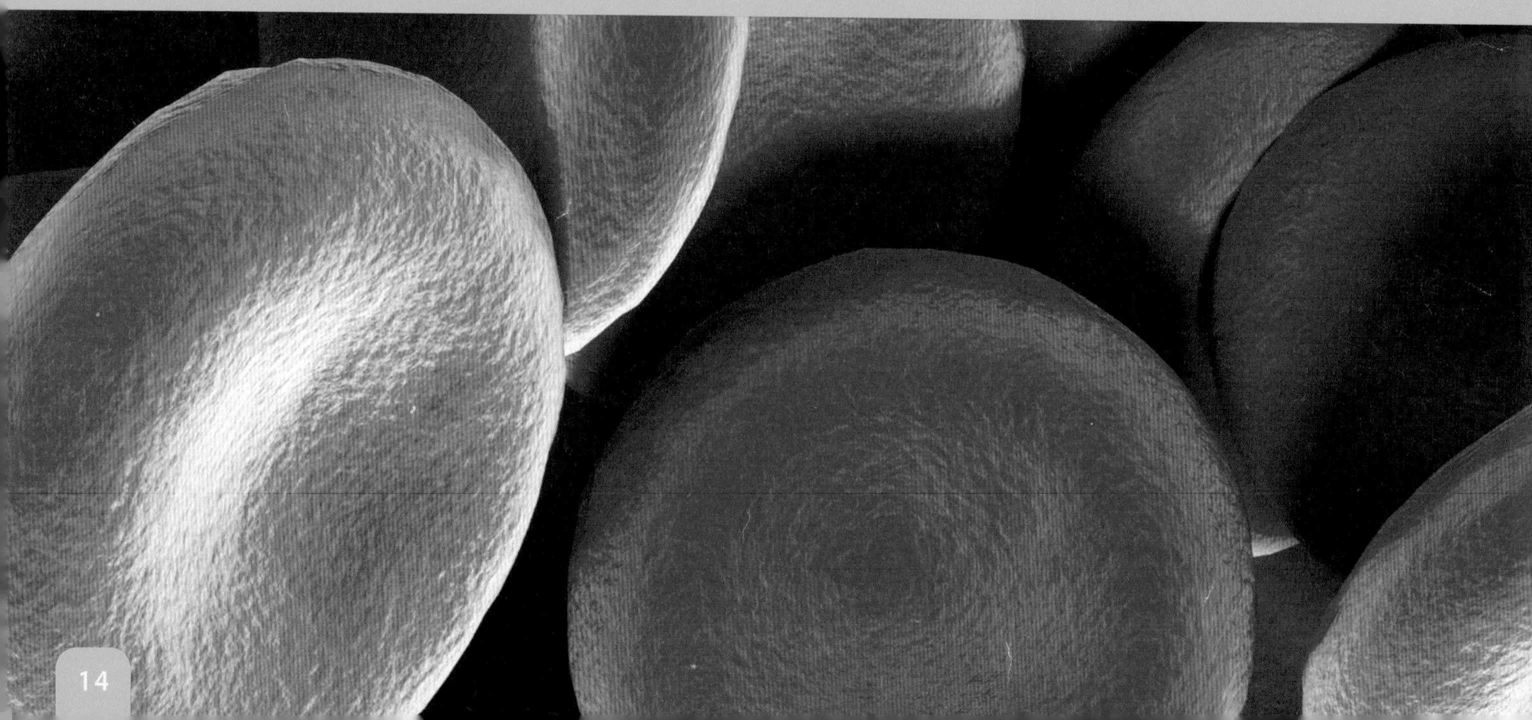

Why study the processes of life?

Look inside the cells of the largest animal or the smallest bacterium and you will find the same things happening. All living things need to be able to process molecules to generate energy, repair damage, and grow new cells. Enzymes speed up these chemical reactions. They need specific conditions to work at their best. Understanding the processes of life helps us to treat diseases and improve food production. It allows us to investigate the workings of all life on Earth.

What you already know

- Genes in the nuclei control how an organism's cells develop and function.
- Genes are instructions for a cell that describe how to make proteins.
- Living organisms are adapted to their environment and compete with one another.
- Nearly all organisms are ultimately dependent on energy from the Sun.
- Plants absorb a small percentage of the Sun's energy for the process of photosynthesis.
- Energy from photosynthesis is stored in the chemicals that make up the plants' cells.

Find out about

- the chemical processes that happen in all living cells
- the way that enzymes speed up chemical reactions
- how the process of respiration releases energy for cells to use
- how photosynthesis captures energy from the Sun and is the start of food chains
- how investigations can explore the relationship between changing variables and outcomes.

The Science

Photosynthesis powers all life on Earth. Respiration breaks down sugars and produces energy for cell processes. Photosynthesis captures energy from sunlight and uses it to combine carbon dioxide and water to make sugars. This energy is passed along the food chain.

Ideas about Science

To investigate the way outcomes are affected when certain factors are changed, it is important to control any other factors that may also affect the outcome.

A Features of all living things

Find out about

- the processes carried out by all living things

All living things are made up of **cells**. Some are just a single cell while others, like you, are made from billions of cells. You may not think you are like single-celled **bacteria** but there are some processes that all living things perform.

Move.

Bacteria and animals may move to find food, escape predators, or find better conditions to grow. Plants are rooted to the spot but grow to find sunlight.

Respire.

Energy is needed to carry out cell processes. Respiration is a series of chemical reactions that release energy from food molecules like sugar. Enzymes allow these reactions to happen in plant, animal, and microbial cells.

Sense.

Organisms need to be able to sense and respond to their surroundings. Plants grow towards sunlight and woodlice run away from it.

Grow.

Bacteria grow and divide to form new bacteria. Plants and animals are made from billions of cells. You have grown from a single fertilised egg cell.

Excrete.

Living cells produce wastes that are toxic if they build up. Wastes are removed by excretion. Carbon dioxide is a waste product of respiration.

Feed.

Living things need a supply of energy. They get this from their food. Plants make their own food during **photosynthesis**.

Reproduce.

All living things eventually die. Reproduction makes new generations that keep the species alive.

Life processes

There is a biological chemical factory in the cytoplasm of every plant cell. It takes small molecules and builds them into large molecules. It uses chemical reactions to build itself, copy itself, and repair itself. It takes large food molecules and breaks these down to release energy in respiration.

All living things are able to do this. Your own cells are performing an amazing set of chemical reactions and processes all the time. All these reactions in living organisms are catalysed by enzymes. Each cell has genes with instructions for making these enzymes.

In the cytoplasm of every one of this plant's cells is a biological chemical factory.

Using life processes: making enzymes

Large vessels, like the fermenter in the picture below, are used to grow bacteria so that their enzymes can be harvested. They contain a nutrient solution, and conditions such as temperature, pH, and oxygen levels are controlled for optimum growth.

As the bacteria grow, they release their enzymes into the nutrient solution. When the nutrients are used up, the solution is filtered to remove the bacteria. The required enzymes can be purified from the solution. They are now available for use in applications such as the food industry, textiles industry, and in the home as biological washing powders.

Bacteria are grown in this fermenter. They make enzymes that are used to make denim jeans look faded.

Like any living organism, bacteria need to take in nutrients. To digest their food, they produce and release enzymes into their surroundings. These break down any food molecules so that they can be absorbed into the microbe.

Key words

- cells
- bacteria
- respiration
- enzymes
- photosynthesis

Questions

1. List the processes that all living things carry out.
2. Which process generates energy for living cells?
3. Suggest ways in which you are able to sense and respond to your surroundings.
4. Suggest some ways that your life would be different if humans had not developed a good understanding of the life processes that happen inside all cells.

B Enzymes

Find out about

- why you cannot live without enzymes
- how enzymes work

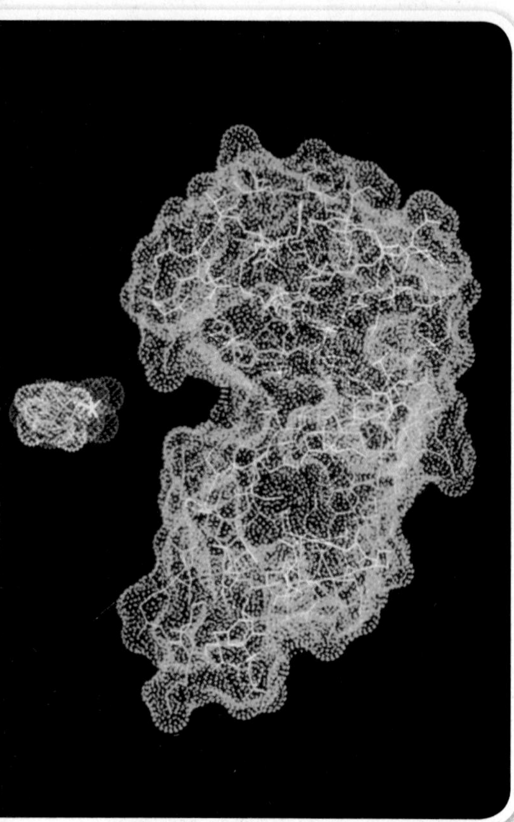

A computer graphic of an enzyme, its active site, and the product of a reaction.

The chemical reactions that take place in cells rely on **enzymes**. Every cell has genes with instructions for making enzymes. Enzymes work best in certain conditions, for example, at a certain temperature. This is an important reason why you need a steady state inside your cells.

What are enzymes?

Enzymes are the **catalysts** that speed up chemical reactions in living organisms. They are **proteins** – large molecules made up of long chains of **amino acids**. The amino acid chains are different in each protein, so they fold up into different shapes. An enzyme's shape is very important to how it works. The sequence of amino acids in each enzyme is determined by instructions in a gene.

How do enzymes work?

Some enzymes break down large molecules into smaller ones. Others join small molecules together. In all cases, the molecules must fit exactly into a part of the enzyme called the **active site**. It is a bit like fitting the right key in a lock. So scientists call the explanation of how an enzyme works the **lock-and-key model**.

1 An enzyme has an active site.

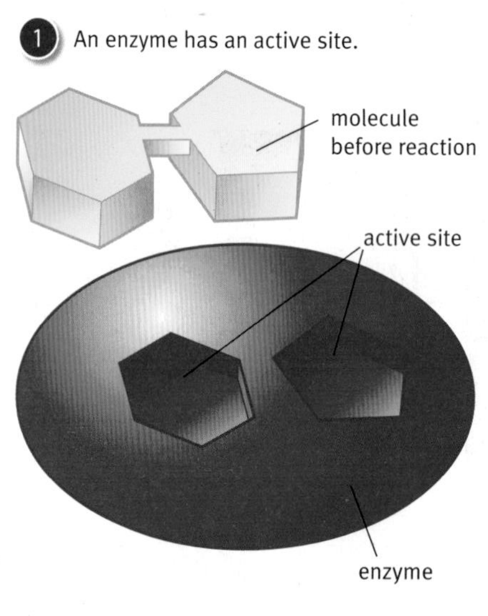

2 The reaction takes place in the active site.

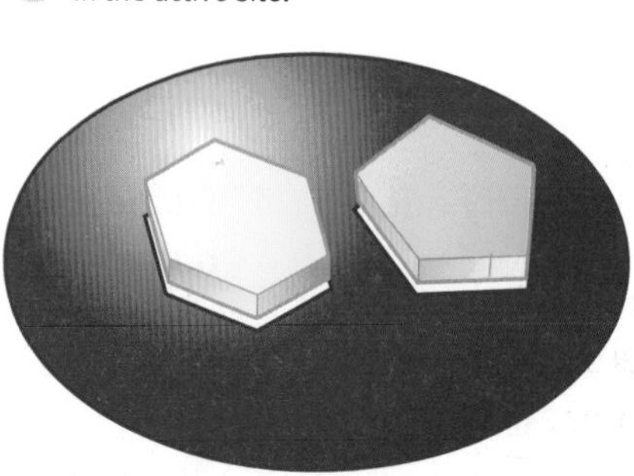

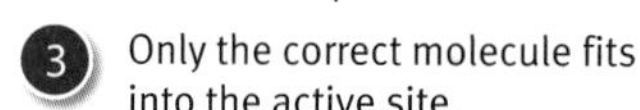

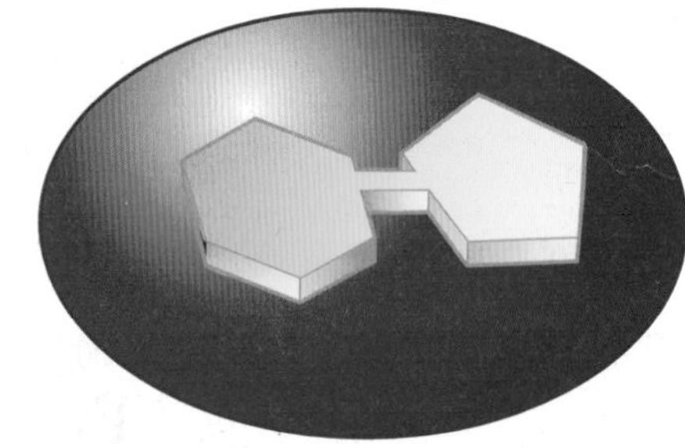

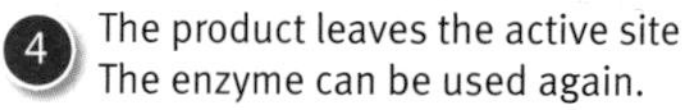

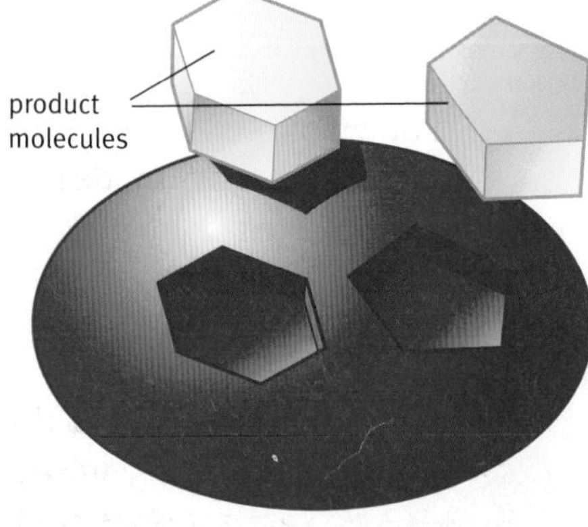

The lock-and-key model of enzyme function. (*Note:* This diagram is schematic. This means that if you were able to see the molecules and enzymes, they would not look like they do here. Enzymes consist of many molecules – see the graphic above left for an idea of scale. However, this is a useful way of picturing, or modelling, how an enzyme works.)

Why do we need enzymes?

At 37 °C, chemical reactions in your body would happen too slowly to keep you alive.

One way of speeding up a reaction is to increase the temperature. A higher body temperature could speed up the chemical reactions in your body. But higher temperatures damage human cells. Also, to keep your body warm you have to release energy from **respiration**. For a higher body temperature you would need a lot more food to fuel respiration.

So, we rely on enzymes to give us the rates of reaction that we need. They can increase rates of reaction by up to 10 000 000 000 times. It is not possible to live without enzymes.

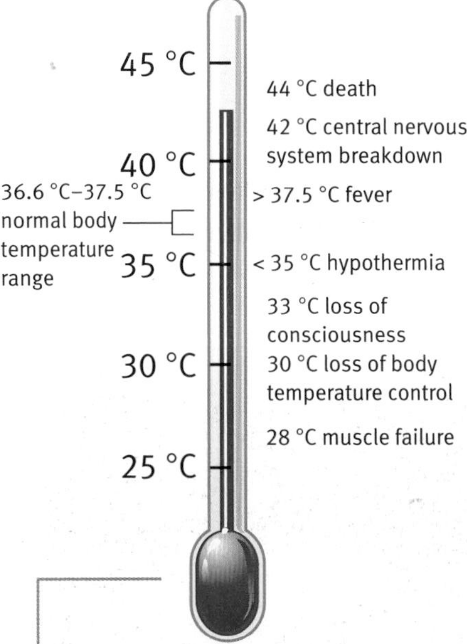

Your core body temperature is about 37 °C, but a small variation either side of this is normal. A core temperature over 42 °C or under 28 °C usually results in death.

Around 80% of the energy from your food is used for keeping warm. If you had to maintain a higher body temperature, you would have to spend a lot more time eating to provide the calories to heat your body.

Shrews have a large surface area for their volume. They lose heat to the environment over their whole body surface. To release enough energy to maintain their body temperature, they have to eat 75% of their body mass in food each day. If not, they die within two to three hours.

Questions

1. Write down:
   a. what enzymes are made of
   b. what enzymes do.
2. Explain how an enzyme works. Use the key words on this page in your answer.
3. The enzyme amylase breaks down **starch** to sugar (maltose); catalase breaks down hydrogen peroxide to water and oxygen. Explain why catalase does not break down starch.

Key words

- ✓ **proteins**
- ✓ **amino acids**
- ✓ **active site**
- ✓ **lock-and-key model**
- ✓ **starch**
- ✓ **catalysts**

Keeping the best conditions for enzymes

Find out about

- how temperature and pH affect enzymes

'Ice fish' such as Antarctic cod are active at 2 °C because their enzymes work best at low temperatures. Most organisms would be dead or very sluggish at 2 °C.

Bacteria living in hot springs have enzymes that withstand high temperatures.

How does temperature affect enzymes?

At low temperatures, enzyme reactions get faster if the temperature is increased.

But above a certain temperature the reaction stops. This is because enzymes are proteins. Higher temperatures change an enzyme's shape so that it no longer works. The diagram below explains what happens using the lock-and-key model.

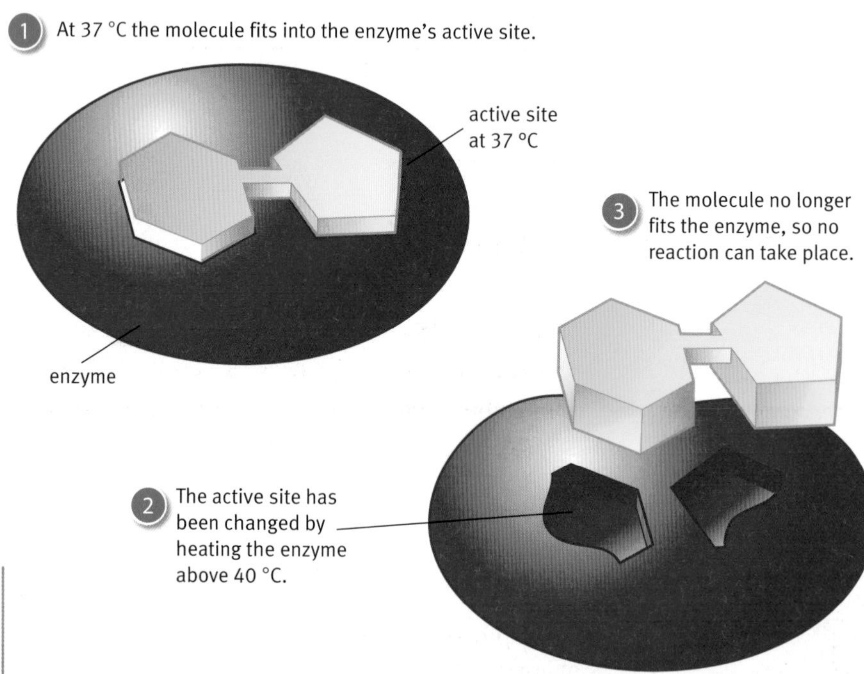

How an enzyme reaction can be stopped by a rise in temperature.

Enzymes are denatured at high temperatures

High temperatures change the shape of an enzyme. They do not destroy it completely. But even when an enzyme cools down, it does not go back to its original shape. Like cooked egg white, the protein cannot be changed back. The enzyme is said to be **denatured**.

Why 37 °C?

The temperature at which an enzyme works best is called its **optimum temperature**. Enzymes in humans work best at around 37 °C. Some organisms have cells and enzymes adapted to different temperatures.

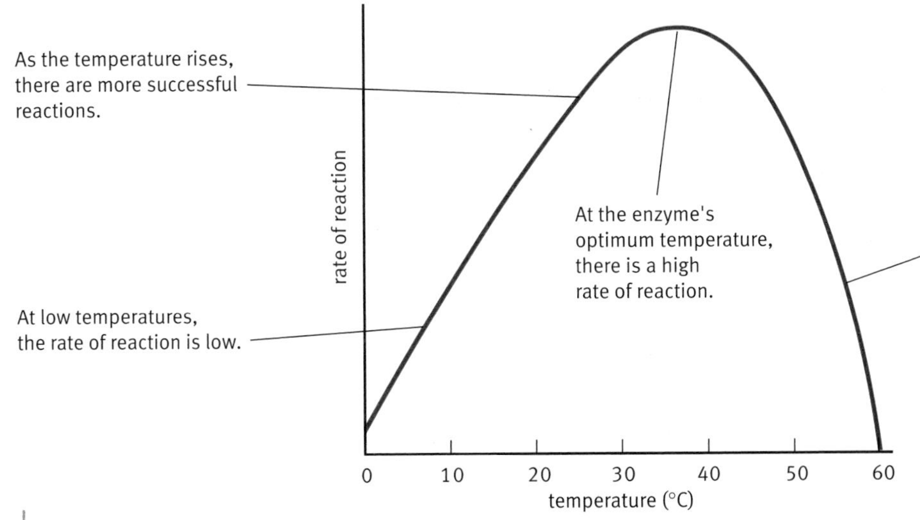

The optimum temperature of an enzyme. The reaction rate is plotted at different temperatures – all other conditions are kept constant.

Key words

- denatured
- optimum temperature

pH also affects enzymes

Proteins including enzymes can be damaged by acids and alkalis. The shape of an enzyme's active site will be changed if bonds holding the protein chains together are broken. The substrate will no longer be able to fit, so the enzyme becomes denatured. Every enzyme has an optimum pH at which it works best.

Enzyme	What it does	Optimum pH
salivary amylase	breaks down starch to sugar (maltose)	4.8
pepsin	breaks down proteins into short chains of amino acids	2.0
catalase	breaks down hydrogen peroxide into water and oxygen	7.6

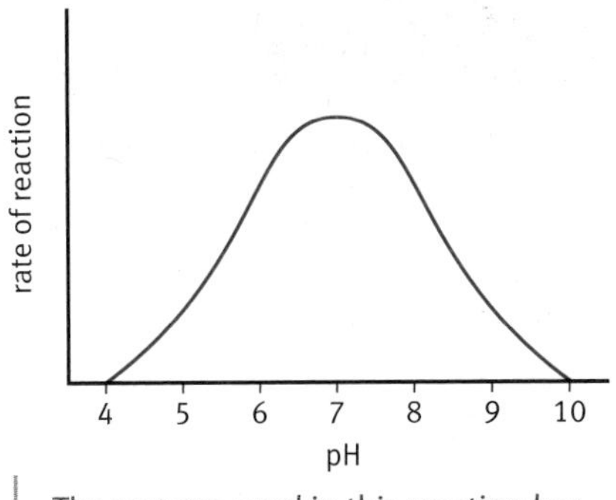

The enzyme used in this reaction has an optimum pH of 7.0.

Questions

1. Explain why increasing the temperature causes enzyme reactions to stop at higher temperatures.
2. What is meant by an enzyme's optimum temperature?
3. Enzymes in food and enzymes from microorganisms such as bacteria and fungi make food decay. Why does food stay fresh in a refrigerator for a few days but for months in a freezer?
4. Starch synthase is an enzyme found in potato tubers, which grow in the UK. The enzyme joins glucose molecules together to store them as starch. Some students compared its rate of reaction at 5 °C, 20 °C, and 45 °C.

   Think about the environment that the potato grows in and suggest what would happen if the students:

   a raised the temperature of the tuber at 5 °C to 20 °C?

   b raised the temperature of the tuber at 20 °C to 45 °C?

D Enzymes at work in plants

Find out about

- how photosynthesis captures energy for life on Earth
- what happens to the glucose made by photosynthesis

Key words

- chlorophyll
- chloroplasts
- cellulose
- vacuole
- cell wall
- cell membrane
- nucleus
- cytoplasm
- glucose

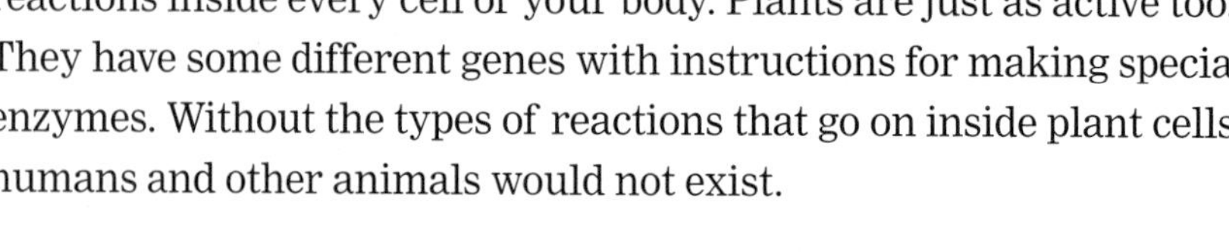

Every second of every day, enzymes catalyse billions of chemical reactions inside every cell of your body. Plants are just as active too. They have some different genes with instructions for making special enzymes. Without the types of reactions that go on inside plant cells, humans and other animals would not exist.

Photosynthesis

Plants capture energy from sunlight. This may look as easy as sunbathing, but the process is one of nature's cleverest tricks. Many complex chemical reactions are involved.

During photosynthesis, plants trap energy from sunlight and use it to make all the molecules they need for growth. These include sugars, starch, enzymes, and **chlorophyll**. These molecules feed other organisms along the food chain. So photosynthesis supplies food for life on Earth.

What happens during photosynthesis?

The chemical equation for photosynthesis is:

$$\underset{\text{carbon dioxide}}{6CO_2} + \underset{\text{water}}{6H_2O} \xrightarrow[\text{chlorophyll}]{\text{light energy}} \underset{\text{glucose}}{C_6H_{12}O_6} + \underset{\text{oxygen}}{6O_2}$$

Do not let the equation mislead you. The reaction does not happen in one go – it has lots of smaller steps. The equation is a convenient way of summing up the process.

A **glucose** molecule is made up of carbon, hydrogen, and oxygen atoms. So glucose is a carbohydrate.

Photosynthesis takes place in **chloroplasts**. They contain a green pigment called chlorophyll. Chlorophyll absorbs light and uses the energy to kick-start photosynthesis.

Energy from light splits water molecules into hydrogen and oxygen atoms. The hydrogen is combined with carbon dioxide from the air to make glucose. The oxygen is released as a waste product. It passes out of the plant into the air. Given enough raw materials, light, and the right temperature, a large tree can make 2000 kg of glucose in a day.

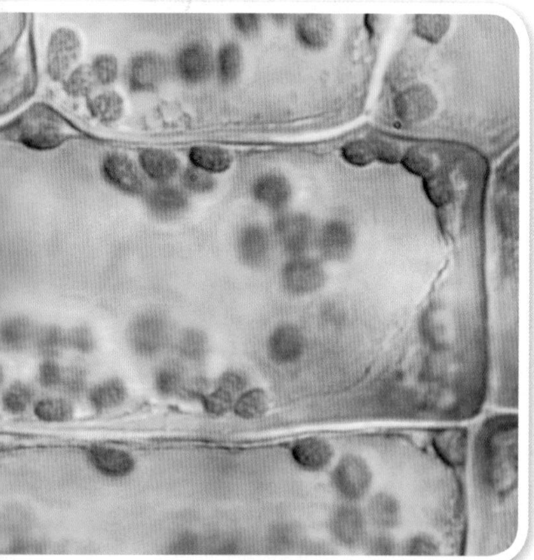

Chloroplasts contain the green pigment chlorophyll and the enzymes that are needed for photosynthesis. (Magnification × 2000.)

Using glucose from photosynthesis

Glucose and oxygen are made by photosynthesis. Glucose can be converted into starch for storage or cellulose to make new cell walls Glucose can also be built up into other molecules such as fats, proteins, and chlorophyll.

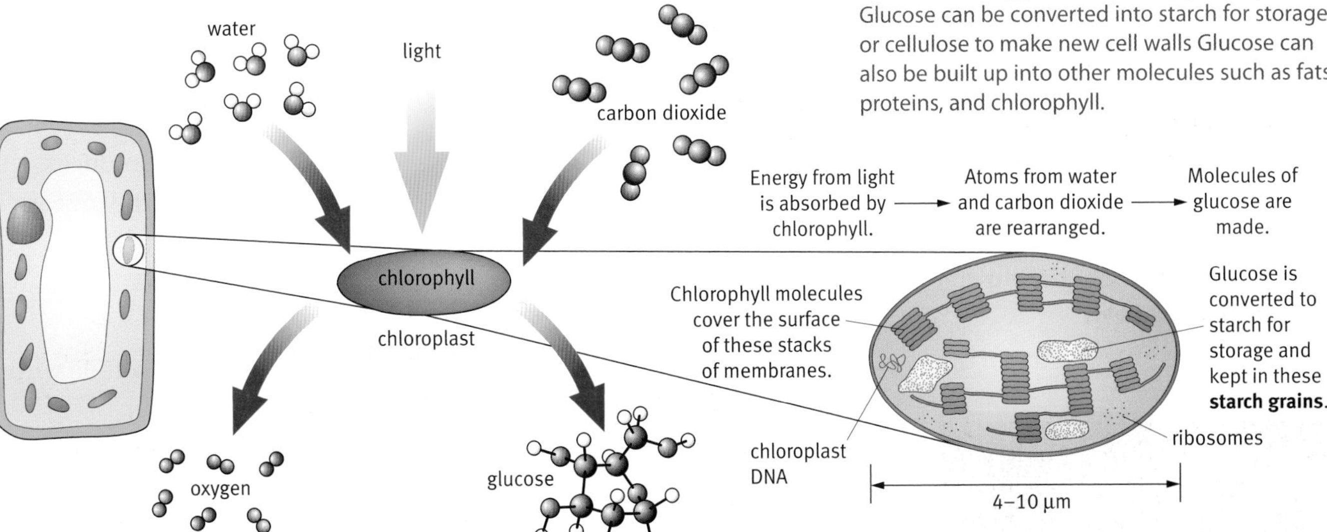

Glucose made during photosynthesis is used by plant cells in three ways.

(1) Making other chemicals needed for cell growth

Glucose is converted into other carbohydrates, as well as fats and proteins. Two important carbohydrates in plants are **cellulose** and starch. Cellulose and starch are both polymers of glucose. They are made up of thousands of glucose molecules linked together.

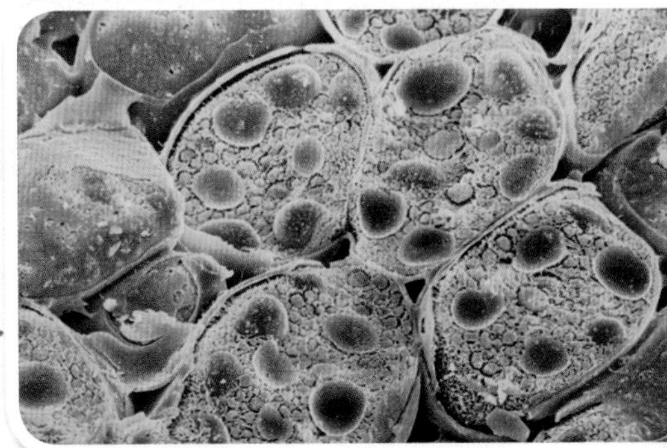

Starch grains in a plant cell store glucose as starch. (Magnification × 200.)

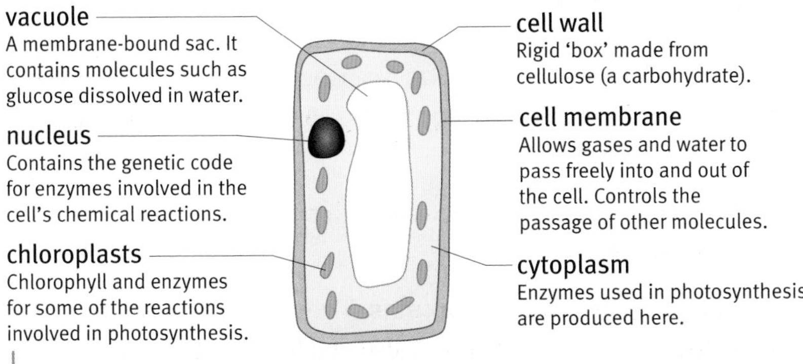

A plant cell contains many different structures involved in photosynthesis.

(2) Storing energy in starch molecules

Any excess glucose is converted into starch. Starch is a storage molecule. Starch can be converted back to glucose when needed. Starch is stored in leaf cells, but some plants have special organs, such as the tubers of a potato, which have cells that are filled with starch.

(3) Releasing energy in respiration

Glucose molecules are broken down by respiration, releasing the energy stored in the molecules. This energy is used to power chemical reactions in the cells, such as converting glucose to cellulose, starch, or proteins.

Questions

1. Write down the word equation that sums up photosynthesis.
2. Draw a diagram to show the flow of chemicals in and out of leaves during photosynthesis.
3. Describe the main stages in making glucose by photosynthesis.
4. Glucose from photosynthesis has three roles in the plant cell. Explain what these are.

Diffusion and gas exchange in plants

Find out about

- how chemicals move in and out of cells
- how carbon dioxide and oxygen move in and out of a leaf during photosynthesis

1 Water just poured onto tea bag

To start with, dissolved molecules from the tea are concentrated close to the tea bag. There are few tea molecules in the rest of the hot water.

2 About 30 seconds later

The tea molecules move from where they are highly concentrated into regions where they are less concentrated.

3 About 2 minutes later

After a few minutes the tea molecules spread evenly throughout the cup.

Molecules move in and out of cells all the time. Cells need a constant supply of raw materials for chemical reactions, and waste needs to be removed.

Photosynthesis in a plant leaf cell uses carbon dioxide and produces waste oxygen. Movement of these molecules takes place by the process of **diffusion**.

Diffusion

Molecules in gases and liquids move about randomly. They collide with each other and change direction. This makes them spread out.

Overall, more molecules move away from where they are concentrated than move the other way. The molecules diffuse from areas of their high concentration to areas of low concentration.

For example, molecules diffuse out of a tea bag when you make a cup of tea. Diffusion is a passive process. It does not need any extra energy.

This swimmer's cells need oxygen and glucose for respiration. They must get rid of carbon dioxide. These molecules move in and out of cells by diffusion.

Supplying photosynthesis

Photosynthesis is a vital process that takes place in leaves using the energy from sunlight. Special structures in leaves allow the chemicals needed for photosynthesis to be delivered and waste oxygen to be removed by diffusion.

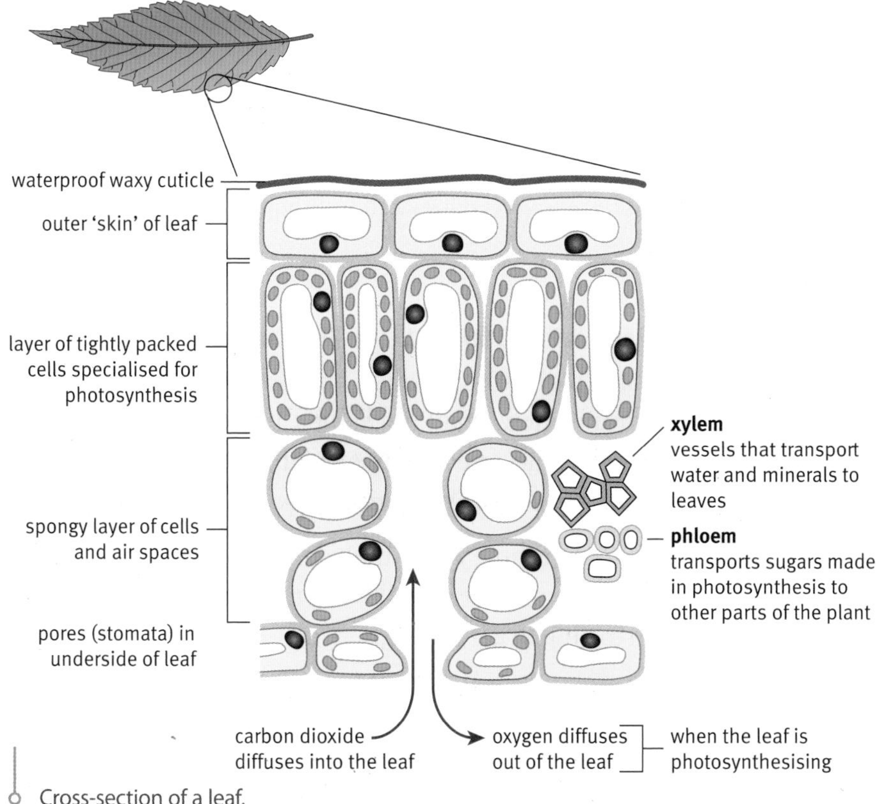

Cross-section of a leaf.

Photosynthesis, diffusion, and gas exchange in leaves

The underside of a leaf contains thousands of tiny holes called **stomata**. These allow carbon dioxide into the leaf and oxygen out. Diffusion drives these gases from high to low concentrations.

	Carbon dioxide	Oxygen
Cells in leaf during photosynthesis	• Carbon dioxide is used in cells for photosynthesis. • Low concentration of carbon dioxide in cells in the leaf. ↑ into leaf: Carbon dioxide diffuses through the stomata and into cells in the leaf to supply photosynthesis.	• Oxygen is produced during photosynthesis. • High concentration of oxygen builds up in cells of the leaf. ↓ out of leaf: Oxygen diffuses out of the cells and through the stomata.
Air surrounding the leaf	• Higher concentration of carbon dioxide in the air.	• Lower concentration of oxygen in the air.

Key word

✓ **diffusion**

Questions

1 Write down a definition for diffusion.

2 Name three chemicals that move in and out of cells by diffusion.

3 Explain how diffusion lets you smell the vinegar from fish and chips on a plate in front of you.

4 Earthworms do not have lungs. They rely on diffusion through their skin to exchange oxygen and carbon dioxide with air spaces in the soil. Use a labelled diagram to explain how diffusion takes oxygen into an earthworm and carbon dioxide out.

F Osmosis

Find out about

- the movement of water molecules by osmosis
- why cells need a steady water balance
- why plant cells store glucose as starch

Water is a vital component of all living things. Without water, the chemical reactions that take place inside cells could not happen.

Osmosis is a specific type of diffusion. It is the process that moves water molecules into and out of cells. The cell membrane is important in osmosis.

Cell membranes are partially permeable

Cell membranes let some molecules through but block others. Tiny channels in the membrane allow small molecules, like water, to travel through them. Larger molecules are too big and cannot get through. Cell membranes are **partially permeable membranes**.

Diagrams **a** and **b** below show how water molecules move during osmosis. In this example the membrane allows water through but the glucose molecules are too big to get through.

Key

partially permeable membrane allows some molecules through and acts as a barrier to others

glucose molecule

water molecule

water molecules associated with glucose molecule (these molecules are not free to move by osmosis)

(*Note:* In these diagrams, the circles represent molecules, not individual atoms. Cell membranes are also made of molecules.)

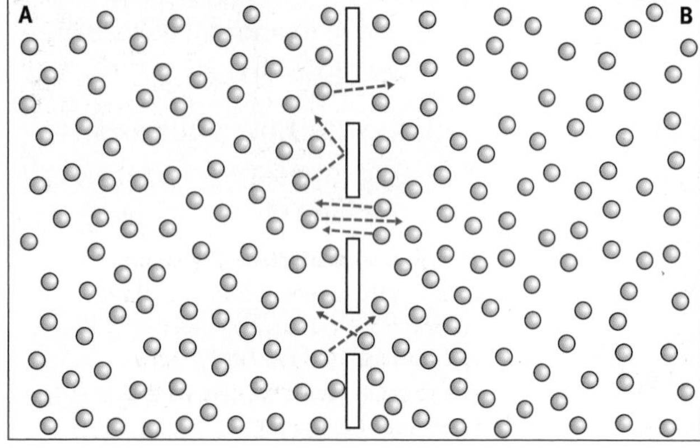

Not to scale

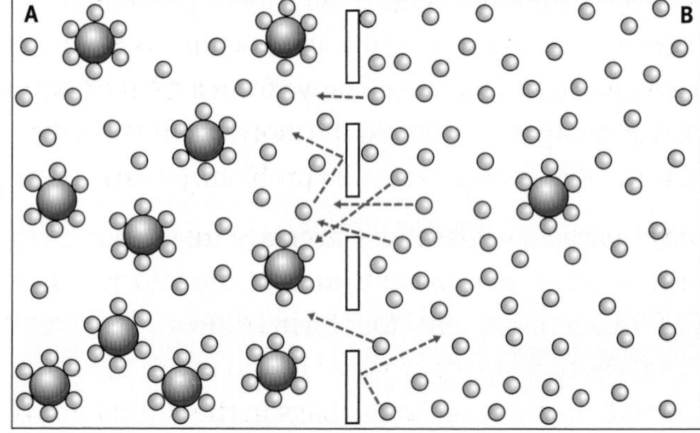

a This membrane is separating water molecules. Water molecules move at random. As many pass from left (A) to right (B) as pass from right (B) to left (A).

b This membrane is separating two glucose solutions. There are more free water molecules and fewer glucose molecules on the right (B). In other words, water molecules that are free to move by osmosis are in higher concentration on the right. So there is overall movement of water from right (B) to left (A).

Key words

- osmosis
- partially permeable membranes
- dilute
- concentrated solution

Movement of water molecules

A solution with a high concentration of water molecules that are free to move is a **dilute** solution. For example, if you make a dilute drink of squash, it has lots of water in it. A more **concentrated solution** of squash would have fewer free water molecules.

More water molecules move away from an area of higher concentration of free water molecules. Think of it as diffusion of water. This overall flow of water from a dilute to a more concentrated solution across a partially permeable membrane is called osmosis.

Osmosis in plant cells

Plants do not have skeletons to give them support. They must keep their structure by having cells that are just the right size and shape. Osmosis is important in this. It drives the uptake of water by plant roots and determines how water passes from one cell to another throughout the whole plant.

If plant cells take in too much water, they bulge and become stretched. Their strong cell wall prevents them from bursting. If plant cells lose too much water, they shrink. Plant cells need to keep just the right amount of water inside their cytoplasm.

Look at the two pictures on the right to see what happens to a plant if it does not have enough water.

The photographs show the same plant. In the lower image, the plant is shown after it has not been watered for 10 days. Notice how the structure has changed as the plant cells have dried.

Glucose is stored as starch

Plants make glucose during photosynthesis. The glucose is transported from the leaves to other cells where it is stored until it is needed for respiration. This poses a problem for the plant cells that store the glucose. They will take in too much water by osmosis. The water would move from a dilute solution surrounding the cells into the more concentrated glucose solution in the cells. To overcome this problem, glucose is stored as starch.

Large carbohydrates like starch are **insoluble**. They have very little effect on the concentration of the solutions in a plant cell. This makes them ideal for storing glucose. The starch does not affect the movement of water in and out of the cells. Starch is kept in small, membrane-bound bags in the cell called **starch grains**.

Questions

1. Explain what is meant by a partially permeable membrane.
2. Write down a definition of osmosis.
3. A student put a raisin (a dried grape) into a glass of water. They noticed that the raisin expanded and swelled. Explain this observation. Include a diagram to show the movement of water.
4. Explain why starch is needed to store glucose in plant cells.
5. Water moves into root cells by osmosis. What does this tell you about the water in the soil?

Minerals from the soil

Find out about

- why plants need minerals
- how minerals are absorbed in the roots by active transport

Nitrates contain this group of atoms:

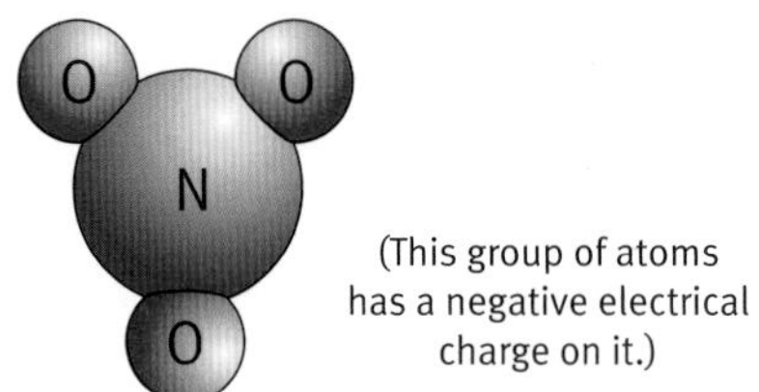

(This group of atoms has a negative electrical charge on it.)

Nitrate ions are found dissolved in soil water, and in rivers and seas.

Plants capture energy from sunlight during photosynthesis. This energy builds glucose for respiration. Glucose also supplies the raw materials needed to make other molecules like proteins, fats, and DNA. These molecules need elements contained in minerals from the soil.

Making proteins needs nitrogen

Proteins are long chains of amino acids. To make amino acids, nitrogen must be combined with carbon, hydrogen, and oxygen atoms from glucose made during photosynthesis.

Most of the Earth's nitrogen is in the air, but plants mainly take in nitrogen from the soil as **nitrate ions**. These nitrates are absorbed by **root hair cells**.

Nitrates are not the only minerals that plants need. For example, they need magnesium to make chlorophyll and phosphates to make DNA. As proteins are used to build cells and make enzymes, nitrates are needed in the largest quantities. Fertilisers contain minerals such as phosphates and nitrates.

Nitrate ions are absorbed by active transport

Plant roots absorb nitrate ions dissolved in water in the soil. Nitrate ions are at a higher concentration inside the root cells than in the surrounding soil. Diffusion would move the nitrate ions out of the roots and into the soil. To overcome this, the cells use a process called **active transport** to pump nitrates from the soil and into the roots.

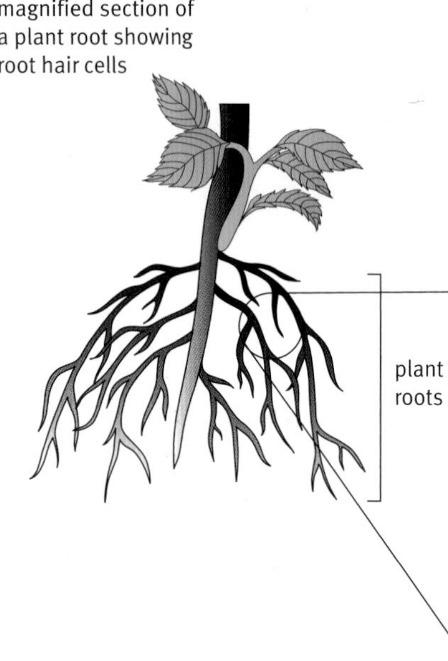

root hair cell

Plant roots have a very large surface area to help them absorb water and minerals. Tiny root hair cells take in minerals, such as nitrates, by active transport.

Another way of getting molecules into cells

When a cell needs to take in molecules that are in higher concentration inside the cell than outside, active transport is used.

In active transport, cells use the energy from respiration to transport molecules across the membrane. An example of active transport is where nitrates are taken into plant roots against their diffusion gradient.

Key words

- nitrate ions
- root hair cells
- active transport
- insoluble
- starch grains

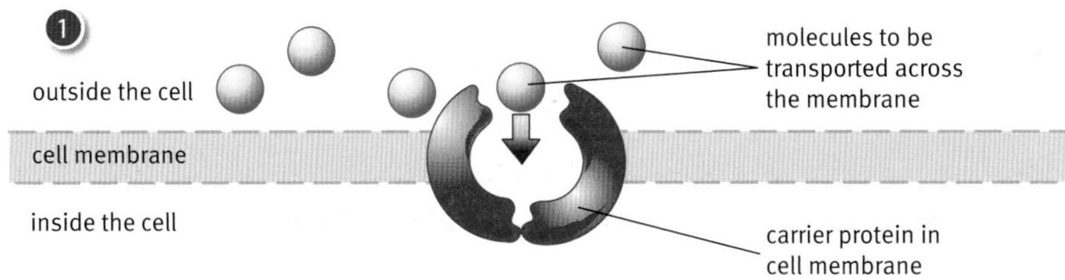

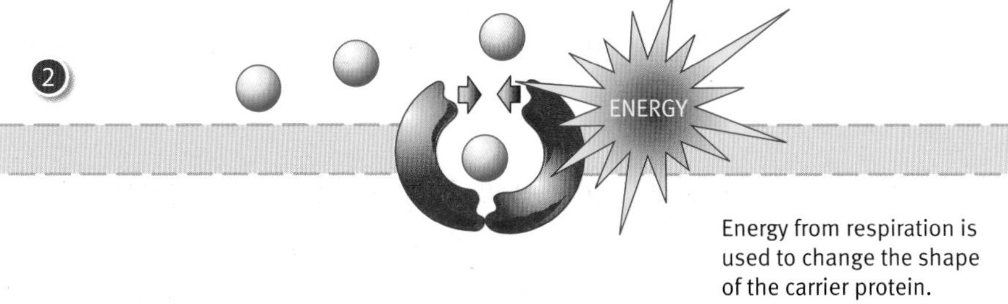

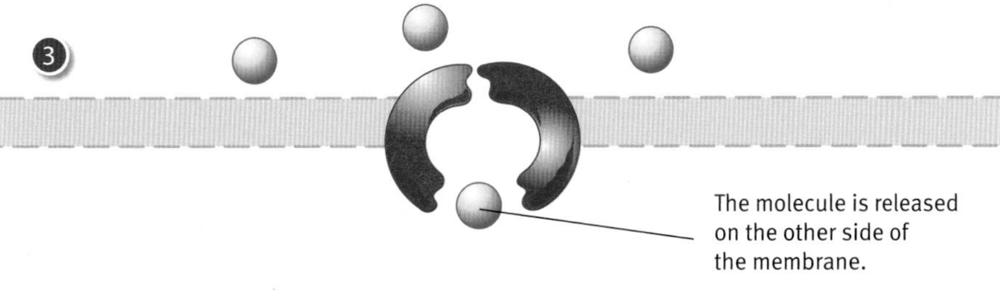

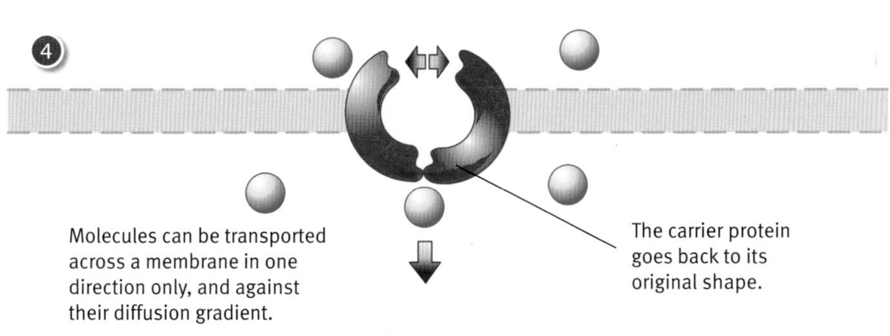

Movement of molecules across a cell membrane by active transport (schematic).

Questions

1. Explain why cells sometimes need to use active transport.
2. Name one chemical that is moved into cells by active transport.
3. Write down two main differences between diffusion and active transport.
4. Why do plant cells need a source of nitrate ions?

The rate of photosynthesis

Find out about

- ✓ what limits the rate of photosynthesis

Intensive tomato farming takes place all year round in this greenhouse.

The conditions inside the greenhouse in the photograph on the left are kept under very careful control. The tomato plants growing here have the optimum conditions for photosynthesis. They are making glucose at their highest rate, so they are growing quickly. All this is planned by the farmer so that the yield from the tomato plants will be as high as possible. Yield is the amount of product the farmer has to sell.

All reactions speed up when the temperature rises, and photosynthesis is no exception. The greenhouse is kept warm at 26 °C. This is the optimum temperature for photosynthesis to take place in these plants. Some plants don't grow in the UK because the temperature is too cold. They are better adapted for life in hot climates. The temperature falls as you climb a mountain. Above a certain height it is too cold for many large plants to photosynthesise effectively.

Faster photosynthesis – light intensity

Other factors have an effect on the rate of photosynthesis. Energy from light drives photosynthesis, so increasing the amount of light a plant receives increases the rate of photosynthesis.

The diagram below shows an experiment to investigate how changing **light intensity** affects the **rate of photosynthesis** in a piece of pondweed. The results from the experiment are shown in the graph. The graph shows that:

- at low light intensities, increasing the amount of light increases the rate of photosynthesis
- at a certain point increasing the amount of light stops having an effect on the rate of photosynthesis.

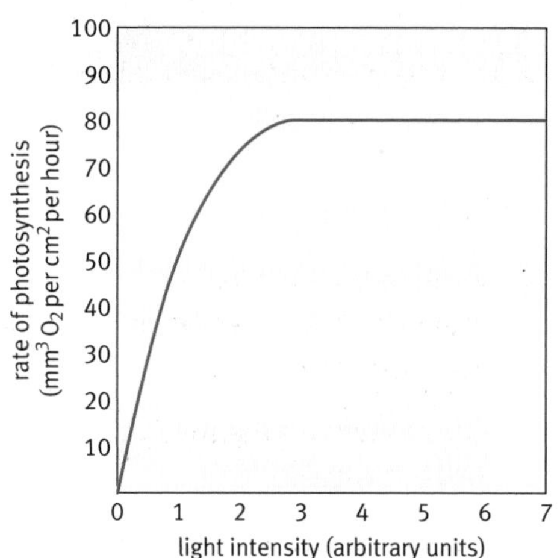

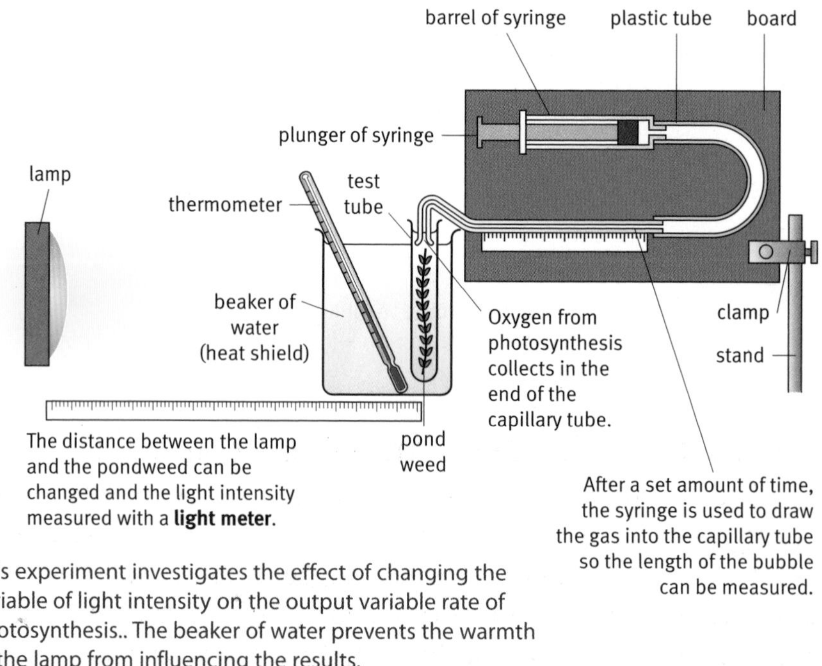

This experiment investigates the effect of changing the variable of light intensity on the output variable rate of photosynthesis.. The beaker of water prevents the warmth of the lamp from influencing the results.

Why does the rate not keep on rising?

Photosynthesis needs more than energy from light. Extra light makes no difference to the rate of photosynthesis if the plant does not have the carbon dioxide, water, or chlorophyll to use the energy from light to the full. The temperature must also be high enough for photosynthesis reactions to speed up. Increasing the light intensity stops having an effect on the rate of photosynthesis because one of these other factors is in short supply. This factor is called the **limiting factor**.

Limiting factors

In a British summer the limiting factor for photosynthesis is often water. Stomata close to prevent water diffusing out of the leaves, and this also has the unavoidable consequence of reducing carbon dioxide diffusion into the leaf.

The graph below shows the effect of increasing light intensity on the rate of photosynthesis at two different carbon dioxide concentrations. At 0.04% CO_2, more light increases the rate of photosynthesis up to a point, until light is no longer the limiting factor. Increasing the CO_2 level to 0.4% makes the rate of photosynthesis higher – CO_2 must have been the limiting factor. But even this graph levels off as another factor becomes in short supply.

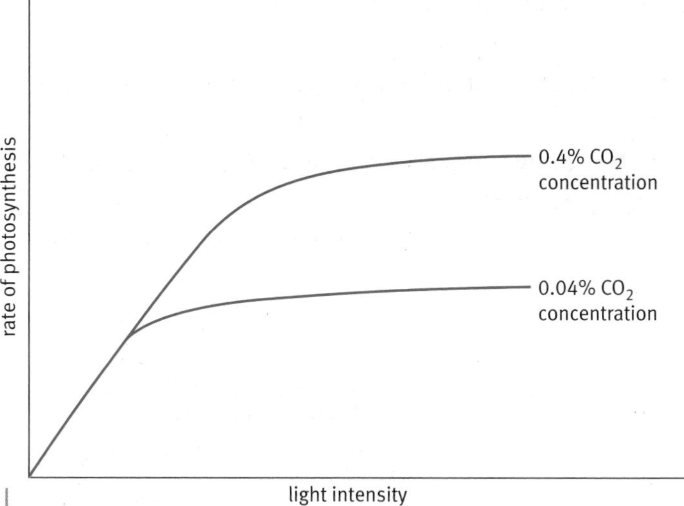

At the higher carbon dioxide concentration, photosynthesis takes place faster. But the rate still levels off. Another factor must be limiting photosynthesis.

Carbon dioxide levels in the greenhouse

Carbon dioxide forms 0.04% of normal air. Levels over about 1% are toxic to plants and animals. The levels in the tomato greenhouse are kept at 0.1%. Raising the concentration higher than this has no effect on the rate of photosynthesis. So it would not be cost effective for the farmer to add more carbon dioxide than this to the greenhouse.

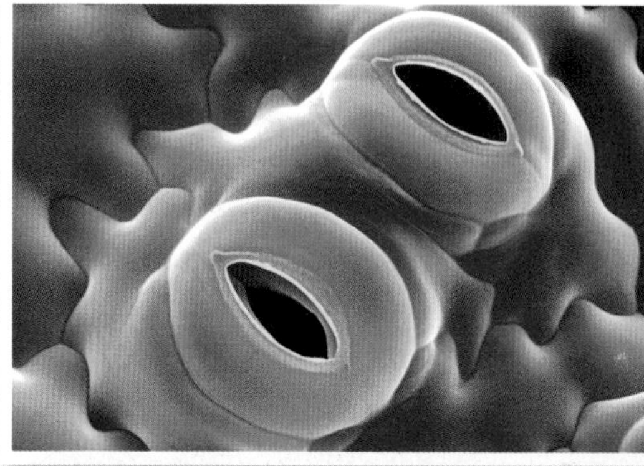

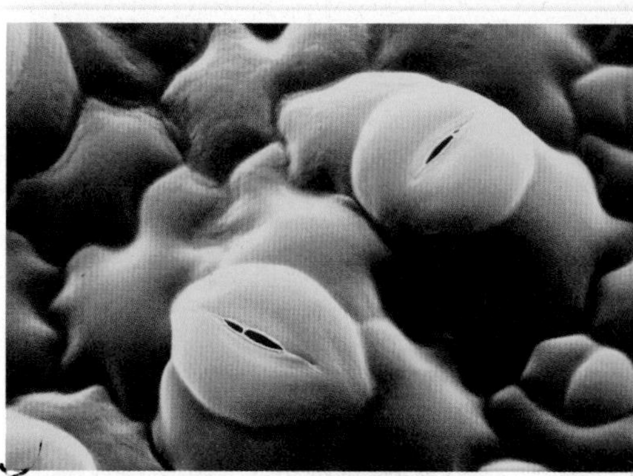

Top: The stomata on the underside of leaves open to allow gases to move in and out of the leaf. *Bottom:* They close to conserve water. (Magnification × 400.)

Key words

- light intensity
- rate of photosynthesis
- light meter
- limiting factor

Questions

1. Write down four factors that can affect the rate of photosynthesis.
2. Explain what is meant by a limiting factor.
3. Suggest a factor that could be limiting bluebells growing on a woodland floor in spring.

Find out about

- how environmental conditions affect the plants that are able to grow
- ways to survey plants in a location

Why do different plants grow in different locations?

Don't plants just grow where we plant them? Well that is the case in gardens and parks, but not in natural ecosystems. The great variety of plants has evolved to take advantage of **habitats** all over the Earth.

Some plants grow well in shade and others need bright light. Some plants need plenty of water and others can survive in deserts. All plants need minerals, water, and light in just the right amounts to be able to grow well.

The shading effect of trees determines the types of plant that can grow and survive in woodland.

Different habitats, different conditions

Woodland is a rich habitat for many plants and animals. Sunlight gets absorbed by tall trees, which hold their leaves up high for maximum photosynthesis. Breaks in the trees allow sunlight to pass through, but most of the woodland floor is in shade. Only specialised plants are able to grow and survive in these conditions.

Key words

- habitat
- samples
- quadrat
- random
- transect

Investigating different habitats

To understand why plants grow in particular locations, factors like soil pH, temperature, light intensity, and the availability of water are measured. Enough individual **samples** must be taken to get a true picture of what the conditions are like.

A square grid, called a **quadrat**, is used to survey the plants in a square metre. The quadrat is placed on the ground. Plants and animals within the quadrat are identified accurately and counted. An identification key can help to name the organisms. The key has descriptions or pictures that can be compared with the specimen. Plant growth is often recorded as percentage (%) cover.

This scientist is using a quadrat to record the plants growing in a square metre.

The positioning of the quadrat in the area being investigated is **random**. This allows reliable comparisons between different locations. Placing quadrats randomly removes bias. Recording the plants and animals in a quadrat involves accurate identification of each species.

Sometimes samples are taken at regular intervals along a straight line called a **transect**. This is useful when looking at how the types of plants change gradually from one area to another, for example, when moving from the shaded part of a wood into an open field. A lightmeter could be used to accurately measure light intensity. The sensor should be held at the same angle and the readings taken one after another so as to compare like with like.

Comparing bluebell growth

The results of a survey of bluebell growth are presented in the table below. Two areas were compared – one in shady woodland and the other in the middle of a field.

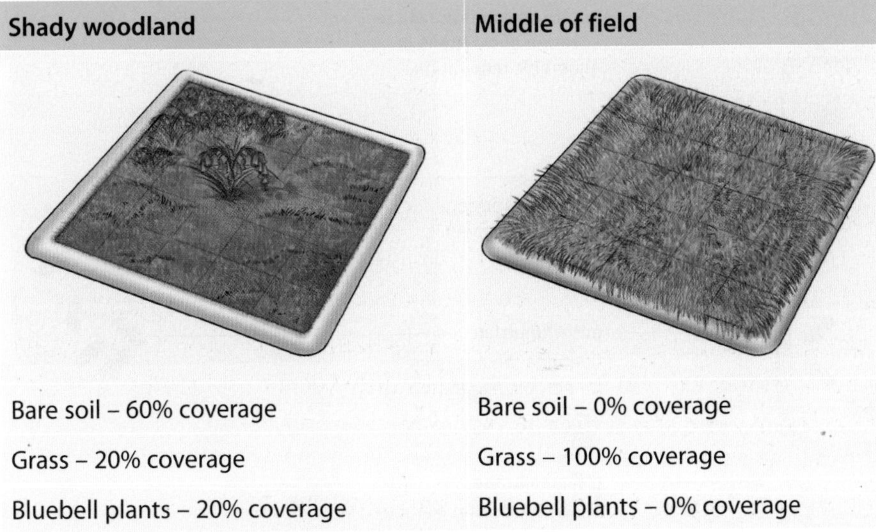

Shady woodland	Middle of field
Bare soil – 60% coverage	Bare soil – 0% coverage
Grass – 20% coverage	Grass – 100% coverage
Bluebell plants – 20% coverage	Bluebell plants – 0% coverage

The findings show that bluebells can survive in the shade. In the field, the bluebells cannot compete with the grass.

Questions

1 What do plants need to be able to grow?

2 What are the different environmental conditions that could affect plant growth?

3 Suggest how you could make sure that plant and animal data recorded at a location gives a reliable indication of the conditions in the area.

4 Cactus plants have thick waxy skins to prevent water loss. They can be grown indoors in the UK. Suggest what would happen if a cactus was planted in a flower bed outside.

J Energy for life

Find out about

- aerobic respiration in plant and animal cells

Photosynthesis combines carbon dioxide and water to make glucose. This captures the energy held in sunlight and converts it into energy held in glucose. Cells release the energy in glucose in a process called respiration. This is a carefully coordinated series of chemical reactions that happen in the cells of all living organisms.

Respiration releases energy in a form that cells can use for processes like active transport, movement, and building molecules used for growth and repair.

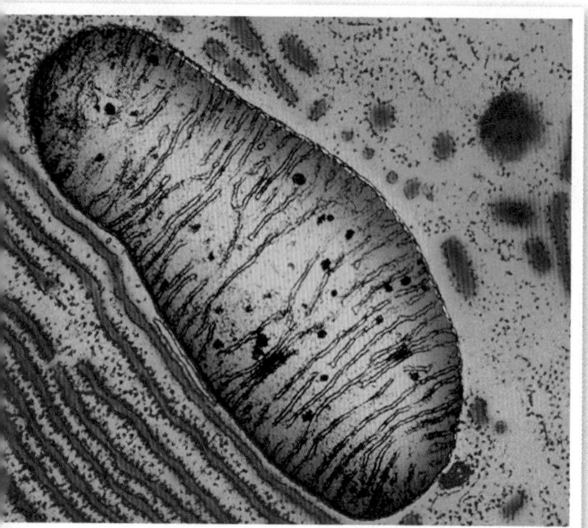

Some of the reactions for respiration take place in the cell cytoplasm, and many happen inside **mitochondria**. This electron micrograph shows a single mitochondrion. (Magnification × 64000.).

Aerobic respiration

Your body is a demanding animal. Billions of cells each carry out thousands of chemical reactions every second to keep you alive. Your cells need a constant supply of energy to drive these reactions. The food you eat provides you with molecules to make new cells. Food is also a store of chemical energy. This is converted by respiration into energy your cells can use.

Most of your energy comes from **aerobic respiration**. During aerobic respiration, glucose from food reacts with oxygen. The reactions release energy from the glucose. Respiration can be summarised by the following equation:

$$\underset{\text{glucose}}{C_6H_{12}O_6} + \underset{\text{oxygen}}{6O_2} \longrightarrow \underset{\text{carbon dioxide}}{6CO_2} + \underset{\text{water}}{6H_2O} \text{ (+ energy released)}$$

Respiration is a long series of reactions. It is summarised by this equation.

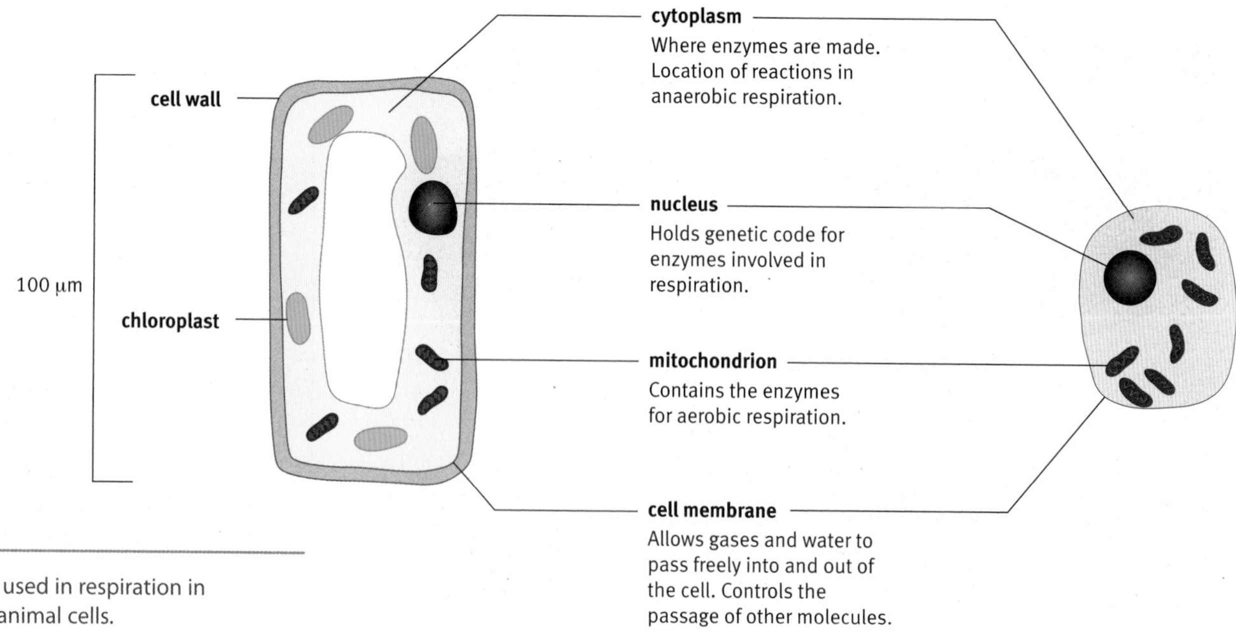

Structures used in respiration in plant and animal cells.

What happens to the energy from respiration?

All respiration releases energy from glucose. This energy will be needed by many processes in the cell.

Key words
- aerobic respiration
- mitochondria
- polymers

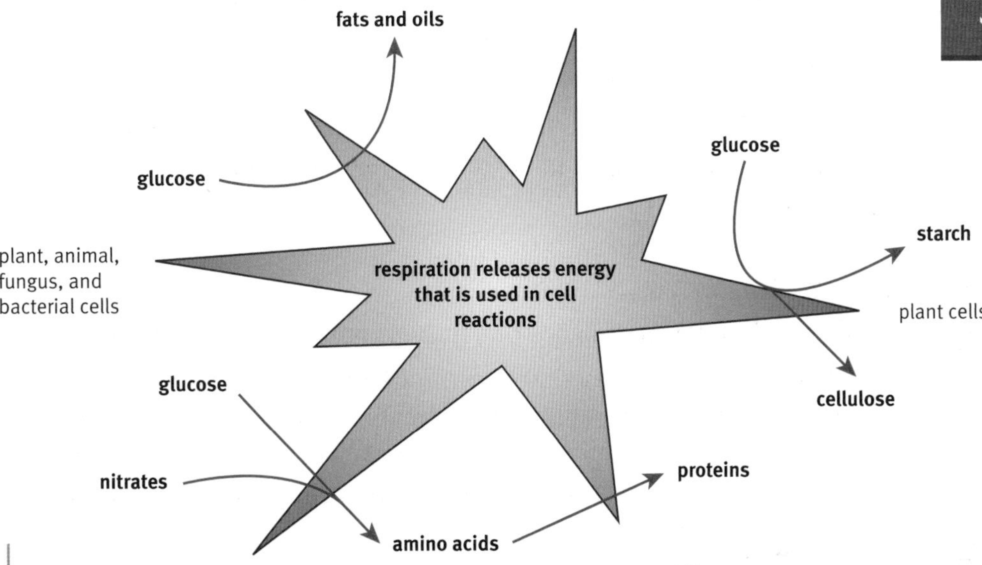

The energy from respiration is used in reactions that make many different molecules. These include **polymers** of glucose (starch and cellulose) and proteins, fats, and oils.

This girl is using energy for many different actions – such as processing information in her brain, moving, maintaining a constant internal environment, and growing and repairing her tissues.

Questions

1 Write down three processes for which your body needs energy.

2 a Write down the word equation for aerobic respiration.
  b Annotate the equation to show where the reactants come from, and what happens to the products.

3 Explain why the uptake of nitrate ions in a plant's roots by active transport requires energy, but the movement of carbon dioxide into leaf cells when they are photosynthesising does not.

K Anaerobic respiration

Find out about

- anaerobic respiration in humans and other organisms

Many animals use a different type of respiration for short bursts of intense energy, for example, when a predator runs after its prey and the prey runs for its life. In these situations the animals' muscles cannot get oxygen quickly enough for aerobic respiration. So they switch to a form of respiration that does not need oxygen. It is called **anaerobic respiration**.

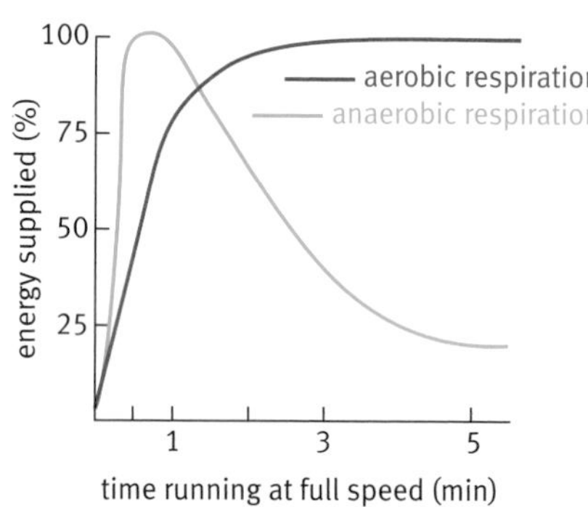

glucose ⟶ lactic acid (+ energy released)

Anaerobic respiration in animals is summarised by this equation.

In animals, anaerobic respiration can only be used for a short period of time. It releases much less energy from each gram of glucose than aerobic respiration. Also, the waste product **lactic acid** is toxic in large amounts. If it builds up in muscles, it makes them feel tired and sore.

Germinating seeds respire anaerobically.

Anaerobic respiration in plants and microorganisms

Other organisms can also use anaerobic respiration when there is limited oxygen available, for example:

- parts of plants, such as roots in waterlogged soils, and germinating seeds
- some microorganisms, such as **yeast** (brewing and baking), lactobacilli (cheese- and yoghurt-making), and bacteria in puncture wounds.

Anaerobic respiration in plants, yeast, and some bacteria produces **ethanol** and carbon dioxide instead of lactic acid.

glucose ⟶ ethanol + carbon dioxide (+ energy released)

Anaerobic respiration in plants and microorganisms is summarised by this equation.

The 100-m sprint takes about 10–11 seconds. The heart and lungs cannot increase oxygen supply to the muscles fast enough, so most of the energy required during the short race comes from anaerobic respiration.

Bacteria and yeast

Bacteria and yeast have been used for thousands of years in the preparation of foods such as bread, cheese, yoghurt, alcoholic drinks, and vinegar. Many of these foods are products of anaerobic respiration.

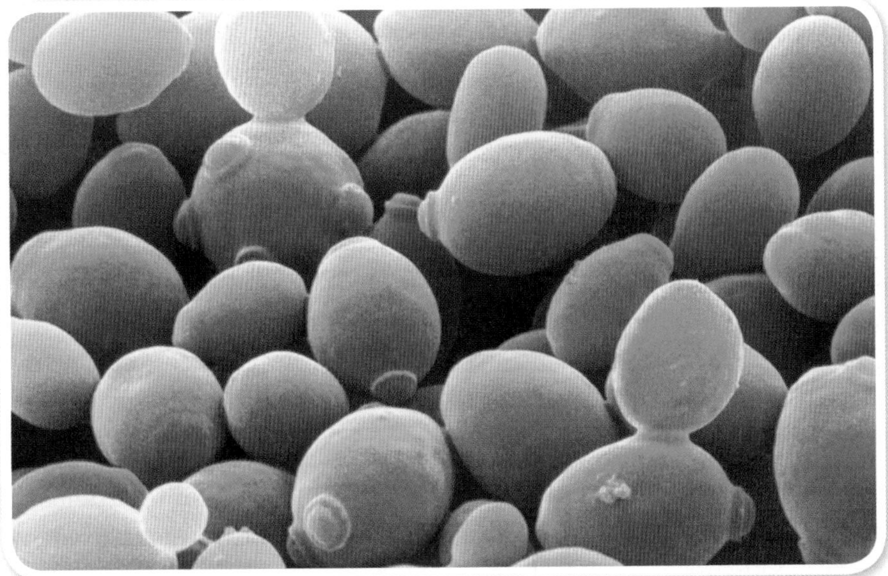

Yeast cells can respire anaerobically until ethanol builds up and becomes too toxic.

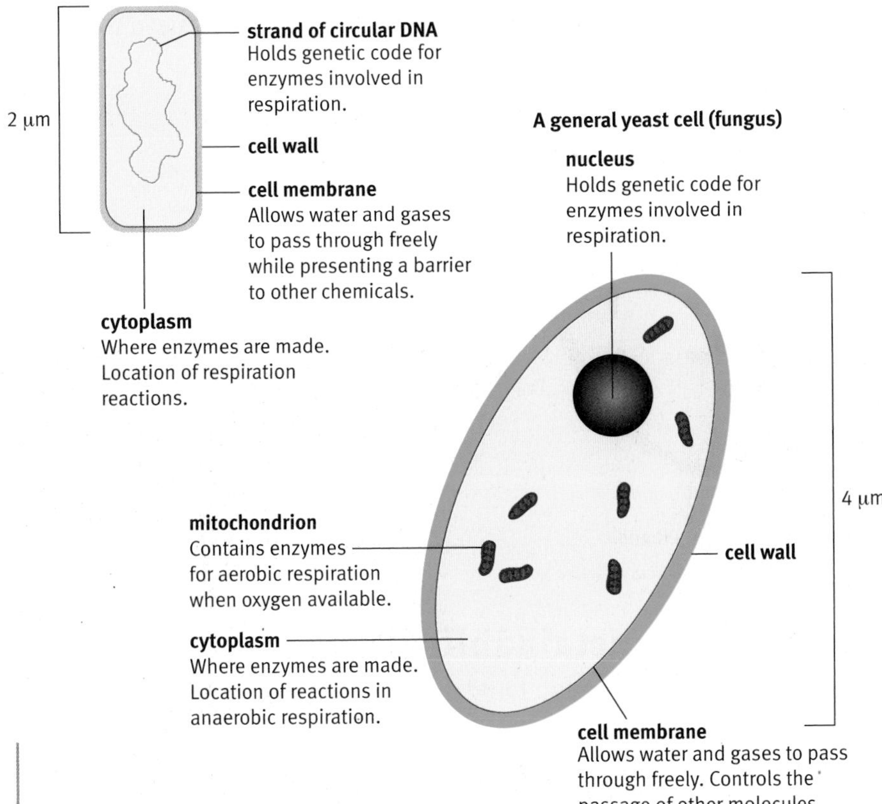

Structure of a typical bacteria and yeast cell.

Key words

- anaerobic respiration
- lactic acid
- yeast
- bacteria
- ethanol

Questions

1. Describe conditions where anaerobic respiration is an advantage to:
   a human beings
   b another organism.
2. Give an example of anaerobic respiration that people draw on to make a useful product.
3. Yeast is used in brewing to make alcoholic drinks. Suggest why the yeast stops growing and dies before it has used up all the sugar in the fermentation solution.

L Useful products from respiration

Find out about

- how bioethanol is made from sugar
- how farm manure can be used to make a valuable fuel

Living things make complex molecules far more efficiently than they can be made in a laboratory. For thousands of years people have harnessed microorganisms to make products such as drinks, bread, yoghurt, and cheese. Microorganisms are added to food ingredients and kept in the right conditions. As they grow, by-products from the microorganisms create the desired product. For example, yeast is used in bread-making. It uses sugars in the flour for respiration and the carbon dioxide it produces makes bread rise.

Carbon dioxide, a by-product of the respiration of yeast, has caused these loaves of bread to rise.

Bioethanol from sugar

Petrol and diesel are used to fuel vehicle engines. They are made from crude oil. **Bioethanol** can also be used to fuel vehicle engines. It is made from the sugars in plant material such as sugar beet, maize, and wheat. Fuel from renewable sources is said to be more **sustainable** and causes less global warming.

Yeast are single-celled organisms. Yeast have been used for thousands of years to brew beers and wine. On a much larger scale, yeast are used to make bioethanol.

Yeast cells take sugars and convert them into ethanol during the process of anaerobic respiration. This process is called **fermentation**. Large fermentation vessels inside the factory contain yeast, sugar, water, and other nutrients. The yeast produce bioethanol that can be used to fuel car engines.

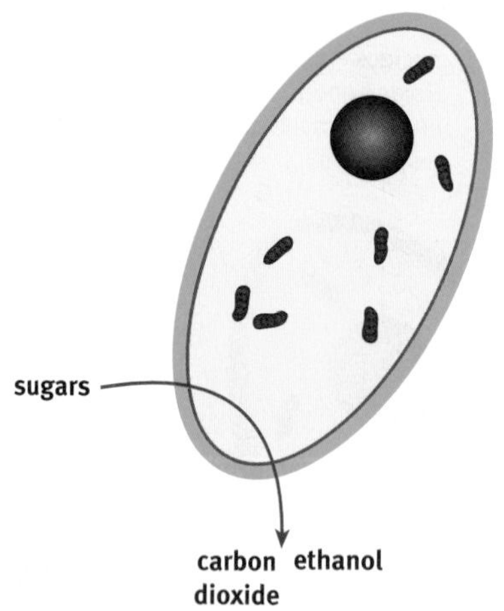

Yeast cells take sugars and convert them into ethanol during the process of fermentation.

This factory produces bioethanol to fuel cars.

Biofuels and sustainability

Biofuels can be controversial. Land may be used to grow crops for fuel instead of food for people. In some areas, forest is cut down so the land can be used to grow crops for fuel. Producers are now developing ways of using non-food crops and algae to end the need to use food crops.

Biogas from waste

Biogas is a fuel obtained from animal manure or human waste. Bacteria break down the organic material in the manure and produce methane gas. This can be used as a fuel to heat buildings and run electricity generators.

Methanogenic bacteria in biogas fermenters produce methane gas during anaerobic respiration.

This chemical factory produces biogas.

Key words

- **bioethanol**
- **sustainable**
- **fermentation**
- **biogas**

Biodigesters typically use manure from farm animals, but some are also used to get biogas from human waste. The biogas contains around 70% methane and 30% carbon dioxide.

Bacteria grow in anaerobic conditions inside the biodigester and produce methane gas, which can be used as a fuel. Manure from animals and human toilets can be collected. The optimum temperature of 36 °C makes this a good option in warm climates.

Biogas is used in some developing countries, such as by smallholders in rural parts of India. It is particularly useful where other sources of energy, such as electricity, are not available.

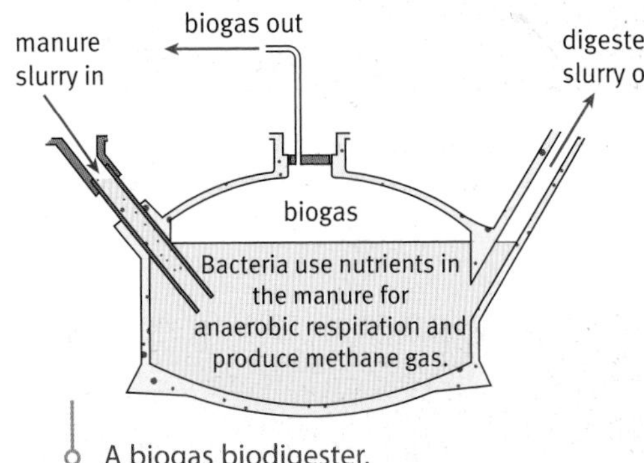

A biogas biodigester.

Questions

1. Why is bioethanol said to be a renewable resource?
2. Explain why producing and using bioethanol contributes less carbon dioxide to the atmosphere than using petrol from crude oil.
3. Describe the process of anaerobic respiration in yeast. Compare it to anaerobic respiration in animal cells.
4. Biogas contains around 70% methane and 30% carbon dioxide. Suggest where the carbon dioxide in the biogas comes from.
5. Draw a flow diagram to show how energy is transferred from sunlight into the methane in biogas. Include details of the processes happening at each stage.

Science Explanations

All living things are made up of cells. Biological processes such as photosynthesis and respiration take place in cells. These processes involve chemical reactions that are speeded up by enzymes.

You should know:

- that some chemical reactions in cells require energy; these reactions include muscle contraction, synthesis of large molecules, and active transport
- that respiration is a series of chemical reactions in plant, animal, and microbial cells, which release energy by breaking down food molecules
- what enzymes are and how they speed up chemical reactions in living organisms
- that cells make enzymes from the instructions carried in our genes
- why the model that describes how enzymes recognise molecules is called the lock-and-key model
- the conditions that enzymes need to work at their optimum rate
- that photosynthesis uses energy from sunlight to make glucose by joining carbon dioxide and water, and releasing oxygen as a waste product
- that chlorophyll absorbs light energy
- that glucose may be converted into other chemicals needed for plant growth
- how to use equipment such as light meters, quadrats, and identification keys to investigate the effect of light on plants
- how to take a transect in fieldwork
- that minerals are taken up by plant roots, and what they are used for in plants
- that diffusion is the passive overall movement of molecules from a region of higher concentration to a region of lower concentration
- how osmosis is the overall movement of water by diffusion from a dilute to a more concentrated solution through a partially permeable membrane
- why active transport, such as the absorption of nitrates by plant roots, requires energy from respiration
- that aerobic respiration breaks down glucose in the presence of oxygen, releasing energy, carbon dioxide, and water; it takes place in animal and plant cells and some microorganisms
- that anaerobic respiration takes place in animal, plant, and some microbial cells
- how energy is released from glucose without oxygen during anaerobic respiration – animal cells form lactic acid, plant cells and yeast produce carbon dioxide and ethanol
- the structures and functions of typical plant, animal, and microbial cells
- how we use the anaerobic respiration of microorganisms to make biogas, bread, and alcohol.

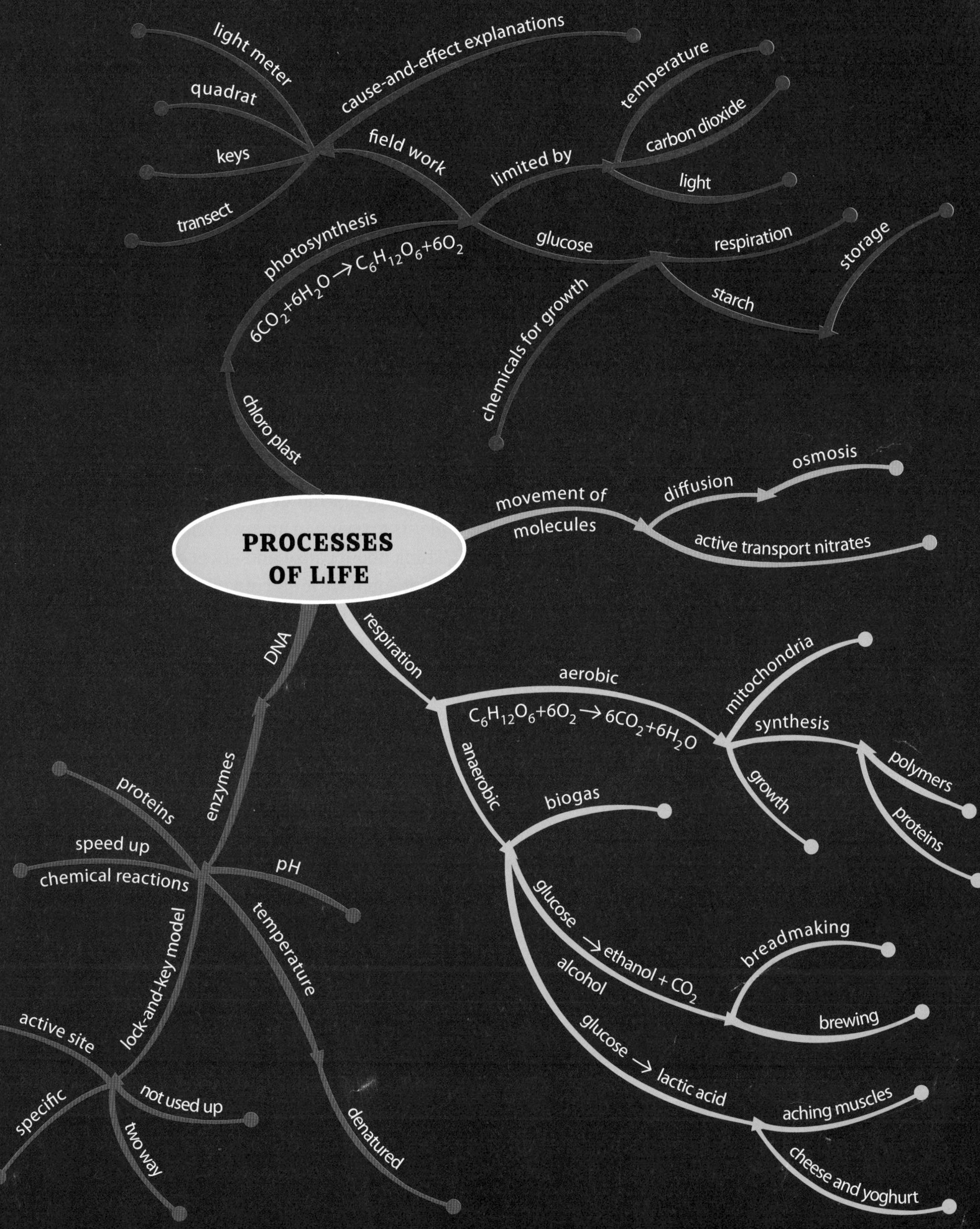
PROCESSES OF LIFE
chloro plast
photosynthesis
$6CO_2 + 6H_2O \rightarrow C_6H_{12}O_6 + 6O_2$
field work
light meter
quadrat
keys
transect
cause-and-effect explanations
limited by
temperature
carbon dioxide
light
glucose
respiration
starch
storage
chemicals for growth
movement of molecules
diffusion
osmosis
active transport nitrates
respiration
aerobic
$C_6H_{12}O_6 + 6O_2 \rightarrow 6CO_2 + 6H_2O$
mitochondria
synthesis
polymers
proteins
growth
anaerobic
biogas
glucose → ethanol + CO_2
alcohol
breadmaking
brewing
glucose → lactic acid
aching muscles
cheese and yoghurt
DNA
enzymes
proteins
speed up
chemical reactions
pH
temperature
denatured
lock-and-key model
active site
specific
two way
not used up

Ideas about Science

This module helps you develop an understanding of life processes. You will also learn more about how scientists explain cause and effect, and how they investigate relationships between factors.

If you take several measurements of the same quantity, these are likely to vary. This may be because:

- you are measuring several individual examples, such as oxygen produced from different samples of pond weed
- the quantity you are measuring is varying, for example, the number of bluebells growing in different areas of a wood
- the limitations of the measuring equipment or because of the way you use the equipment.

The best estimate of a true value of a quantity is the mean of several measurements. The true value lies in the spread of values in a set of repeat measurements.

- A measurement may be an outlier if it lies outside the range of the other values in a set of repeat measurements.
- When comparing information on plants growing in different places, a difference between their means is likely to be real if their ranges do not overlap.
- A correlation shows a link between a factor and an outcome, for example, as light intensity increases, the rate of photosynthesis increases.
- A correlation does not always mean that the factor being changed causes the outcome.

Scientists often think about processes in terms of factors that may affect an outcome (or outcome variable). When you investigate the relationship between a factor and an outcome, it is important to control all the other factors that you think might affect the outcome of the investigation. This is called a 'fair test'.

You should be able to:

- identify input variables for photosynthesis
- explain how temperature and pH affect enzyme reactions.

You also need to explain why it is necessary to control all of these variables in an investigation and consider how you would perform a fair test.

When you plan an investigation you need to identify the likely effect of a factor on an outcome. You also need to know that if you don't control the variables in an investigation, the investigation will be flawed.

You should understand why plants grow in particular places. You could measure factors like soil pH, temperature, light intensity, and water. Compare this data with what you know about the plants growing in an area, and what plants need. Think about cause-and-effect explanations for what you have found out.

- Enough individual samples must be taken to get a true picture of the conditions.

You should be able to find ways of collecting reproducible data and be able to explain why repeating measurements leads to a better estimate of the quantity measured. Here are two examples:

- A light meter can be used to accurately measure light intensity. The sensor should be held at the same angle for each reading. Time of day and shading from trees or bushes should be taken into account.
- One way of sampling using quadrats is to position quadrats randomly in the area being measured. This removes bias. Recording the quadrat data involves accurate identification of each species.

Review Questions

1 This question is about enzymes.

a What are enzymes?
Your answer should include:
- what type of chemical they are
- what they do.

b Enzymes and some molecules fit together. Which model explains this?
Choose the correct answer.

enzyme-and-molecule model
lock-and-key model
puzzle-shaped model

c Which of the following does not affect biological enzymes?

i light
ii pH
iii temperature

2 Alex draws a set of diagrams to show what happens during a reaction involving an enzyme.
He makes a mistake and draws one incorrect stage.

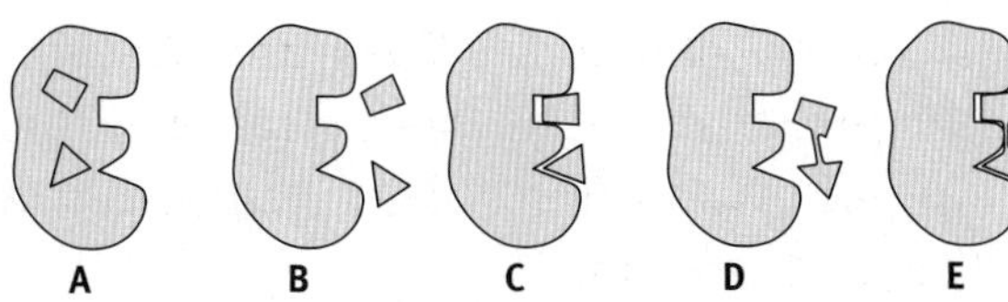

The stages are not drawn in the correct order.
Put the stages in the correct order.
The last one has been done for you.

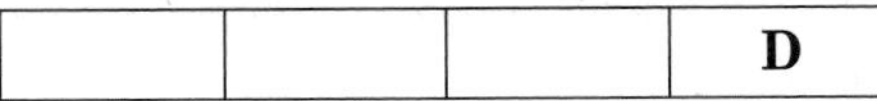

3 Photosynthesis takes place in green plants.

a Describe three factors that limit the rate of photosynthesis.

b The rate at which photosynthesis takes place can be measured. One way to do this is to count the number of bubbles given off by a water plant in one minute.

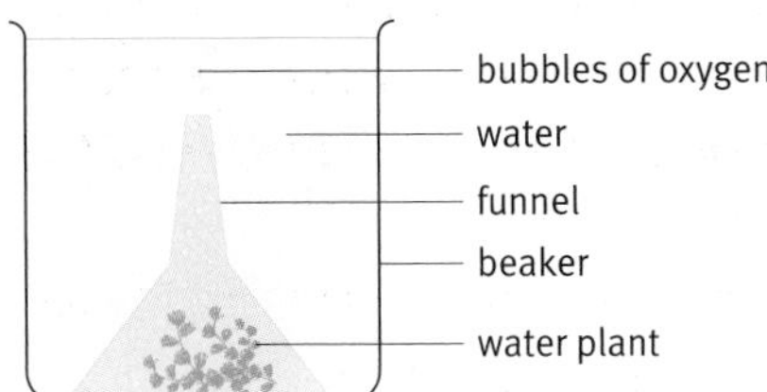

Explain why trying to count the number of bubbles in one minute may not give an accurate value for the rate of photosynthesis.

4 Steve is an athlete. He runs marathons. As he runs he releases energy both aerobically and anaerobically.
Explain **how** and **why** Steve uses both of these types of respiration when he runs a marathon.

5 Molecules move across cell membranes by diffusion, osmosis, and active transport.
Describe and explain the differences between these three different processes.

C4 Chemical patterns

Why study chemical patterns?

The periodic table is important in chemistry because it helps to make sense of the mass of information about all the elements and their compounds. The table offers a framework that can give meaning to all the known facts about properties and reactions.

What you already know

- Elements are made up of just one type of atom.
- A molecule is a group of atoms chemically joined together.
- During a chemical reaction atoms are conserved and mass is conserved.
- The properties of the reactants and products of chemical changes are different.
- There are patterns in the way chemicals react.
- Acids neutralise alkalis to form salts.
- When an acid reacts with a metal carbonate, the products are a salt, water, and carbon dioxide.
- Chlorine can be used to kill microorganisms.
- A chemical change caused by the flow of an electric current is called electrolysis.

Find out about

- the chemistry of some very reactive elements
- the patterns in the periodic table
- how scientists can learn about the insides of atoms
- the use of atomic theory to explain the properties of chemicals
- the ways in which atoms become charged and turn into ions.

The Science

Chemists use ideas about atomic structure to explain the periodic table and the properties of elements. Spectroscopy led to the discovery of new elements. Spectroscopy is now used to study chemicals and their reactions.

Ideas about Science

For a scientific explanation to be accepted and become a theory, it must be tested by other scientists to check it is supported by evidence and can be used to make predictions. Atomic theory explains patterns in the periodic table. Ionic theory explains electrolysis. Without ionic theory, a method of extracting aluminium metal would not have been developed.

The periodic table

Find out about

- relative masses of atoms
- the development of the periodic table
- groups and periods

A scientist in the 1800s explaining his latest research. Those who first tried to link the properties of elements to their relative atomic masses weren't taken seriously by other scientists. Their explanations didn't include all the known elements.

Germanium – one of Mendeleev's missing elements. He used his version of the periodic table to predict that the missing element would be a grey metal that would form a white oxide with a high melting point. An element was discovered in 1886 that fitted into the space left by Mendeleev. Its properties were very similar to those he predicted.

Looking for patterns

Around the beginning of the 1800s, only about 30 elements were known, but by the end of that century almost all of the stable elements found on Earth had been discovered.

Finding so many elements, with such a wide range of properties, encouraged chemists to look for patterns, for example, patterns between the properties of elements and the masses of their atoms.

Relative atomic masses

In those days scientists could not measure the actual masses of atoms so they compared the masses of atoms with the mass of the lightest one, hydrogen. This is called the **relative atomic mass**.

In the early 1800s, Johann Döbereiner, a German scientist, noticed that there were several groups where three elements have similar properties (for example, calcium, strontium, and barium). For each group, the relative atomic mass of the middle element was the mean of the relative atomic masses of the other two.

Almost 50 years later an English chemist, John Newlands, arranged the known elements in order of their relative atomic masses. He saw that every eighth element had similarities – like an octave in music. But this seemed only to work for the first 16 known elements. This made other scientists reluctant to accept Newlands' ideas – they did not account for all the data.

Elements in order

Dmitri Mendeleev, a Russian scientist, produced patterns with real meaning when he lined up elements in order of their relative atomic mass. Mendeleev's inspiration was to realise that not all of the elements had been discovered. He left gaps for missing elements to produce a sensible pattern, and predicted what the properties of these missing elements would be.

With the elements in order of relative atomic mass, Mendeleev spotted that at intervals along the line there were elements with similar properties. Using elements known today, for example, you can see that the third, eleventh, and nineteenth elements (lithium, sodium, and potassium) are very similar.

Periodicity

A repeating pattern of any kind is a **periodic** pattern. The table of the elements gets its name from the repeating, or periodic, patterns you see when the elements are in order.

The periodic table now

In the periodic table, the elements are arranged in rows, one above the other. Each row is a **period**. The most obvious repeating pattern is from metals on the left to non-metals on the right. Every period starts with a very reactive metal in Group 1 and ends with an unreactive gas in Group 8. Elements with similar properties fall into a column. Each column is a **group** of similar elements.

Key words

- ✔ relative atomic mass
- ✔ periodic
- ✔ period
- ✔ group

group number

period number	1	2											3	4	5	6	7	0
1							1 H hydrogen 1											4 He helium 2
2	7 Li lithium 3	9 Be beryllium 4											11 B boron 5	12 C carbon 6	14 N nitrogen 7	16 O oxygen 8	19 F fluorine 9	20 Ne neon 10
3	23 Na sodium 11	24 Mg magnesium 12											27 Al aluminium 13	28 Si silicon 14	31 P phosphorus 15	32 S sulfur 16	35.5 Cl chlorine 17	40 Ar argon 18
4	39 K potassium 19	40 Ca calcium 20	45 Sc scandium 21	48 Ti titanium 22	51 V vanadium 23	52 Cr chromium 24	55 Mn manganese 25	56 Fe iron 26	59 Co cobalt 27	59 Ni nickel 28	63.5 Cu copper 29	65 Zn zinc 30	70 Ga gallium 31	73 Ge germanium 32	75 As arsenic 33	79 Se selenium 34	80 Br bromine 35	84 Kr krypton 36
5	86 Rb rubidium 37	88 Sr strontium 38	89 Y yttrium 39	91 Zr zirconium 40	93 Nb niobium 41	96 Mo molybdenum 42	98 Tc technetium 43	101 Ru ruthenium 44	103 Rh rhodium 45	106 Pd palladium 46	108 Ag silver 47	112 Cd cadmium 48	115 In indium 49	119 Sn tin 50	122 Sb antimony 51	128 Te tellurium 52	127 I iodine 53	131 Xe xenon 54
6	133 Cs caesium 55	137 Ba barium 56	57–71	178 Hf hafnium 72	181 Ta tantalum 73	184 W tungsten 74	186 Re rhenium 75	190 Os osmium 76	192 Ir iridium 77	195 Pt platinum 78	197 Au gold 79	201 Hg mercury 80	204 Tl thallium 81	207 Pb lead 82	209 Bi bismuth 83	209 Po polonium 84	210 At astatine 85	222 Rn radon 86
7	223 Fr francium 87	226 Ra radium 88	89–103	261 Rf rutherfordium 104	262 Db dubnium 105	266 Sg seaborgium 106	264 Bh bohrium 107	277 Hs hassium 108	268 Mt meitnerium 109	271 Ds darmstadtium 110	272 Rg roentgenium 111							

lanthanoid metals	139 La lanthanum 57	140 Ce cerium 58	141 Pr praseodymium 59	144 Nd neodymium 60	145 Pm promethium 61	150 Sm samarium 62	152 Eu europium 63	157 Gd gadolinium 64	159 Tb terbium 65	162 Dy dysprosium 66	165 Ho holmium 67	167 Er erbium 68	169 Tm thulium 69	173 Yb ytterbium 70	175 Lu lutetium 71
actinoid metals	227 Ac actinium 89	232 Th thorium 90	231 Pa protactinium 91	238 U uranium 92	237 Np neptunium 93	242 Pu plutonium 94	243 Am americium 95	247 Cm curium 96	247 Bk berkelium 97	249 Cf californium 98	254 Es einsteinium 99	253 Fm fermium 100	256 Md mendelevium 101	254 No nobelium 102	257 Lr lawrencium 103

other non metals · noble gases · halogens · transition metals · alkaline earth metals · alkali metals

Key

relative atomic mass (1), symbol (H), name (hydrogen), proton number (1)

symbol: **black** = solid, **blue** = liquid, **red** = gas, **white** = synthetically prepared

The periodic table. Over three-quarters of the elements are metals. They lie to the left of the table.

Questions

1 In the periodic table, identify and name:
   a a liquid in Group 7
   b a Group 1 metal that does not occur naturally
   c a gaseous element with properties similar to sulfur
   d a solid element similar to chlorine
   e a liquid metal with properties similar to zinc.

2 How many times heavier is:
   a a magnesium atom than a carbon atom?
   b a sulfur atom than a helium atom?

3 Explain how Mendeleev could predict the properties of the unknown element germanium from what was known about other elements such as silicon and tin.

4 Give two reasons why scientists accepted Mendeleev's method of arranging the elements.

B The alkali metals

Find out about

- group 1 metals
- reactions with water and chlorine
- similarities and differences between Group 1 elements

The metals in Group 1 of the periodic table are very reactive. They are so reactive that they have to be kept under oil to stop them reacting with oxygen or moisture in the air.

Chemists call these elements the **alkali metals** because they react with water to form alkaline solutions. It is the compounds of these metals that are alkalis – not the metals themselves.

There are six elements in Group 1. Two of them, rubidium and caesium, are so reactive and rare that you are unlikely to see anything of them except on video. A third, francium, is highly radioactive. Its atoms are so unstable that it does not occur naturally. As a result, the study of Group 1 usually concentrates on lithium (Li), sodium (Na), and potassium (K).

The alkali metals are corrosive and highly flammable. For this reason forceps should be used when handling these metals and goggles must be worn.

Cutting a lump of sodium to show a fresh, shiny surface of the metal.

Strange metals

Most metals are hard and strong. The alkali metals are unusual because you can cut them with a knife. Cutting them helps to show up one of their most obvious metallic properties: they are very shiny. However, they **tarnish** quickly in the air – the shiny surface becomes dull with the formation of a layer of oxide. Group 1 elements, like other metals, are good conductors of electricity.

Most metals are dense and have a high melting point. Again the alkali metals are odd: they float on water and melt on very gentle heating.

Pellets of sodium hydroxide. The traditional name is caustic soda. Caustic substances are corrosive. They attack skin. Alkalis such as NaOH are more damaging to skin and eyes than many acids.

Reactions with water

The reactions with water should be carried out behind a safety screen, because the reactions can be violent and particles can be thrown out from the surface of the water. Drop a small piece of grey lithium into water and it floats, fizzes gently, and disappears as it forms lithium hydroxide (LiOH). It dissolves, making the solution alkaline. It is possible to collect the gas and use a burning splint to show that it is hydrogen.

$$\text{lithium} + \text{water} \longrightarrow \text{lithium hydroxide} + \text{hydrogen}$$

The reaction of sodium with water is more exciting. The reaction gives out enough energy to melt the sodium, which skates around on the surface of the water. It fizzes as hydrogen is formed. Like lithium, the sodium forms its hydroxide (sodium hydroxide, NaOH), which dissolves to give an alkaline solution.

The reaction of potassium with water is very violent. The hydrogen given off catches fire at once, and the metal moves around the surface of

Key words

- alkali metal
- tarnish
- crystalline

the water quickly. The result is an alkaline solution of potassium hydroxide (KOH). All the alkalic metals form hydroxides with the formula MOH, where M is the symbol for the metal.

Reactions with chlorine

Hot sodium burns in chlorine gas with a bright yellow flame. It produces clouds of sodium chloride crystals (NaCl). This is everyday table salt, used to flavour food.

The other alkali metals react with chlorine in a similar way. Lithium produces lithium chloride (LiCl). Potassium produces potassium chloride (KCl). Like everyday salt, these compounds are also colourless, **crystalline** solids that dissolve in water.

Chemists use the term **salt** to cover all the compounds of metals with non-metals. So the chlorides of lithium, sodium, and potassium are all salts.

Trends

The alkali metals are all very similar, but they are not identical. Their reactions with water show that there is a **trend** in their **chemical properties**. The alkali metals increase in reactivity down the group from lithium to sodium to potassium. There are also trends in their **physical properties**.

	Melting point (°C)	Boiling point (°C)	Density (g/cm^3)
lithium, Li	181	1342	0.53
sodium, Na	98	883	0.97
potassium, K	63	760	0.86

Physical properties of lithium, sodium, and potassium.

Lithium, sodium, and potassium all have a low density. The melting points and boiling points decrease down the group.

Compounds of the alkali metals

The compounds of the alkali metals are very different to the elements. The elements are dangerously reactive. But chlorides of sodium and potassium, for example, have a vital role to play in our blood and in the way in which our nerves work.

Many compounds of alkali metals are soluble in water. Soluble sodium compounds, in particular, make up a number of common everyday chemicals including sodium hydroxide (in oven cleaners), sodium hypochlorite (in bleach), and sodium hydrogencarbonate (as the bicarbonate of soda in antacid remedies and baking powder).

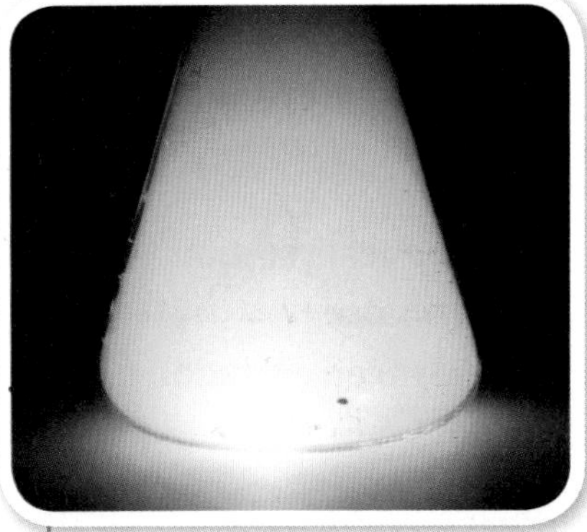

Sodium burning in chlorine gas.

Key words

- salt
- trends
- physical properties
- chemical properties

Questions

1 Predict these properties of rubidium (Rb).
   a Its melting point
   b What happens to a freshly cut surface of the metal in the air?
   c What happens if you drop a small piece of rubidium onto water?

2 For the hydroxide of rubidium predict:
   a its formula
   b whether or not it is soluble in water.

3 For the chloride of caesium (Cs) predict:
   a its colour
   b its formula
   c whether or not it is soluble in water.

4 Suggest and explain the precautions necessary when potassium reacts with water.

Chemical equations

Find out about

- chemical symbols
- formulae
- balanced equations

Equations are important because they do for chemists what recipes do for cooks. They allow chemists to work out how much of the starting materials to mix together and how much of the products they will get.

Chemical models

In a **chemical change** there is no change in mass because the number of each type of atom stays the same. The atoms rearrange, but no new ones appear and no atoms are destroyed during a chemical reaction.

Hydrogen burns in oxygen to form **molecules** of water. The models in the diagram below show what happens. In each water molecule there is only one oxygen atom. So one oxygen molecule (= two red atoms in the model) reacts with two molecules of hydrogen (each molecule = two white atoms in the model) to make two water molecules. There are equal numbers of hydrogen atoms and oxygen atoms on each side of the arrow. This is a model of a **chemical equation**.

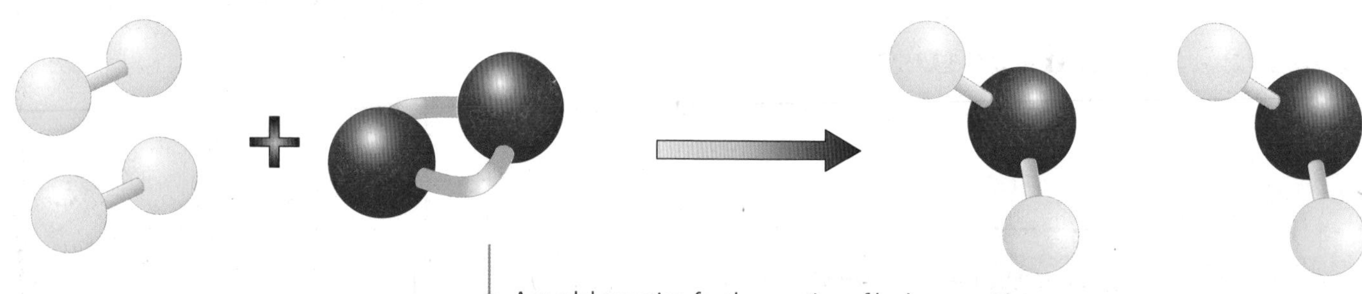

A model equation for the reaction of hydrogen with oxygen.

oxygen atom
hydrogen atom

The reactants and the products of a reaction 'balance'.

Chemical symbols

Using models to describe every reaction would be very tiresome. Instead, chemists write symbol equations to show the numbers and arrangements of the atoms in the reactants and the products.

When written in symbols, the equation in the figure above becomes:

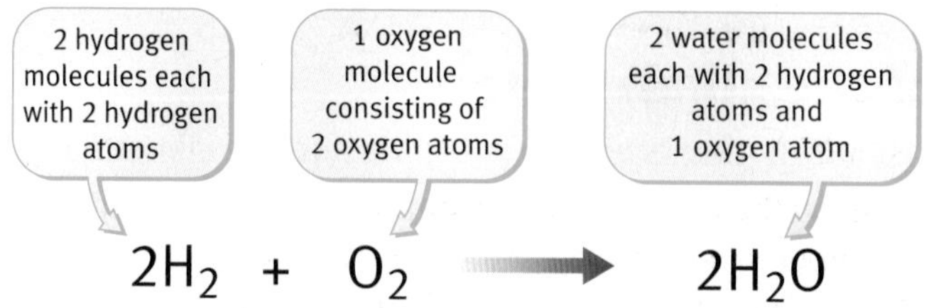

The equation is balanced because it has the same number of atoms of each type on the left and the right of the arrow. Because no atoms have been gained or lost the reactants have the same mass as the products.

Formulae

You can't write an equation unless you first know:

- all the starting chemicals (the reactants)
- everything that is formed during the change (the products).

When writing an equation, you have to write down the correct chemical **formulae** for the reactants and products. Chemists have worked these out by experiment, and you can look them up in data tables.

If the element or compound is molecular, you write the formula for the molecule in the equation. This applies to most non-metals (O_2, H_2, Cl_2) and most compounds of non-metals with non-metals (H_2O, HCl, NH_3).

Not all elements and compounds consist of molecules. For all metals, and for the few non-metals that are not molecular (C, Si), you just write the symbol for a single atom.

The compounds of metals with non-metals are also not molecular. For these compounds you write the simplest formula for the compound, such as LiOH, NaCl, or K_2CO_3.

Writing balanced equations

Follow the four steps shown in the margin to write **balanced equations**.

Worked example

Write a balanced equation to show the reaction of natural gas (methane, CH_4) with oxygen.

Step 1 Describe the reaction in words:

methane + oxygen ⟶ carbon dioxide + water

Step 2 Write down the formulae for the reactants and products:

$$CH_4 + O_2 \longrightarrow CO_2 + H_2O$$

Step 3 Balance the equation.

You must not change any of the formulae. You balance the equation by writing numbers in front of the formulae. These numbers then refer to the whole formula.

$$CH_4 + 2O_2 \longrightarrow CO_2 + 2H_2O$$

Step 4 Add state symbols.

State symbols usually show the states of the elements and compounds at room temperature and pressure. The chemicals in an equation may be solid (s), liquid (l), gaseous (g), or dissolved in water (aq, for aqueous).

$$CH_4(g) + 2O_2(g) \longrightarrow CO_2(g) + 2H_2O(l)$$

Key words

- chemical change
- molecules
- chemical equation
- formulae
- balanced equation

RULES FOR WRITING BALANCED EQUATIONS

STEP 1 Write down a word equation.

STEP 2 Underneath, write down the correct formula for each reactant and product.

STEP 3 Balance the equation, if necessary, by putting numbers in front of the formulae.

STEP 4 Add state symbols.

NEVER change the formula of a compound or element to balance the equation.

Question

1 Write balanced symbol equations for these reactions of the alkali metals:

a sodium with water

b potassium with water

c sodium with chlorine

d lithium with chlor

D The halogens

Find out about

- ✔ Group 7 elements
- ✔ halogen molecules
- ✔ similarities and differences between Group 7 elements

Crystals of the mineral fluorite (calcium fluoride).

Salt formers

The Group 7 elements are all very reactive non-metals. They include fluorine, chlorine, bromine, and iodine. Group 7 elements are also known as the **halogens**. The name 'halo-gen' means 'salt-former'. These elements form salts when they combine with metals. Examples include everyday 'salt' itself, which occurs as the mineral halite (NaCl), and fluorite (CaF_2). A type of fluorite called Blue John is found in Derbyshire and is used in jewellery.

The halogens are interesting because of their vigorous chemistry. As elements they are hazardous because they are so reactive. This is also why they are not found existing freely in nature, but occur as compounds with metals.

You must wear eye protection when working with any of the halogens. You should use chlorine and bromine in a fume cupboard because they are **toxic** gases. You must wear chemical-resistant gloves when using liquid bromine, which is **corrosive**. It is not normally possible to study fluorine because it is so dangerously reactive.

Non-metal patterns

Like most non-metals, the halogens are molecular. They each consist of **diatomic** molecules with the atoms joined in pairs: Cl_2. Br_2 and I_2. They have low melting and boiling points because the forces between the molecules are weak, and so it is easy to separate them and turn the halogens into gases. The melting points and boiling points of the halogens increase down the group.

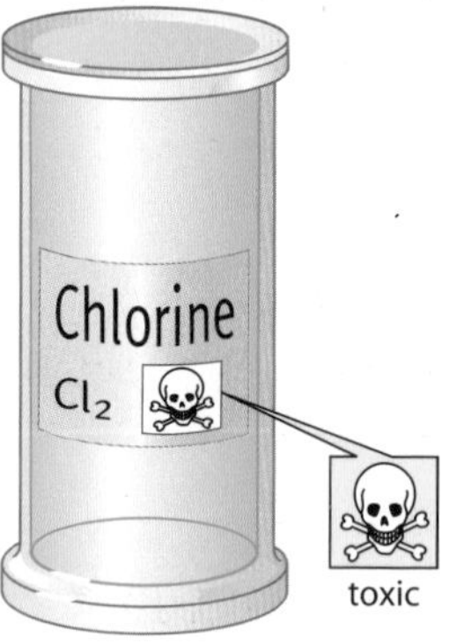

- dense, pale-green gas
- smelly and poisonous
- occurs as chlorides, especially sodium chloride in the sea
- melting point −101 °C
- boiling point −35 °C

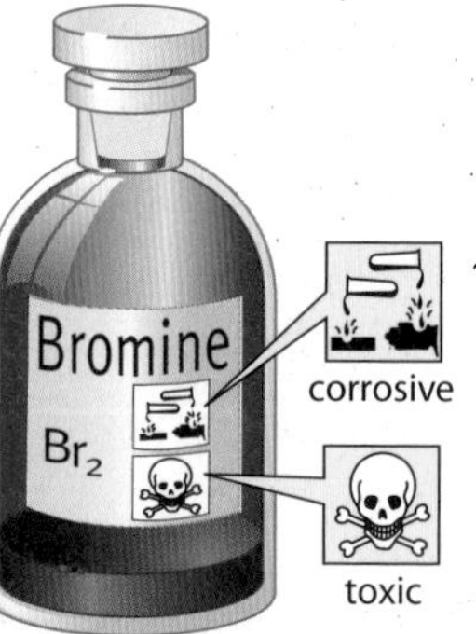

- deep red liquid with red–brown vapour
- smelly and poisonous
- occurs as bromides, especially magnesium bromide in the sea
- melting point −7 °C
- boiling point 59 °C

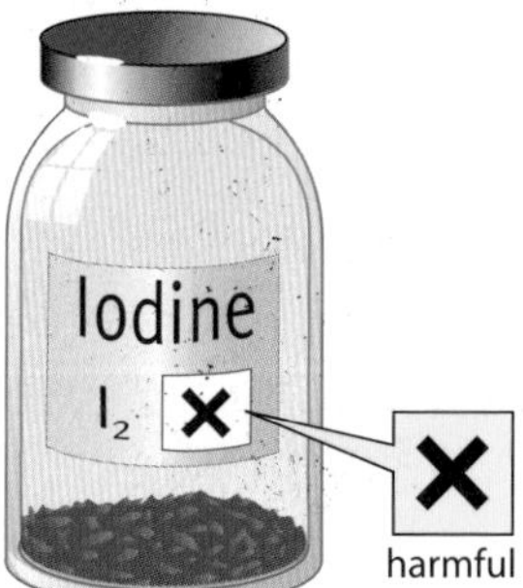

- grey solid with purple vapour
- smelly and harmful
- occurs as iodides and iodates in some rocks and in seaweed
- melting point 114 °C
- boiling point 184 °C

Key words

- ✔ halogens
- ✔ diatomic
- ✔ toxic
- ✔ corrosive
- ✔ bleach
- ✔ displacement reactions
- ✔ harmful

Halogen patterns

All the halogens can harm living things. They can all kill bacteria. Domestic **bleach** is a solution of chlorine in sodium hydroxide, sold to disinfect worktops and toilets. Iodine solution can be used to prevent infection of wounds. These uses illustrate the decreasing reactivity of halogens down the group. Chlorine is much too reactive for its solution to be used to treat wounds.

The reactions of the halogens with an alkali metal such as sodium show the same decrease in reactivity down the group. The reactions with another metal, iron, show the same trend. Hot iron glows brightly in chlorine gas. The product is iron chloride ($FeCl_3$), a rust-brown solid. Iron also glows when heated in bromine vapour, but less brightly. Iron does not even glow when heated in iodine vapour.

The trend in reactivity of the halogens is also shown by **displacement reactions**. For example, if a pale-green solution of chlorine is added to a colourless solution of sodium bromide, the solution immediately turns red because bromine has been formed. Chlorine is more reactive than bromine, so it displaces bromine from its salt, sodium bromide:

$$Cl_2(aq) + 2NaBr(aq) \longrightarrow 2NaCl(aq) + Br_2(aq)$$

In the same way, a solution of bromine will displace iodine from sodium iodide.

Hot iron in a jar of chlorine.

Practical importance

Compounds of the halogens are very important in everyday life.

The chemical industry turns everyday salt (NaCl) into chlorine (and sodium or sodium hydroxide). The chlorine is used to make plastics such as PVC. It is also used in water treatment to stop the spread of diseases. But some chlorine compounds, such as chlorofluorocarbons (CFCs), which were used as refrigerants in the past, have damaged the ozone layer in the upper atmosphere.

Most of the bromine we use comes from the sea. Liquid bromine is extremely corrosive. However, the chemical industry makes important bromine compounds including medical drugs, and pesticides to protect crops.

Traces of iodine compounds are essential in our diet. Where there is little or no natural iodine, it is usual to add potassium iodide to everyday table salt or to drinking water. This prevents disease of the thyroid, a gland in the neck. Iodine and its compounds are starting materials for the manufacture of medicines and dyes.

Questions

1 Write balanced equations for the reactions of:
   - a iron with chlorine to form $FeCl_3$
   - b potassium with bromine
   - c lithium with iodine.

2 Fluorine (F_2) is the first element in Group 7.
   - a Predict the effect of passing a stream of fluorine over iron. Write an equation for the reaction.
   - b What would you expect to see if fluorine gas is passed into a solution of sodium bromide? Write an equation for the reaction.
   - c Astatine is a halogen and is below iodine in the periodic table. Would you expects its melting point to be higher or lower than the melting point of iodine?

The discovery of helium

Find out about

- flame colours
- line spectra
- discovery of the elements

Robert Bunsen (1811–1899), who discovered the flame colours of elements with the help of his new burner.

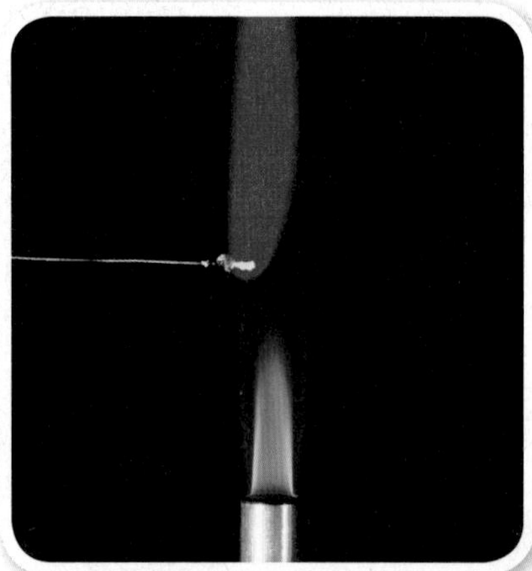

The bright red flame produced by lithium compounds. The compounds of other elements also produce colours in a flame:

- sodium – bright yellow
- potassium – lilac
- calcium – orange–red
- barium – green

A new burner for chemistry

Robert Bunsen moved to the University of Heidelberg in Germany in 1852. Before taking up the job as professor of chemistry, he insisted on having new laboratories. He also demanded gas piping to bring fuel from the gas works, which had just opened to light the city streets.

Existing burners produced smoky and yellow flames. Bunsen wanted something better. In 1855 he invented the type of burner that is still used today in laboratories all over the world.

The great advantage of Bunsen's burner is that it can be adjusted to give an almost invisible flame. Bunsen used his burner to blow glass. He noticed that whenever he held a glass tube in a colourless flame, the flame turned yellow.

Flame colours

Soon Bunsen was experimenting with different chemicals, which he held in the flame at the end of a platinum wire. He found that different chemicals produced characteristic **flame colours**.

Bunsen thought that this might lead to a new method of chemical analysis, but he soon realised that it seemed only to work for pure compounds. It was hard to make any sense of flames from mixtures. So he mentioned his problem to Gustav Kirchhoff, who was the professor of physics.

Flame spectra

'My advice as a physicist,' said Kirchhoff, 'is to look not at the colour of the flames, but at their spectra.'

Kirchhoff built a spectroscope, by putting a glass prism into a wooden box and inserting two telescopes at an angle. Light from a flame entered through one telescope. It was split into a spectrum by the prism and then viewed with the second telescope.

Bunsen and Kirchhoff soon found that each element has its own characteristic spectrum when its light passes through a prism. Each spectrum consists of a set of lines. With their spectroscope, they were able to record the **line spectra** of many elements.

Seen using spectroscopy, cadmium's main lines are red, green, and blue (many fainter lines aren't visible in the photograph). The lines give cadmium its unique fingerprint.

Using **spectroscopy**, Bunsen discovered two new elements in the waters of Durkheim Spa. He based their names on the colours of their spectra. He called them caesium and rubidium, from the Latin for 'sky blue' and 'dark red'.

A Sun element

In 1868 there was a total eclipse of the Sun. Normally, the blinding light from the centre of the Sun makes it impossible to see the much fainter light from the hot gases around the edges of the star. During an eclipse, the Moon hides the whole bright disc of the Sun but not the much fainter light from the hot gases around the edges. This makes it possible to study the light from these gases.

Pierre Janssen, a French astronomer, took very careful observations of the Sun's spectrum during the 1868 eclipse. In the spectrum of the light, he saw a yellow line where no yellow line was expected to be.

Excited by these observations, both Janssen and an English astronomer, Joseph Lockyer, developed new methods to study the light from the Sun's gases. They worked independently, but both came to the same conclusion. There must be an unknown element in the Sun, producing the unexpected yellow line in the spectrum. Janssen and Lockyer published their findings at almost the same time – a coincidence that led to their becoming good friends.

The new element was called 'helium' from the Greek word *helios*, meaning Sun. Both astronomers were still alive in 1895 when William Ramsay, a British chemist, used spectroscopy to discover helium on Earth. The gas came from boiling up a rock containing uranium with acid.

A solar eclipse in 1868 helped scientists to discover helium. During an eclipse it is possible to study the spectra of the light from the hot gases around the edges of the Sun.

WARNING!

Never look directly at the Sun, even during an eclipse. You can damage your eyes or even be blinded!

Key words

- flame colour
- line spectra
- spectroscopy

Questions

1 Why was it important for Bunsen to have a burner with a colourless flame?

2 Why is it not possible to analyse chemical mixtures simply by looking at their flame colours?

3 Use your knowledge of Group 1 chemistry to suggest an explanation for the fact that rubidium and caesium were not discovered until the technique of spectroscopy was developed.

4 In the story of Janssen and Lockyer, identify a statement that is data and a statement that is an explanation of the data.

5 Janssen and Lockyer discovered helium at about the same time. Explain why this meant that other scientists were more likely to accept their findings.

F Atomic structure

Find out about

- atomic theory
- the nuclear model of the atom
- protons, neutrons, and electrons

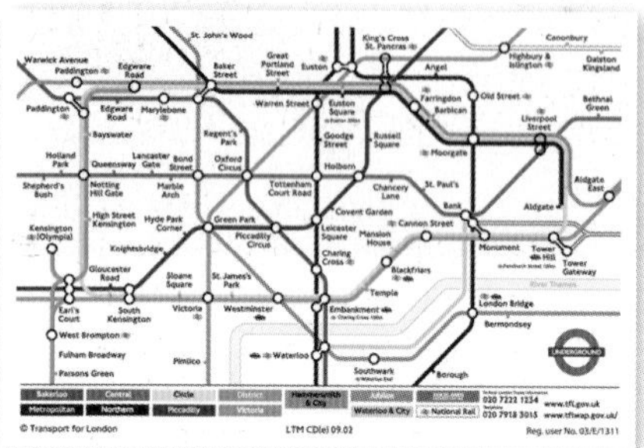

Part of the map of the London Underground.

Atomic models

A picture, or model, of an atom can be used to understand how atoms join together to form compounds and how atoms rearrange during chemical reactions.

Scientists use different models to solve different problems. There is not one model that is 'true'. Each model can represent only a part of what we know about atoms.

It is like using maps to travel through London. The usual map of the Underground is a very useful guide for getting from one tube station to another. It is 'true' in that it shows how the lines and stations connect, but it can't solve all of a traveller's problems. The map doesn't show how the tube stations relate to roads and buildings on the surface. For that you need a street map.

Atomic models from 1800 to the present. The diameter of an atom is about ten million times smaller than a millimetre. These diagrams are distorted. For atoms at this scale, the nuclei would be invisibly small.

1804
Dalton's solid atom.

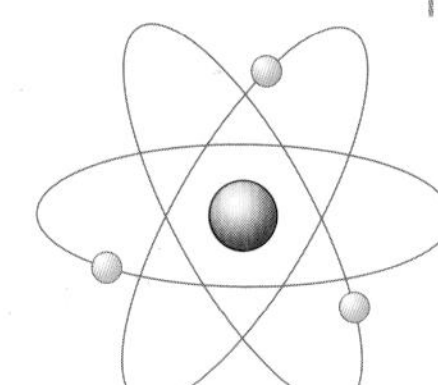

1913
The Bohr–Rutherford 'Solar System' atom, in which electrons orbit round a very small nucleus.

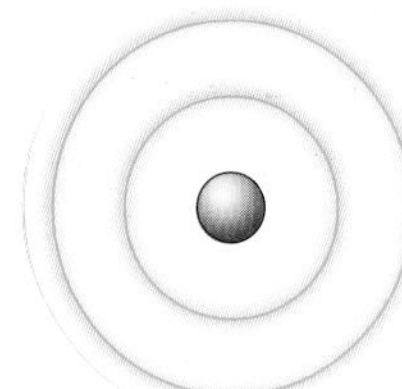

1924
A model of the atom in which the electrons are no longer treated as particles but pictured as occupying energy levels, which give rise to regions of negative charge around the nucleus (charge clouds).

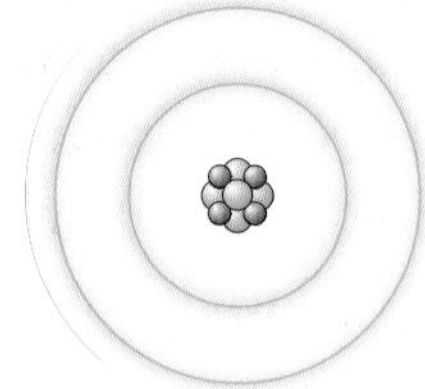

1932
The atom in which the nucleus is built up from neutrons as well as protons.

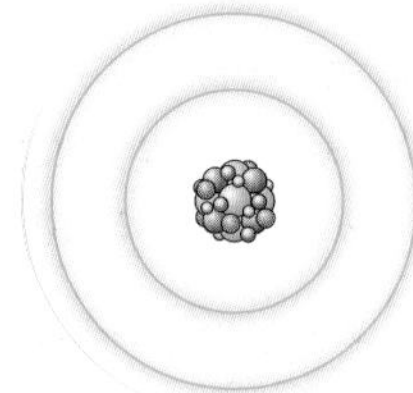

2000+
The present-day atom in which the nucleus is built up from many kinds of particles.

Dalton's atomic theory

The story of our modern thinking about atomic structure began with John Dalton. In 1803 John Dalton was studying mixtures of gases and their properties. He did not just summarise the data he collected, but used his creativity to think up an explanation that would account for the data. In Dalton's theory, everything is made of atoms that cannot be broken down. The very word 'atom' means 'indivisible'.

The main ideas in Dalton's theory still apply to chemistry today. So far as chemistry is concerned, each element does have its own kind of atom, and the atoms of different elements differ in mass. The idea that

Key words

- nucleus
- protons
- neutrons
- electrons
- proton number
- sub-atomic particles

equations must balance is based on Dalton's view that atoms are not created or destroyed during chemical changes.

Even so, Dalton's theory is limited. It cannot explain the pattern of elements in the periodic table. Nor can it explain how atoms join together in elements and compounds. A better explanation was needed that would account for most, if not all, the data available.

The Large Hadron Collider at CERN in Switzerland is designed to collide beams of protons at very high energy. It was first switched on in 2008. After fixing some initial problems, CERN engineers have gradually increased the proton energy. When it is fully working, it will probe more deeply into matter than ever before.

Inside the atom

In time it became clear that atoms are not solid, indivisible spheres. From the middle of the nineteenth century, scientists began to find ways of exploring the insides of atoms. Still today, scientists are spending vast sums of money to build particle accelerators (atom-smashers), which work at higher and higher energies. They hope to discover more about the fine structure of atoms.

It is possible to explain much more about the chemistry of elements and compounds with the help of a model of atomic structure that includes **sub-atomic particles**, the particles that make up atoms.

A model for chemistry

In your study of chemistry you will be using an atomic model that dates back to 1932, when James Chadwick discovered the neutron. In this model the mass of the atom is concentrated in a tiny, central **nucleus**. The nucleus consists of **protons** and **neutrons**. The protons have a positive electric charge. Neutrons are uncharged.

Around the nucleus are the **electrons**. The electrons are negatively charged. The mass of an electron is so small that it can often be ignored. In an atom, the number of electrons equals the number of protons in the nucleus (the **proton number**). This means that the total negative charge equals the total positive charge, and overall an atom is uncharged.

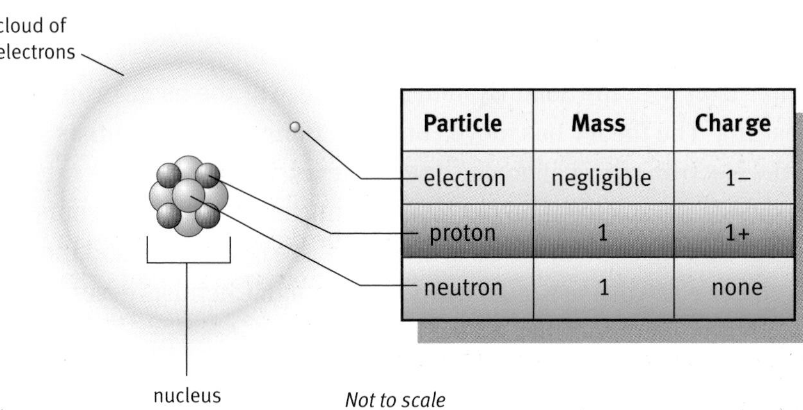

Particle	Mass	Charge
electron	negligible	1−
proton	1	1+
neutron	1	none

All atoms consist of these three basic particles. The nucleus of an atom is very, very small. The diameter of an atom is about ten million million times greater than the diameter of its nucleus.

Question

1 With the help of the periodic table on page 47 work out:

a the element with one more proton in its nucleus than a chlorine atom

b the element with one proton fewer in its nucleus than a neon atom

c the number of protons in a sodium atom

d the number of electrons in a bromine atom

e the size of the positive charge on the nucleus of a fluorine atom

f the total negative charge on the electrons in a potassium atom.

Find out about

- evidence for energy levels
- electrons in shells
- electron arrangements

Electrons in orbits

In 1913, the Danish scientist Niels Bohr came up with an explanation for the line spectra from atoms. He made a close study of the spectrum from hydrogen.

In the Bohr model for atoms, the electrons orbit the nucleus as the planets orbit the Sun. Bohr's idea was that heating atoms gives them energy. This forces the electrons to move to higher-energy orbits further from the nucleus. These electrons then drop back from outer orbits to inner orbits. They give out light energy as they do so. Each energy jump corresponds to a particular colour in the spectrum. The bigger the jump, the nearer the line to the blue end of the spectrum. Only certain energy jumps are possible, so the spectrum consists of a series of lines.

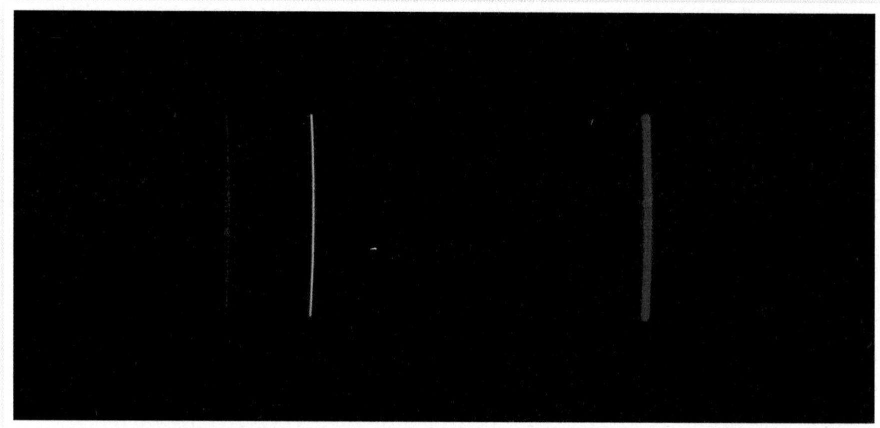

The line spectrum of hydrogen. Atomic theory can explain why this spectrum is a series of lines.

Bohr was able to use his theory to calculate sizes of the energy jumps. He could then deduce the energy levels of electrons in the various orbits.

Electrons in shells

The comparison with the Solar System and the use of the term 'orbit' can be misleading. The theory has moved on since Bohr's time. Scientists still picture the electrons at a particular **energy level**. However, in the modern theory the electrons do not orbit the nucleus like planets round the Sun. All that theory can tell us is that there are regions around the nucleus where electrons are most likely to be found. Chemists describe these regions as 'clouds' of negative charge.

Think of each electron cloud as a **shell** around the nucleus. Each shell is one of the regions in space where there can be electrons. The shells only exist if there are electrons in them. Electrons in the same shell have the same energy.

Key words

- energy level
- shell
- electron arrangement

Electron arrangements

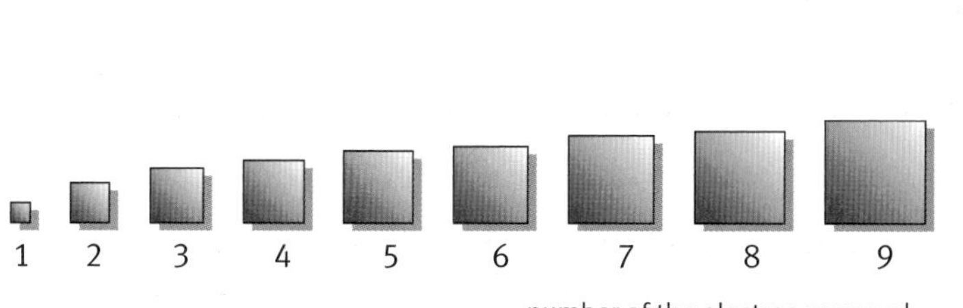

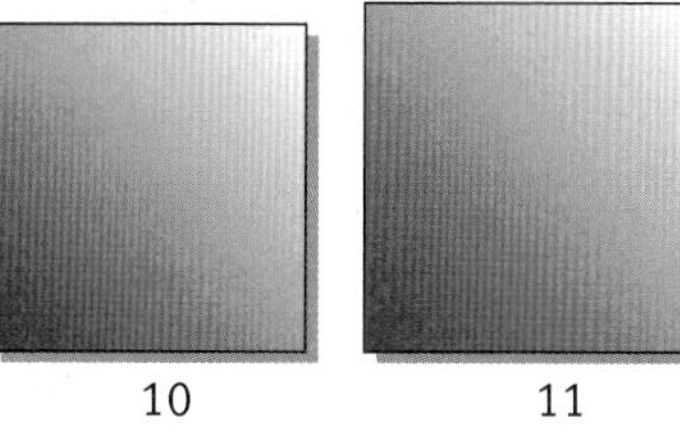

The areas of the squares are in proportion to the amount of energy needed to remove the electrons one by one from a sodium atom.

Each electron shell can contain only a limited number of electrons. The innermost shell with the lowest energy fills first. When it is full, the electrons go into the next shell. Evidence for this theory comes not only from spectra, but also from measurements of the energy needed to remove electrons from atoms.

There are 11 electrons in a sodium atom. Scientists have measured the quantities of energy needed to remove these electrons one by one. The values are represented by the areas of the squares in the picture above. It is quite easy to remove the first electron. The next eight are more difficult to remove. Finally it becomes really hard to remove the last two electrons, which are held very powerfully because they are in the shell closest to the nucleus.

This supports the idea that the electrons in a sodium atom are arranged in three shells, as shown in the diagram on the right. The diagram shows common representations of the **electron arrangement** of the element.

The first shell that is closest to the nucleus can hold up to two electrons. The second shell can hold eight. Once the second shell holds eight electrons, the third shell starts to fill.

If there are more electrons, they occupy further shells. After the first 20 elements, the arrangements become increasingly complex as the shells hold more electrons and the energy differences between shells get smaller.

Na
electron (e^-)
nucleus
electron shells

This diagram can be abbreviated to:

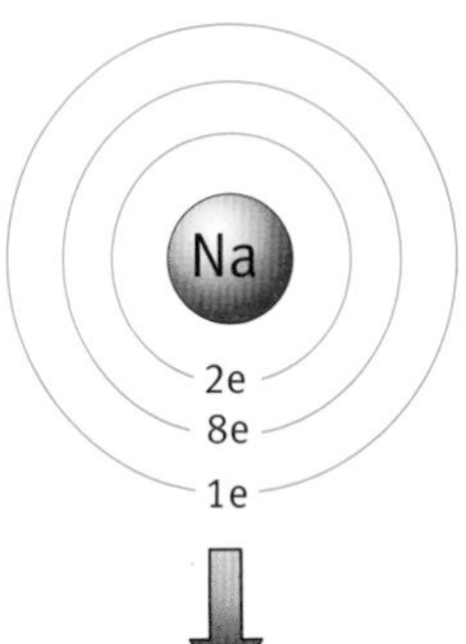

or even more simply to:
Na: 2e.8e.1e or 2.8.1

Two-dimensional representations of the electrons in shells in a sodium atom.

Questions

1. Draw diagrams to show the electrons in shells for these atoms:
   a beryllium **b** oxygen **c** magnesium.
   Refer to the periodic table on page 47 for the proton numbers, and therefore the number of electrons, in each atom.
2. How does the diagram at the top of this page support the electron arrangements of a sodium atom shown in the diagram on the right?

H Electrons and the periodic table

Find out about

- atomic structure and periods
- electron arrangements and groups
- explaining similarities and differences between the elements

The periodic table then and now

Scientists discovered electrons in 1897, nearly 30 years after Mendeleev published his first periodic table. Mendeleev knew nothing about atomic structure, and he used the relative masses of atoms to put the elements in order.

A modern periodic table shows the elements in order of proton number, which is also the number of electrons in an atom. One convincing piece of evidence for the 'shell model' of atomic structure is that it can help to explain the patterns in the periodic table.

Periods

The diagram below shows the connection between the horizontal rows of the periodic table and the structure of atoms. From one atom to the next, the proton number increases by one and the number of electrons increases by one. So the electron shells fill up progressively from one atom to the next.

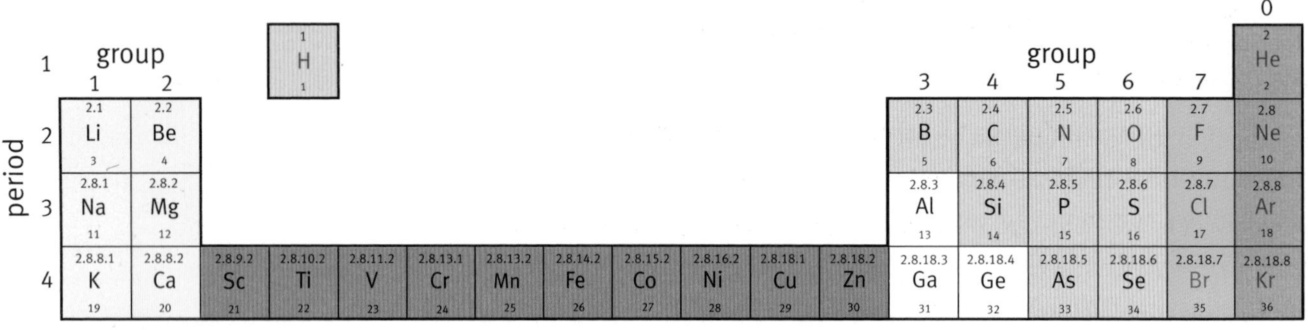

Key

2.4 — number of electrons in each shell
C — symbol
6 — proton number

Electron arrangements for the first 20 elements in the periodic table.

The first period, from hydrogen to helium, corresponds to filling the first shell. The second shell fills across the second period, from lithium (2.1) to neon (2.8). Eight electrons go into the third shell, from sodium (2.8.1) to argon (2.8.8), and then the fourth shell starts to fill from potassium to calcium.

In fact the third shell can hold up to 18 electrons. This shell is completed from scandium to zinc, before the fourth shell continues to fill from gallium to krypton. This accounts for the appearance of the block of transition metals in the middle of the table. Why this happens cannot be explained by the simple theory described here. You will find out the explanation if you go on to a more advanced chemistry course.

Groups

When atoms react, it is the electrons in their outer shells that get involved as chemical bonds break and new chemicals form. It turns out that elements have similar properties if they have the same number and arrangement of electrons in the outer shells of their atoms.

Three of the alkali metals appear in the diagram on the right. You can see that they each have one electron in the outer shell of their atoms. This is the case for the other alkali metals too. This helps to account for the similarities in their chemical properties.

The alkali metals are not all the same because their atoms differ in the number of inner full shells. A sodium atom has two inner filled shells, so it is larger than a lithium atom, and its outer electron is further away from the nucleus. As a result, the two metals have similar, but not identical, physical and chemical properties.

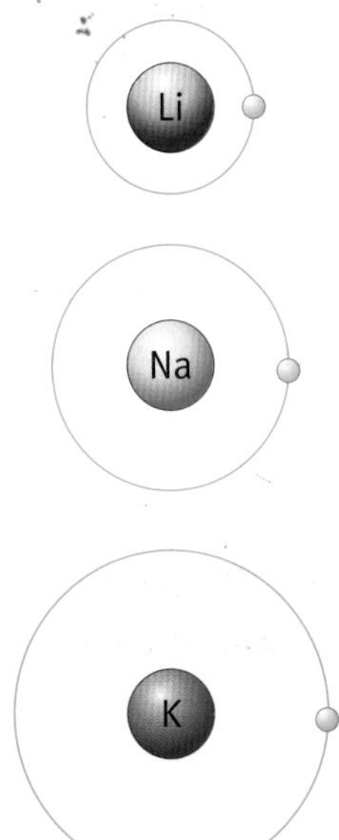

The trend in the size of the atoms of Group 1 elements reflects the increasing number of full, inner electron shells down the group. Only the outer shells are shown here.

Metals and non-metals

Elements with only one or two electrons in the outer shell are metals, with the exception of hydrogen and helium. Elements with more electrons in the outer shell are generally non-metals, though there are exceptions to this, such as aluminium, tin, and lead. The halogens are non-metal elements with seven electrons in the outer shell.

At the end of each period there is a noble gas. This is a group of very unreactive elements. The first member of the group is helium.

The term 'noble' has been used by alchemists and chemists for hundreds of years to describe elements that will not react easily. The chemical nobility stand apart from the hurly-burly of everyday reactions.

Questions

1 Explain the meaning of this statement: the electron arrangement of chlorine is (2.8.7).

2 a What are the electron arrangements of the elements beryllium, magnesium, and calcium?

b In which group of the periodic table do these three elements appear?

c Are these elements metals or non-metals?

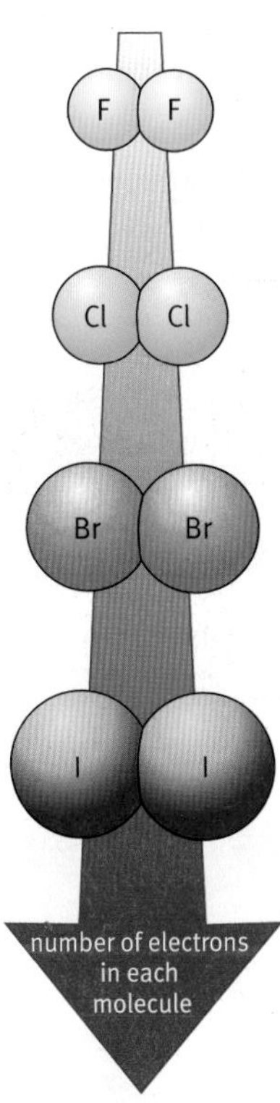

The trend in the size of the molecules of Group 7 elements reflects the increasing number of full, inner electron shells in the atoms down the group.

Find out about

- salts
- properties of salts
- electricity and salts

Why are salts so different from their elements?

Compounds of metals with non-metals are salts. Chemists can explain the differences between a salt and its elements by studying what happens to the atoms and molecules as they react. A good example is the reaction between two very reactive elements to make the everyday table salt you can safely sprinkle on food.

A chemical reaction in pictures: sodium and chlorine react to make sodium chloride.

Sodium chloride crystals. Sodium chloride is soluble in water. The chemical industry uses an electric current to convert sodium chloride solution into chlorine, hydrogen, and sodium hydroxide.

Crystals of the mineral galena, which is an ore of lead. Galena consists of insoluble lead sulfide.

Crystals of the mineral pyrite. Pyrite consists of insoluble iron sulfide.

Salts

Salts such as sodium chloride are crystalline. The crystals of sodium chloride are shaped like cubes. So are the crystals of calcium fluoride shown in Section D.

Salts have much higher melting and boiling points than chemicals such as chlorine and bromine, which are made up of small molecules.

Chemical	Formula	Melting point (°C)	Boiling point (°C)
sodium	Na	98	890
chlorine	Cl_2	−101	−34
sodium chloride	NaCl	808	1465
potassium	K	63	766
bromine	Br_2	−7	58
potassium bromide	KBr	730	1435

Sodium chloride is an example of a salt that is soluble in water. There are many other examples of soluble salts, including most of the compounds of alkali metals with halogens.

Some salts are insoluble in water. Lithium fluoride is an example of a salt that is only very slightly soluble in water. Many minerals consist of insoluble salts. Fluorite (CaF_2) is one example. Others are galena (PbS) and the brassy-looking pyrite (FeS_2), sometimes called fool's gold.

Molten salts and electricity

The apparatus on the right is used to investigate whether or not chemicals conduct electricity. The crucible contains some white powdered solid. This is zinc chloride.

At first there is no reading on the meter, showing that the solid does not conduct electricity. This is true of all compounds of metals with non-metals; they do not conduct electricity when solid.

Heating the crucible melts the zinc chloride. As soon as the compound is **molten**, there is a reading on the meter. This shows that a current is flowing round the circuit. As a liquid, the compound is a conductor. That is not all. The electric current causes the compound to decompose chemically. The most obvious change is the bubbling around the positive electrode. The gas produced is chlorine.

After a while, it is possible to show that zinc has formed at the negative electrode. This is done by switching off the current, allowing the crucible to cool, and adding pure water to dissolve any remaining zinc chloride. Shiny pieces of zinc metal remain. So the electric current splits the compound into its elements: zinc and chlorine.

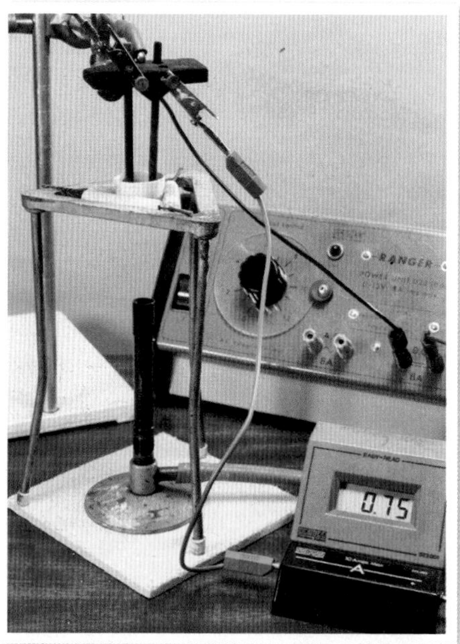

The crucible contains zinc chloride. The carbon rods dipping into the crucible are the electrodes. A current begins to flow in the circuit when the zinc chloride is hot enough to melt.

Salts in solution and electricity

Soluble salts also conduct electricity. This can be studied using the apparatus shown on the right. There are changes at the electrodes when an electric current flows.

The presence of water has an effect on the chemicals produced when a salt solution conducts electricity. The products are not always the same as the elements in the compound.

This apparatus is used to study the changes at the electrodes when a solution of a salt conducts electricity. In this example, the flow of an electric current is producing gases at the electrodes.

Questions

1 Draw up a table to compare the physical properties of sodium, chlorine, and sodium chloride.

2 Refer to the table of data on the left. Which of the chemicals is a liquid:
  a at room temperature?
  b at the boiling point of water?
  c at 1000 °C?

3 Draw a two-dimensional line diagram and circuit diagram to represent the apparatus used to show that zinc chloride conducts electricity when hot enough to melt.

Key word

✔ molten

Ionic theory

Find out about

- ions
- ionic compounds
- explaining properties of salts

Michael Faraday lectured at the Royal Institution. He started the Christmas lectures, which continue today in the same lecture theatre.

Key words

- ions
- electrolysis

Questions

1 What big idea occurred to Faraday that enabled him to explain how solutions of salts and molten salts conduct electricity?

2 Why was it important for Faraday to communicate his ideas to other scientists?

Electrolysis

An electric current can split a salt into its elements, or other products, if it is molten or dissolved in water. This process is called **electrolysis**. The term 'electro-lysis' is based on two Greek words that mean 'electricity-splitting'.

The discovery of electrolysis was very important in the history of chemistry. Electrolysis made it possible to split up compounds, which previously no-one could decompose. In 1807 and 1808, an English chemist, Humphry Davy, used electrolysis to isolate for the first time the elements potassium, sodium, barium, strontium, calcium, and magnesium.

Faraday's theory

Michael Faraday worked with Humphry Davy. He began as an assistant, then established himself as a leading scientist in his own right. In 1833 he began to study the effects of electricity on chemicals. He had to think creatively to come up with an explanation that would account for his observations.

Faraday decided that compounds that can be decomposed by electrolysis must contain electrically charged particles. Since opposite electrical charges attract each other, he could imagine the negative electrode attracting positively charged particles and the positive electrode attracting negatively charged particles.

The charged particles move towards the electrodes. When they reach the electrodes, they turn back into atoms. This accounts for the chemical changes that decompose a compound during electrolysis.

Faraday named the moving, charged particles **ions**, from a Greek word meaning 'to go'.

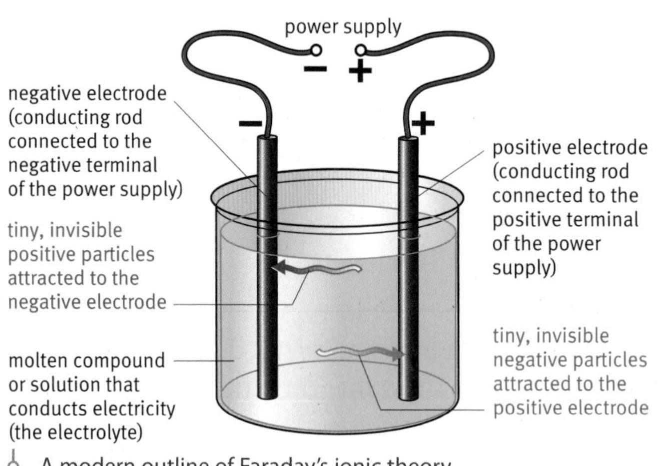

A modern outline of Faraday's ionic theory.

Explaining electrolysis

Chemists continue to use ionic theory to explain electrolysis. According to the theory, salts such as sodium chloride consist of ions.

Sodium chloride is made up of sodium ions and chloride ions. Sodium ions, Na^+, are positively charged. The chloride ions, Cl^-, carry a negative charge. These oppositely charged ions attract each other.

A crystal of sodium chloride consists of millions and millions of Na^+ and Cl^- ions closely packed together. In the solid, these ions cannot move towards the electrodes, and so the compound cannot conduct electricity. The ions can move when sodium chloride is hot enough to melt or when it is dissolved in water.

During electrolysis, the negative electrode attracts the **positive ions**. The positive electrode attracts the **negative ions**. When the ions reach the electrodes, they lose their charges and turn back into atoms.

Metals form positive ions, and non-metals generally form negative ions.

Elements and compounds

Ionic theory can help to explain why compounds are so different from their elements. Sodium atoms and chlorine molecules are dangerously reactive. Sodium chloride is safe because its ions are much less reactive.

Questions

3 Why do solid compounds made of ions not conduct electricity?

4 Chemists sometimes call the negative electrode the cathode. Cations are the ions that move towards the cathode. What is the charge on a cation? Which type of element forms cations? Give an example of a cation.

5 Chemists sometimes call the positive electrode the anode. Anions are the ions that move towards the anode. What is the charge on an anion? Which type of element forms anions? Give an example of an anion.

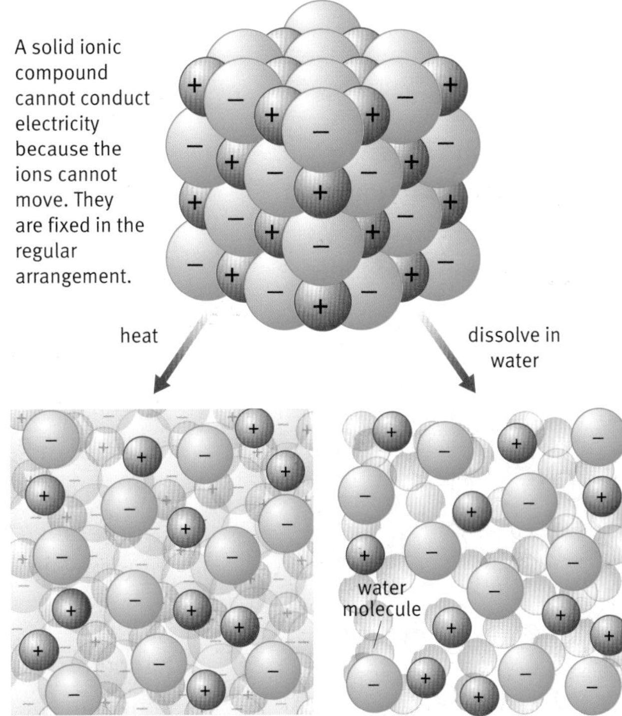

When an ionic compound is heated strongly, the ions move so much that they can no longer stay in the regular arrangement. The solid melts. Because the ions can now move around independently, the molten compound conducts electricity.

When an ionic compound has dissolved, it can conduct electricity because its ions can move independently among the water molecules.

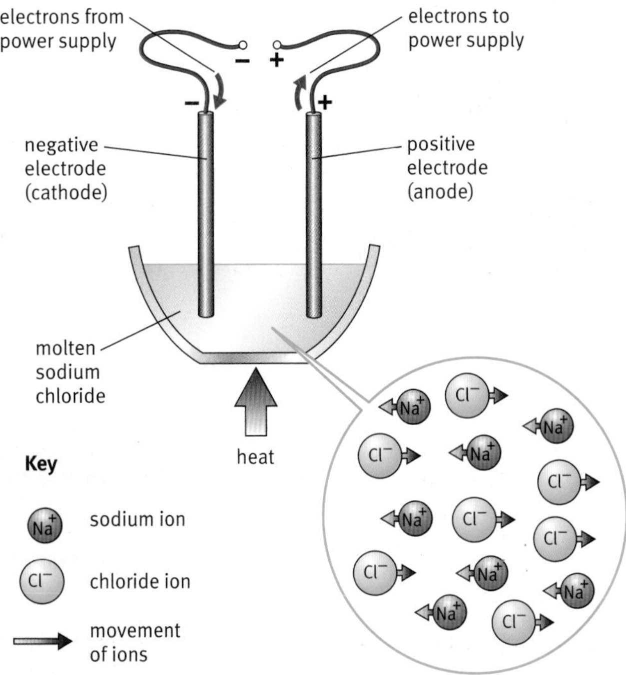

Sodium chloride conducts when molten because its ions can move towards the electrodes.

Ionic theory and atomic structure

Find out about

- atoms and ions
- electron arrangements of ions
- the formulae of ionic compounds

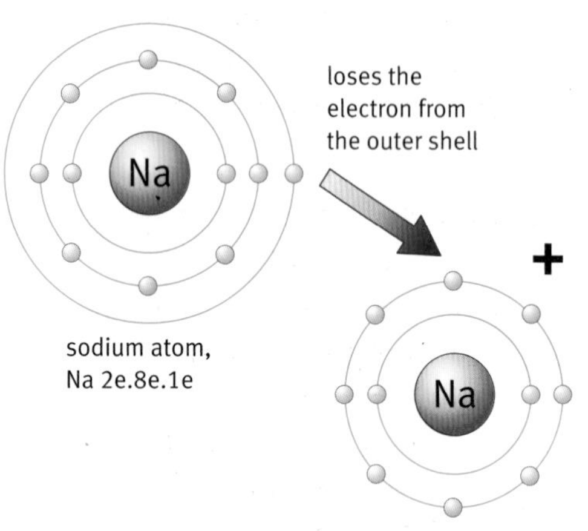

A sodium atom turns into a positive ion when it loses a negatively charged electron.

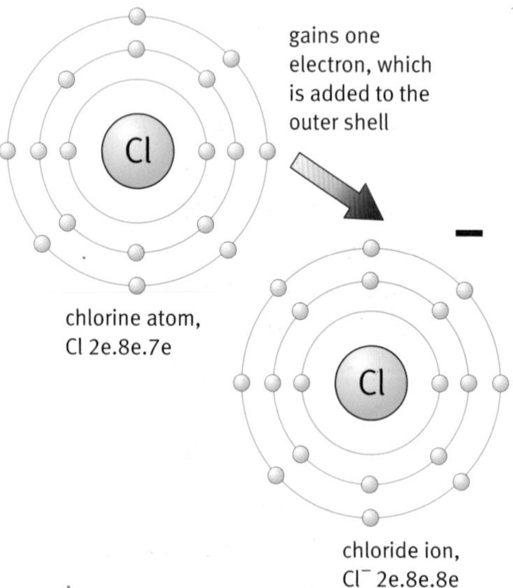

A chlorine atom turns into a negative ion by gaining an extra negatively charged electron.

Atoms into ions

Faraday could not explain how atoms turn into ions because he was working long before anyone knew anything about the details of atomic structure. Today, chemists can use the shell model for electrons in atoms to show how atoms become electrically charged.

The metals on the left-hand side of the periodic table form ions by losing the few electrons in their outer shell. This leaves more protons than electrons, and so the ions are positively charged.

All the metals in Group 1 have one electron in the outer shell. The diagram in Section G shows that removing the first electron from a sodium atom needs relatively little energy. The same is true for the other Group 1 metals, so they all form ions with a 1+ charge: Li^+, Na^+, and K^+, for example.

Chlorine gas consists of Cl_2 molecules. But it is easier to see what happens when chloride ions form by looking at one atom at a time, as shown in the diagram on the left. As each chlorine atom turns into an ion, it gains one electron and becomes negatively charged, Cl^-.

Electron arrangements of ions

Notice that when sodium and chlorine atoms turn into ions, they end up with the same electron arrangement as the nearest noble gas in the periodic table. This is generally true for simple ions of the first 20 or so elements in the periodic table. An explanation of why this is so involves an analysis of the energy changes that occur when metals react with non-metals. This is something you will study if you go on to a more advanced chemistry course.

Ions into atoms

Electrolysis turns ions back into atoms. Metal ions are positively charged, so they are attracted to the negative electrode. It is a flow of electrons from the battery into this electrode that makes it negative. Positive metal ions gain electrons from the negative electrode and turn back into atoms.

Non-metal ions are negatively charged, so they are attracted to the positive electrode. This electrode is positive because electrons flow out of it to the battery. Negative ions give up electrons to the positive electrode and turn back into atoms.

Formulae of ionic compounds

The formula of sodium chloride is NaCl because there is one sodium ion (Na^+) for every chloride ion (Cl^-). There are no molecules in everday table salt, only ions.

Not all ions have single positive or negative charges like sodium and chlorine. The formula of lead bromide is $PbBr_2$. In this compound there are two bromide ions for every lead ion. All compounds are overall electrically neutral, so the charge on a lead ion must be twice that on a bromide ion. A bromide ion, like a chloride ion, has a single negative charge (Br^-), so a lead ion must have a double positive charge (Pb^{2+}).

Compound	Ions present		Formula
	Positive	Negative	
magnesium oxide	Mg^{2+}	O^{2-}	MgO
calcium chloride	Ca^{2+}	Cl^- Cl^-	$CaCl_2$
aluminium oxide	Al^{3+} Al^{3+}	O^{2-} O^{2-} O^{2-}	Al_2O_3

Examples of formulae of ionic compounds.

Ions in the periodic table

The charges of simple ions show a periodic pattern. You can see this from the diagram below, showing the ionic symbols in the periodic table. Many of the transition metals in the middle block of the table can form more than one type of ion. Iron, for example, can form Fe^{2+} and Fe^{3+} ions, while copper can exist as Cu^+ and Cu^{2+} ions. Why this should be so is something you will study if you go on to a more advanced chemistry course.

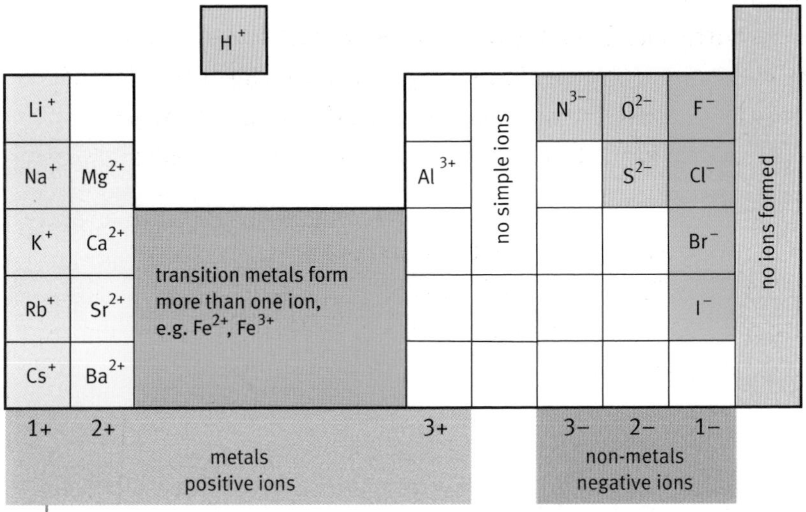

H^+
Li^+
Na^+ Mg^{2+}
K^+ Ca^{2+}
Rb^+ Sr^{2+}
Cs^+ Ba^{2+}
transition metals form more than one ion, e.g. Fe^{2+}, Fe^{3+}
Al^{3+}
no simple ions
N^{3-} O^{2-} F^-
S^{2-} Cl^-
Br^-
I^-
no ions formed
1+ 2+ 3+
metals positive ions
3− 2− 1−
non-metals negative ions

Simple ions in the periodic table.

Questions

1 Draw diagrams to show the number and arrangement of electrons in a lithium atom and in a lithium ion. What is the charge on a lithium ion?

2 Draw diagrams to show the number and arrangement of electrons in a fluorine atom and in a fluoride ion. What is the charge on a fluoride ion?

3 Write down the electron arrangements of:
   a a fluoride ion (nucleus with 9 protons and 10 neutrons)
   b a neon atom (nucleus with 10 protons and 10 neutrons)
   c a sodium ion (nucleus with 11 protons and 12 neutrons).

   In what ways are a fluoride ion, a neon atom, and a sodium ion the same? How do they differ?

4 With the help of the table of ions, work out the formulae of these ionic compounds:
   a potassium iodide
   b calcium bromide
   c aluminium chloride
   d magnesium nitride
   e aluminium sulfide.

5 Use the table of ions to work out the charge on the following ions:
   a copper (Cu) in $CuCl_2$
   b zinc (Zn) in ZnO
   c iron (Fe) in Fe_2O_3.

Chemical species

Find out about

- atoms, molecules, and ions
- chemical species

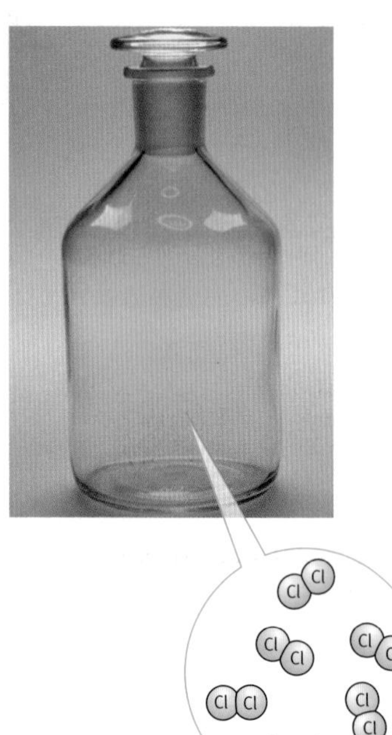

Chlorine gas consists of chlorine molecules. Chlorine molecules are chemically very reactive.

In this module you have met the idea that the same element can take different chemical forms with distinct properties. Chemists describe these different forms as **chemical species**.

Species of chlorine

Chlorine has three simple species: atom, molecule, and ion. Each of these species of chlorine has distinct properties. Chlorine atoms (Cl) do not normally exist on their own. They rapidly pair up to form chlorine molecules (Cl_2). However, ultraviolet radiation can split chlorine (and chlorine compounds) into atoms. This is what happens to CFCs such as CCl_3F when they get into the upper atmosphere. In the full glare of the Sun's radiation, the molecules break up into atoms. The highly reactive free chlorine atoms rapidly destroy ozone. Lowering the concentration of ozone creates the so-called 'hole' in the ozone layer.

Chlorine gas at room temperature consists of chlorine molecules (Cl_2). These are very reactive, as illustrated by the chemistry of the alkali metals and halogens described in Section D. The chlorine molecules are reactive enough to do damage to human tissues, so the gas is given the label 'toxic'.

Chloride ions are quite different. They occur in compounds such as sodium chloride and magnesium chloride. Chloride ions in these salts are essential to life and occur in all living tissues. Chloride ions are chemically active in many ways, but they are not as reactive and harmful as the atoms or molecules of the element.

Polar stratospheric cloud formations over the Arctic (seen as thin orange and brown layers). Ice particles in these clouds provide a surface for the chemical reactions that release reactive chlorine atoms, which can then destroy ozone.

There are more complex species of chlorine, with the element joined to other atoms. This includes molecules that contain chlorine and other elements, such as tetrachloromethane (CCl_4).

Chlorine dioxide is oxidising, toxic, and corrosive. As a liquid it can be explosive. Its properties are different to those of chlorine, which is toxic, and oxygen, which is oxidising.

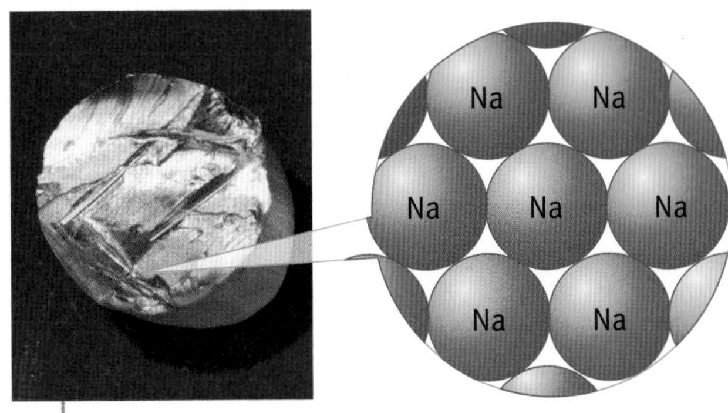

Sodium metal consists of sodium atoms. Sodium atoms are chemically very reactive.

Species of sodium

There are only two species of sodium: atom and ion. The atoms in sodium metal are chemically very reactive. In the presence of other chemicals, the sodium atoms react to produce compounds containing sodium ions.

Sodium combines with chlorine to produce the ionic compound sodium chloride. This is made up of two chemical species: Na^+ and Cl^-. These two ions are quite unreactive. Sodium chloride is soluble in water, but its solution has a neutral pH. Water does not react with the ions.

When sodium reacts with water, it produces another ionic compound: sodium hydroxide, containing Na^+ and OH^-. The sodium hydroxide dissolves in the water to give a solution that is very alkaline. It is the hydroxide ions that make a solution of sodium hydroxide alkaline, not the sodium ions.

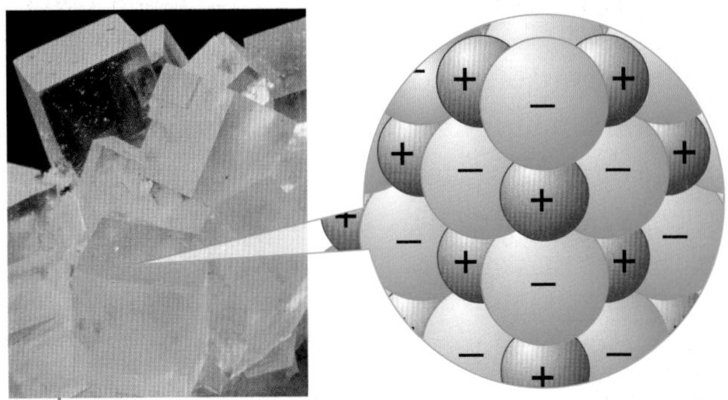

Sodium chloride consists of sodium ions and chloride ions. These ions are not very reactive.

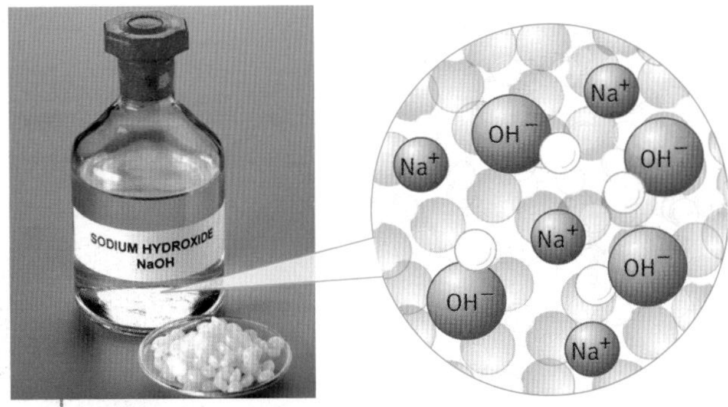

Sodium hydroxide is a strong alkali. It consists of sodium ions and hydroxide ions. In solution, the ions move around separately, mixed with water molecules.

Questions

1 Use the idea of chemical species to explain why the properties of sodium chloride are very different from the properties of its elements sodium and chlorine.

2 Give the name and formulae of all chemical species in:
  a potassium
  b bromine
  c potassium bromide.

3 a Identify four distinct chemical species in unpolluted air, giving their names and formulae.
  b Identify three more chemical species present in the polluted air of a busy city street.

Science Explanations

Chemists have identified patterns and come up with theories to make sense of the world, and to explain how roughly 100 elements can give rise to such a huge variety of chemical compounds.

You should know:

- that the chemists' model of the atom has a tiny central nucleus, containing protons and neutrons, which is surrounded by electrons
- that the number of electrons is equal to the number of protons and that the electrons in an atom have definite energies
- how the arrangement of electrons in shells is determined by the way that the electron shell with the lowest energy fills first until full, then the next shell starts to fill, and so on
- that the chemistry of an element is largely determined by the number and arrangement of the electrons in its atoms
- that elements in the periodic table are lined up in order of their proton numbers so that there are repeating patterns across each row (period) in the table
- that each column in the periodic table consists of a group of related elements that have similar chemistries because they have the same number of electrons in their outer shells
- that there are trends in the properties of the elements down a group because of the increasing number of inner full shells
- that Group 1 elements are the alkali metals, which react with moist air, water, and chlorine, becoming more reactive down the group
- that Group 7 elements are the halogens made up of diatomic molecules
- why there is a trend in the physical state of the halogens at room temperature
- that the halogens become less reactive down the group
- that chemists use word equations and symbol equations to describe reactions
- why safety precautions are important when working with hazardous chemicals such as Group 1 metals, alkalis, and the halogens
- how ionic compounds form when metals react with non-metals such that metal atoms lose electrons while the non-metal atoms gain electrons
- how a regular lattice of ions gives rise to the shape of the crystals of an ionic compound
- that the properties of an ionic compound are the properties of its ions, which behave in a different way to the atoms or molecules of its elements
- that ionic compounds conduct electricity when molten or when dissolved in water, because the ions are charged and they are able to move around in the liquid.

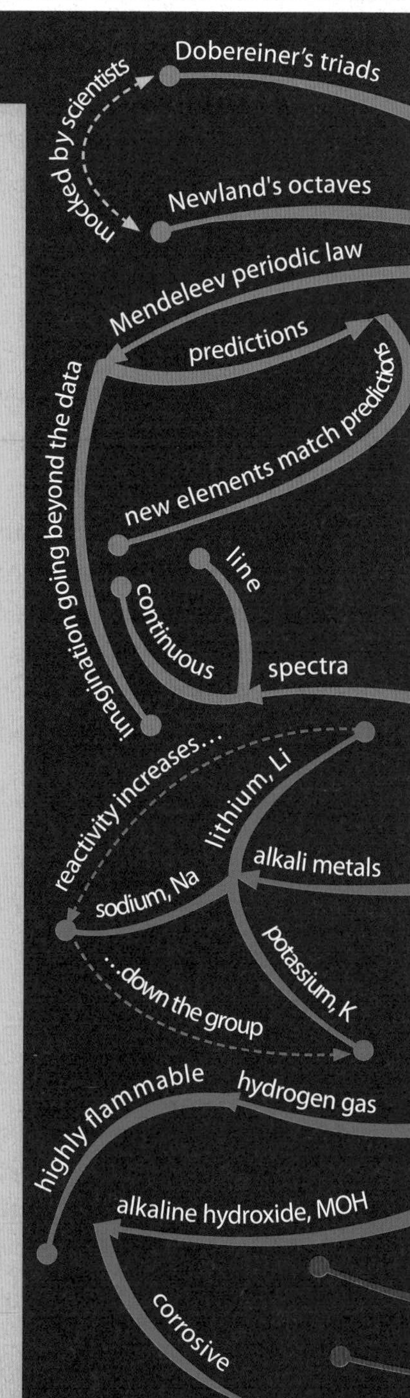

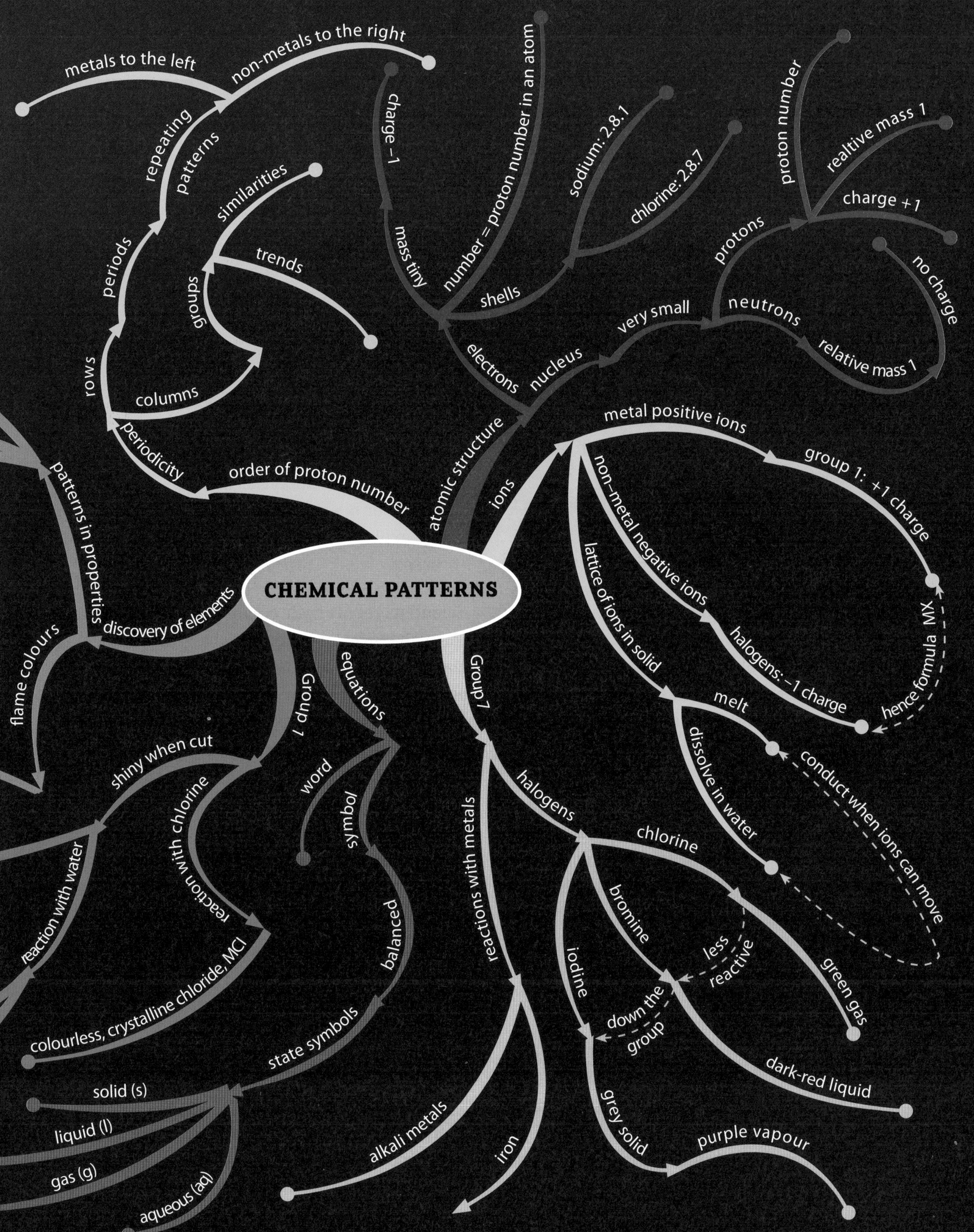
CHEMICAL PATTERNS
order of proton number
periodicity
rows
periods
repeating patterns
metals to the left
non-metals to the right
columns
groups
similarities
trends
atomic structure
electrons
mass tiny
charge –1
number = proton number in an atom
shells
sodium: 2.8.1
chlorine: 2.8.7
nucleus
very small
protons
proton number
realtive mass 1
charge +1
neutrons
no charge
relative mass 1
ions
metal positive ions
group 1: +1 charge
non-metal negative ions
halogens: –1 charge
hence formula MX
lattice of ions in solid
melt
dissolve in water
conduct when ions can move
Group 7
halogens
chlorine
green gas
bromine
dark-red liquid
iodine
grey solid
purple vapour
less reactive
down the group
reactions with metals
alkali metals
iron
equations
word
symbol
balanced
state symbols
solid (s)
liquid (l)
gas (g)
aqueous (aq)
Group 1
shiny when cut
reaction with chlorine
colourless, crystalline chloride, MCl
reaction with water
discovery of elements
patterns in properties
flame colours

Ideas about Science

Scientific explanations are based on data but they go beyond the data and are distinct from them. An explanation has to be thought up creatively to account for the data. A new explanation may explain a range of phenomena not previously thought to be linked. The explanation should also allow predictions to be made about new situations or examples.

In the context of the discovery of new elements and the development of the periodic table you should be able to:

- give an account of scientific work and distinguish statements that report data from statements of explanatory ideas (hypotheses, explanations, and theories)
- recognise that an explanation may be incorrect even if the data is correct
- identify where creative thinking is involved in the development of an explanation, such as Mendeleev's insight that he had to leave gaps for undiscovered elements
- recognise data or observations that are accounted for by (or conflict with) an explanation, such as the data from spectra that could be explained by the existence of new elements
- give good reasons for accepting or rejecting a proposed scientific explanation, such as the way that the shell model of atomic structure accounts for the arrangement of elements in the periodic table
- identify the best of given scientific explanations for a phenomenon, such as Mendeleev's use of his periodic table to predict the existence of unknown elements
- understand that when a prediction agrees with an observation this increases confidence in the explanation on which the prediction is based, but does not prove it is correct. For example, Mendeleev correctly predicted the properties of missing elements in his periodic table

- understand that when a prediction disagrees with an observation this indicates that one or the other is wrong and decreases confidence in the explanation on which the prediction is based.

Scientists report their claims to other scientists through conferences and journals. Scientific claims are only accepted once they have been evaluated critically by other scientists. Scientists are usually sceptical about claims that cannot be repeated by anyone else and about unexpected findings until they have been replicated (by themselves) or reproduced (by someone else). You should be able to:

- broadly outline the peer review process, in which new scientific claims are evaluated by other scientists
- recognise that new scientific claims that have not yet been evaluated by the scientific community are less reliable than well-established ones; an example is the lack of acceptance of early attempts to find connections between the chemical properties of the elements and their relative atomic mass
- identify the fact that a finding has not been reproduced by another scientist as a reason for questioning a scientific claim.

Review Questions

1 A teacher reacts sodium with chlorine. The product is sodium chloride, NaCl.

a The sodium bottle has two hazard symbols.

i Give the meaning of each symbol.

ii List two safety precautions the teacher must take when using sodium. Link each to a hazard.

b Write a balanced symbol equation for the reaction of sodium with chlorine, including state symbols.

c Sodium chloride is made up of two types of ion.

i Give the formulae of the two ions.

ii State the number of protons and electrons in each ion.

iii Draw diagrams to show the electron arrangements of each ion.

2 The table shows part of the periodic table, with only a few symbols included.

period \ group	1	2	3	4	5	6	7	0
1								He
2						O	F	
3	Na	Mg	Al	Si		S	Cl	
4				Ge			Br	
5								
6	Cs							

a Using only the elements in the table, write down symbols for:

i a metal that floats on water

ii an element with similar properties to silicon (Si)

iii three elements that have molecules made up of two atoms

iv the most reactive metal.

b Predict the state of fluorine, F, at room temperature. Give a reason for your prediction.

3 The table shows the properties of three halogens.

Halogen	Boiling point (°C)	Formula of compound with iron
chlorine	–34.7	$FeCl_3$
bromine	58.8	$FeBr_3$
iodine	184	

a **i** Describe the trend in boiling points.

ii Predict the formula of iron iodide.

iii The formula of a bromide ion is Br^-. Work out the charge on the iron ion.

b Describe the pattern of reactivity of the halogens going down the group. Give examples to illustrate your answer. Use ideas about electron arrangement and the formation of ions to help you explain the pattern.

c Give an example to show how the model of atomic structure with electrons in shells can explain the similarities between the halogens.

P4 Explaining motion

Why study motion?

Humans have always been interested in how things move and why they move the way they do. Motion is such an obvious part of our everyday lives that we cannot really claim to know very much about the natural world if we cannot explain and predict how objects move.

What you already know

- Speed is calculated by the distance covered divided by time.
- An unbalanced force changes the motion of an object.
- The weight of an object is due to the gravitational force between the Earth and the object.
- The air resistance on a moving object depends on its shape and its speed.

Find out about

- forces always arising from an interaction between two objects
- friction and reaction of surfaces
- instantaneous and average speed, velocity, and acceleration
- the idea of momentum, and how the momentum of an object changes when a force acts on it
- everyday examples of motion, including the principles on which traffic-safety measures are based
- gravitational potential energy and kinetic energy.

The Science

One tantalising thought has always driven people who have studied motion – is it possible that every example of motion we observe can be explained by a few simple rules (or laws) that apply to everything? Remarkably the answer is 'yes'. And these laws are so exact and precise that they can be used to predict the motion of an object very accurately.

Ideas about Science

A scientific explanation cannot be deduced by just looking at the data – it needs someone to think creatively to explain the observations. Many people had tried to describe and explain how things move. To write his laws of motion Isaac Newton built on the ideas of those who had come before. But making the link between an apple falling on the ground and the Moon orbiting the earth required a leap of imagination.

A Forces and interactions

Find out about

- ✓ how forces arise when two objects interact
- ✓ contact and action-at-a-distance forces

What makes things move the way they do? And what makes things stop, or stay still? To start something moving, we have to push or pull it. To stop a moving object, we have to exert a **force** on it, against its direction of motion.

Where do forces come from?

Look at the photograph of a firework rocket exploding. The burning sparks move rapidly out from the centre, where the chemicals in the rocket have exploded. Notice that the starburst is symmetrical. For every moving spark, another spark moves in the exact opposite direction. This tells us something very important about forces. They always come in pairs.

The chemical reaction inside the firework produces forces that send the burning fragments out equally in all directions, producing a sphere of sparks.

Let's consider a simpler example, where there are two moving objects. Sophie and Sam stand in the centre of an ice rink. What happens if Sophie gives Sam a gentle push?

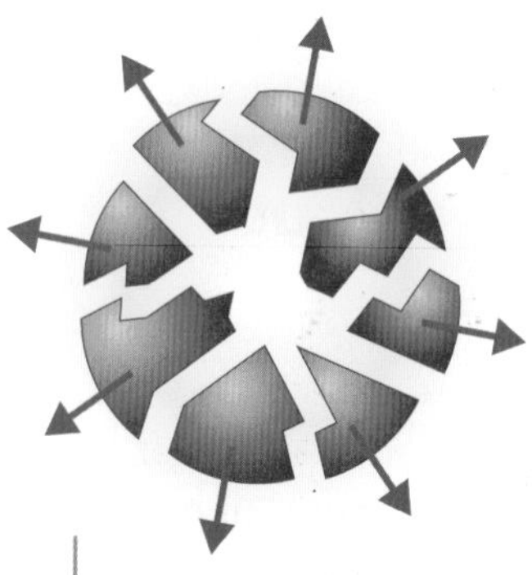

An exploding firework rocket. For every moving spark, there is another spark moving in exactly the opposite direction.

The answer is that both of them will move. When Sophie exerts a force on Sam by pushing him, she experiences a force herself. Sam exerts a force on Sophie – not by pushing but just by being there as an obstacle to push against. The same is true if they stand a distance apart holding a rope. If either one pulls, both will start to move together.

This tells us something very important about forces:

- Forces always arise from an **interaction** between two objects. So forces always come in pairs. The two forces in an **interaction pair** are:
  - equal in size
  - opposite in direction.

This is always true. And it does not depend on the size or strength of the two people involved. Another important thing to notice is that

- the two forces act on different objects.

In this example, one force of the pair acts on Sam and the other on Sophie.

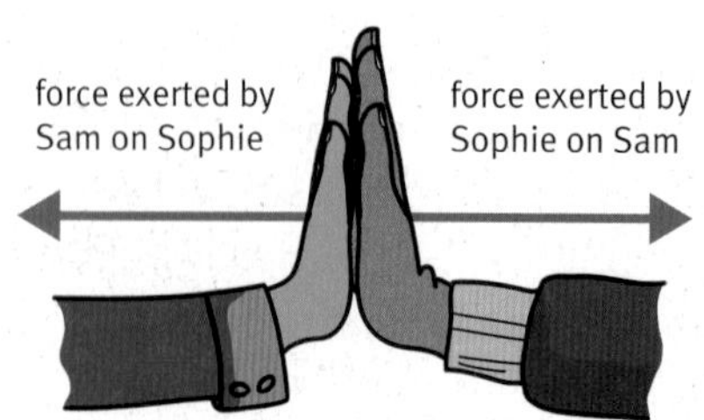

Forces always arise in pairs. Here Sophie pushes Sam and also experiences a force herself in return.

Two kinds of interaction

Forces arise from interactions between pairs of objects. When they are caused by two objects touching, we call them **contact forces**. Contact forces exist only while the objects are actually touching. As soon as the objects separate, the forces stop. The two objects may, of course, keep moving. We will come back to this later in this chapter.

There is a second kind of interaction between objects, called **action-at-a-distance**. One common example is magnetism. The two ring magnets in the diagram are repelling each other. Both threads are at an angle – because both magnets experience a force. The same is true when magnets attract. If you hold a fridge magnet close to the door, the magnet experiences a force towards the door – and the fridge experiences a force towards the magnet.

Both the fridge magnet and the door experience a force. Action-at-a-distance forces always come in pairs.

Gravity is another example of action-at-a-distance. An apple falls from a tree because of the force exerted on it by the Earth, pulling it downwards. But gravity is an attraction between two objects. The apple also exerts an equal and opposite force on the Earth! This does not have any visible effect, however, because the Earth is so massive.

Unlike contact forces, action-at-a distance forces act all the time, even when the two interacting objects are apart. They get weaker as the distance between the objects increases.

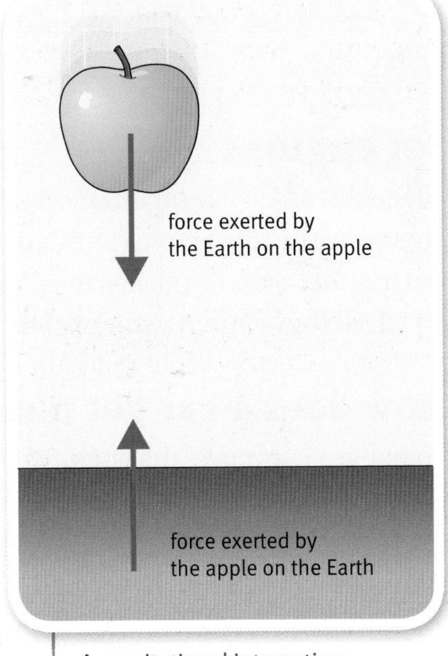

A gravitational interaction – again forces always arise in pairs.

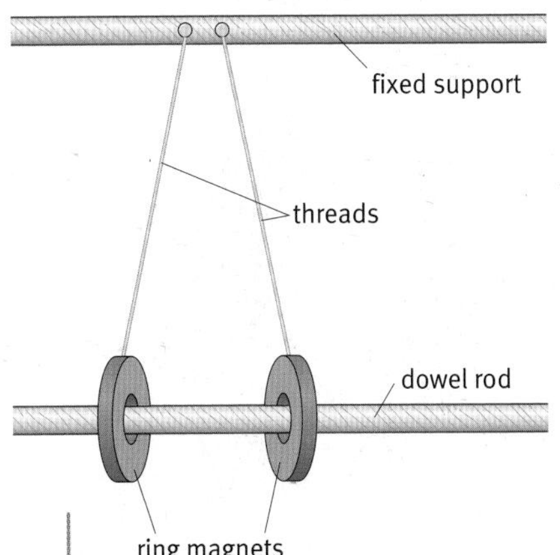

These two ring magnets are repelling each other. Notice that both magnets are being pushed aside.

Questions

1. List four examples of interaction pairs of forces mentioned on these pages.
2. What three things are always true about interaction pairs?
3. How could you modify the apparatus in the top diagram to show that *attraction* forces between magnets also arise in pairs? Sketch how you would set it up, and write down what you would expect to see.

Key words

- **force**
- **interaction**
- **interaction pair**
- **contact forces**
- **action at a distance**

Getting moving

Find out about

- the forces that enable people and vehicles to get moving
- rockets and jet engines

The start of the longest journey humans have made so far – to the Moon. A huge force is needed to push a rocket like this upwards. It is provided by the hot exhaust gases, which are formed by burning the fuel.

To get something moving, an interaction pair of forces is always involved. Here are some examples.

Rockets

As its fuel burns, a rocket pushes out hot gases from its base. The rocket exerts a large force downwards on these gases. The other half of this interaction pair is the force exerted on the rocket by the escaping gases. This pushes the rocket upwards.

The photograph on the left shows one of the most famous rocket launches: the *Apollo 11* mission to land the first humans on the Moon. The interaction pair of forces is shown on the photo.

Rockets carry with them everything they need to make the burning gases they push out. This means that they can work anywhere, including empty space.

Jet engines

Jet engines also use an interaction pair of forces. Air is drawn into the engine and pushed out at high speed from the back. The other force of the interaction pair pushes the engine forward. Jet engines need to draw air in, so they cannot work in space.

How does a car get moving?

To make a car move, the engine has to make the wheels turn. This causes a forward force on the car. To understand how, think first about a car trying to start on ice. If the ice is very slippery, the wheels will just spin. The car will not move at all. The spinning wheels produce no forward force on the car. Now imagine a car on a muddy track. The rally car below is throwing up a shower of mud as it tries to get going.

As it rotates, the wheels exert a force backwards on the ground – with dramatic results in this case!

You can see that there is an interaction between the wheels and the ground. The wheels are causing a backwards force on the ground surface. This makes the mud fly backwards. Mud, however, moves when the force is quite small. The other force of the interaction pair is the forward force on the car. It is equal in size. So it is also small – and not big enough to get the car moving.

Now imagine a good surface and good tyres, which do not slip. Again, the engine makes the wheel turn. It pushes back on the road. Friction between the wheel and the road surface stops it slipping so it exerts a very large force backwards on the road surface. So the other force of the interaction pair is the same size. This large forward force gets the car moving.

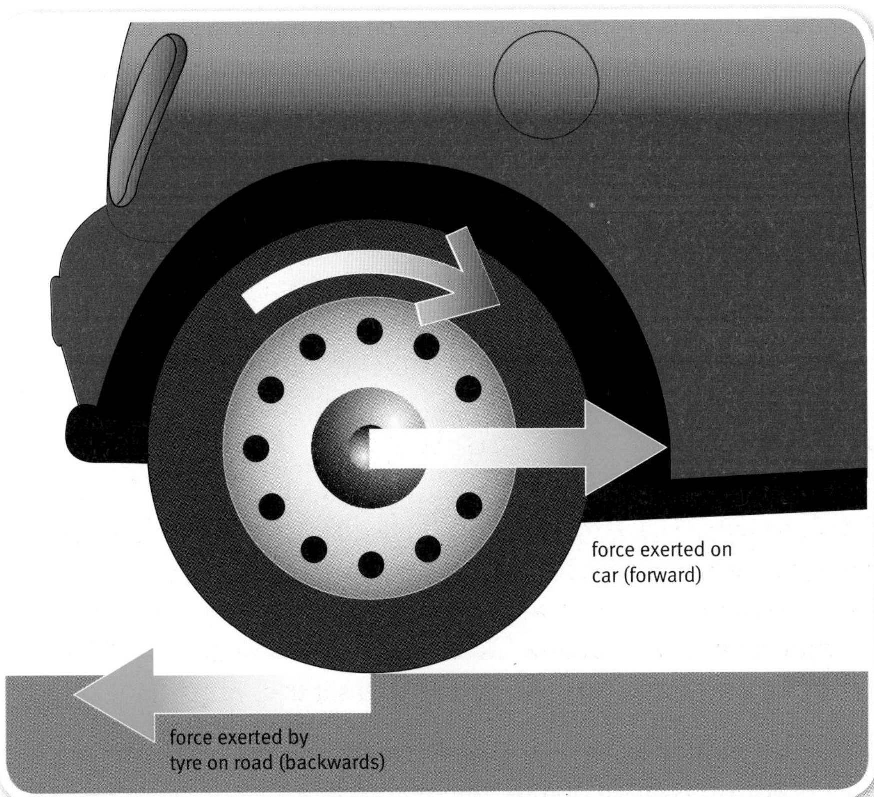

If the tyre grips the road and does not spin, the second force of the interaction pair results in a large forward force on the axle. This pushes the car forward.

Walking

When you walk, you push back on the ground with each foot in turn. The ground then pushes you forward. You are not usually aware of this. When you walk across a floor, it does not feel as though you are pushing backwards on it. You only become aware of the importance of this interaction when the surface is slippery – for example, when you try to walk on an icy surface, where there is not enough friction. Because you cannot push it back, it is unable to move you forward.

Questions

1 Jet engines are suitable for aircraft but not for travel in space. Explain why. How do rockets overcome this problem?

2 A boat propeller pushes water backwards when it spins round. Use the ideas on these pages to write a short paragraph explaining how this makes the boat move forward. Draw a diagram and label the main forces involved, to illustrate your explanation.

3 Sketch a matchstick figure walking. Mark and label the interaction pair of forces on the foot in contact with the ground.

Icy surfaces are difficult to walk on. Your foot cannot get a grip to push back on the surface, so the surface does not push you forwards.

C Friction

Find out about

- friction and what causes it
- how to add the forces acting on an object

Friction is the interaction between two surfaces when they slide over each other – or when they are pushed to try to make them slide over each other. There is a friction force on *both* objects involved. As always, the two forces of the interaction pair are equal in size and opposite in direction. The friction force on each object acts in the direction that would help to prevent it sliding over the other.

Friction enables cars and people to get moving. If friction didn't exist, walking would be impossible and wheeled vehicles could not move. But what causes friction and how does it work?

The bumps and hollows on each surface cause a force, as one object slides (or tries to slide) across the other.

What causes friction?

Friction is caused by the roughness of the sliding surfaces. Even surfaces that seem smooth have quite large humps and hollows if you look at them under a microscope.

So, at a microscopic level, two 'smooth' surfaces in contact are not really touching everywhere. And some of the bumps on one surface will fit into hollows on the other. As one slides across the other, these bumps collide. This causes a force that resists the sliding.

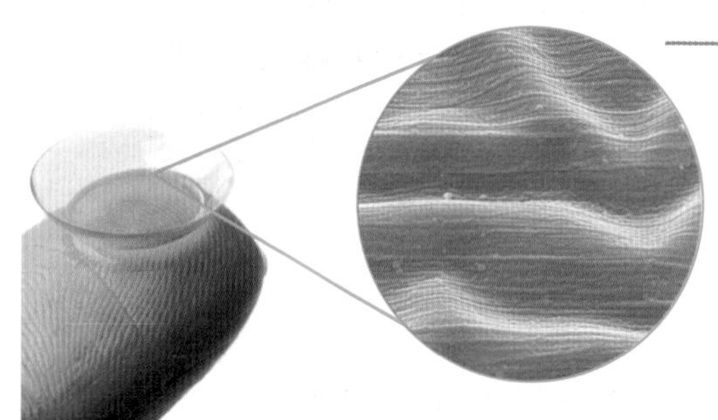

Even surfaces that appear smooth are really quite rough. At the microscopic level they have humps and hollows. The photograph on the right shows the surface of a contact lens magnified 1080 times.

Friction between an object and the floor results in a pair of opposite forces. The arrows show the direction of the forces exerted on the foot and the floor. In reality, the bumps and hollows in the two surfaces are very much smaller.

When you walk, it's as if your shoes have corrugated soles – and the floor is also corrugated – on the microscopic level. The diagram on the left exaggerates this in order to show the effect more clearly. Pushing back with your foot will cause a backwards force on the ground, and a forwards force on your foot.

Questions

1 List three everyday situations in which we try to reduce friction, and three where we try to make friction as large as possible.

2 Use the ideas on these pages to write short explanations of the following observations:

a We can reduce the friction between two surfaces by putting oil on them.

b It is easier to push a box across the floor when it is empty than when it is full.

Friction – a responsive force

What size is the friction force between two surfaces, and what does it depend on? Think about the forces involved as Jeff tries to push a large box along a level floor.

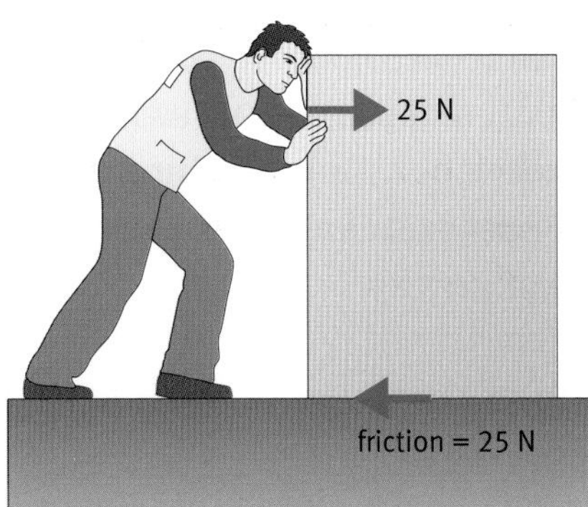

1 Jeff pushes the box with a force of 25 N, to try to slide it along. It does not move. The friction force exerted by the floor on the box is 25 N. This exactly balances Jeff's push.

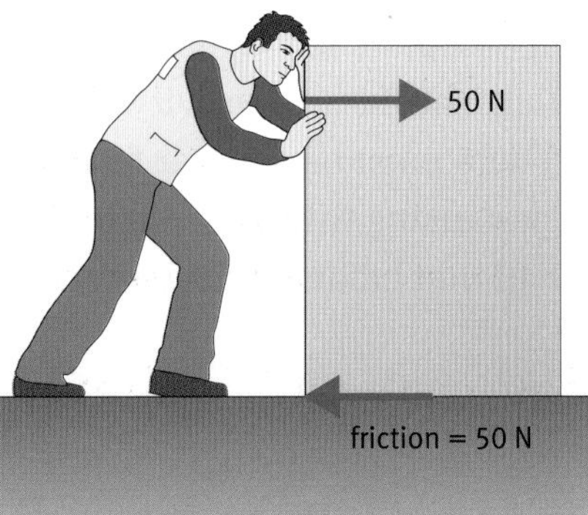

2 Jeff then pushes harder, with a force of 50 N. The box still does not move. The friction force exerted by the floor on the box is now 50 N. Again, this balances Jeff's push.

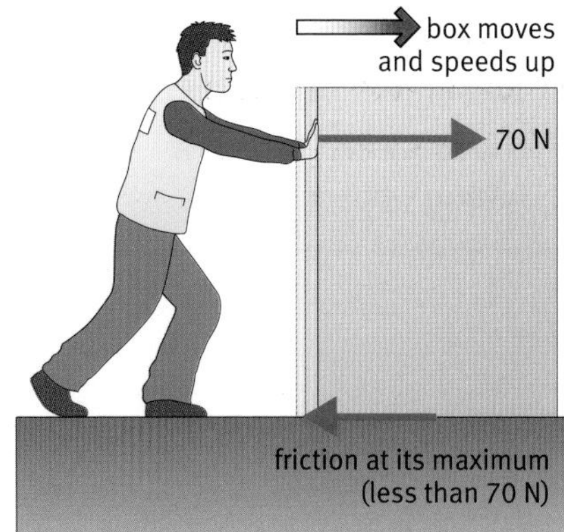

3 Jeff pushes harder still, exerting a force of 70 N. The box starts to move. 70 N is bigger than the maximum friction force for this box and floor surface.

The size of the friction force depends on the size of the external force applied by Jeff, up to a certain limit. If Jeff's push is below this limit, friction matches it exactly. The two forces cancel each other out and the box doesn't move. But if Jeff's force is above this limit, the box will move. So what determines this limit? It depends on the weight of the box, and on the roughness of the two surfaces in contact.

Adding forces

The discussion above used an idea that may seem obvious:

- If there is a force acting on an object, but it is not moving, then there must be another force balancing (or cancelling out) the first one.

If the forces acting on an object balance each other, we say they add to zero. Adding several forces that act on the same object is straightforward. But you must take the direction of each force into account. The sum of all the forces acting on an object is called the **resultant force**. The diagrams on the right show some examples.

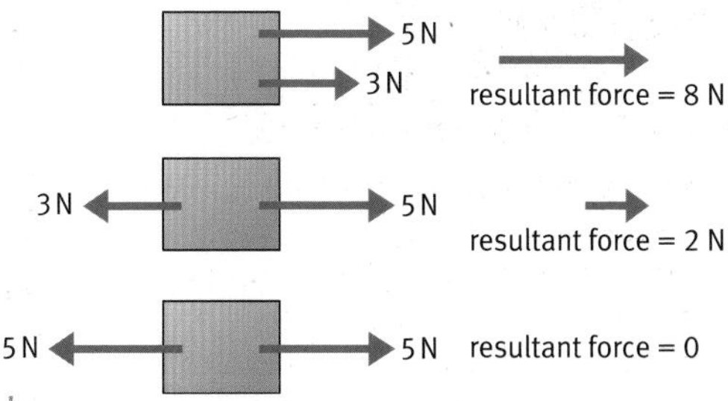

To find the resultant force acting on an object, you add the separate forces. You must take account of their directions.

Question

3 Sketch the first diagram of the workman, Jeff, pushing the box. Mark and label the forces acting *on Jeff* as he pushes. Use the length of each force arrow to indicate the size of the force.

Key words

- friction
- resultant force

D Vertical forces

Find out about

- how a surface exerts a reaction force on any object that presses on it
- the forces acting on a falling object

If you hold a tennis ball at arm's length and let it go, it immediately starts to move downwards. There is a force acting on the tennis ball. This force is the pull exerted on it by the Earth. It is due to the interaction known as gravity.

But if you put a tennis ball on a table so that it does not roll about, it does not fall. The force of gravity has not suddenly stopped or been switched off. There must be another force that cancels it out. The only thing that can be causing this is the table. The table must exert an upward force on the ball that balances the downward force of gravity.

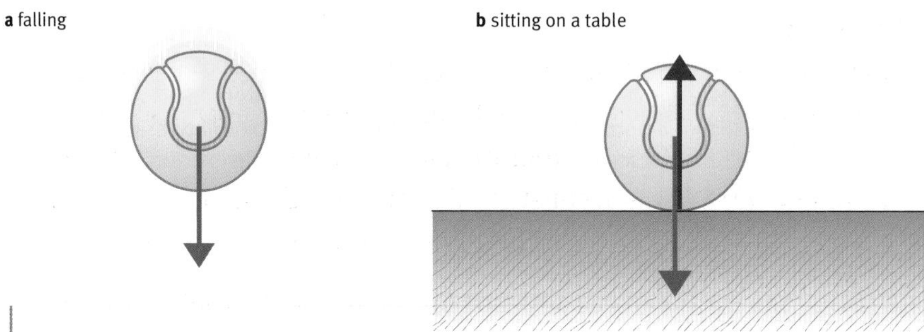

The forces acting on a tennis ball **a** falling and **b** sitting on a table.

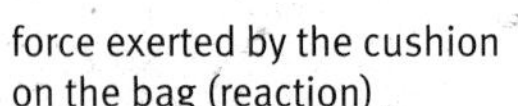

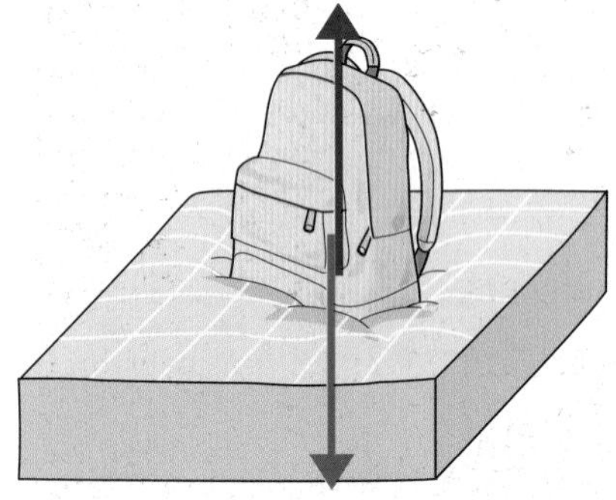

The bag squeezes the foam until the upward force of the springy foam on the bag exactly balances the downward gravity force on the bag.

How can a table exert a force?

Although it may seem strange, tables can and do exert forces. To understand how, imagine an object, like a school bag, sitting on the foam cushion of a sofa. The bag presses down on the foam, squashing it a bit. Because foam is springy, it then pushes upwards on the bag, just like a spring. Like a spring, the more it is squeezed, the harder it pushes back. So the bag sinks into the foam until the push of the foam on it exactly balances the downward pull of gravity on it.

The same thing happens, though on a much smaller scale, when the bag sits on a table top. A table top is not so easily squeezed as a foam cushion. But it *can* be squashed. This is not visible to the naked eye, however. We call this upward force that a surface exerts when something presses on it the **reaction** of the surface.

The size of the reaction force depends on the downward force that is causing it. The surface distorts just enough to make the reaction force balance the downward force on the object due to gravity. The resultant force on it is zero.

There is, of course, a limit to this. If the downward force exerted on a table is bigger than it can take, the table top will break! But up to this limit, the reaction matches the downward force.

Walls can push too!

Any surface can exert a reaction force, not just horizontal ones. The diagram on the right shows Deborah, a roller-skater. When she pushes on the wall, it pushes back on her. She immediately starts to move backwards. Deborah's push squashes the wall where her hands touch it. Although we cannot see any distortion, this part of the wall is compressed. Like a spring, it exerts an equal force back on Deborah's hands.

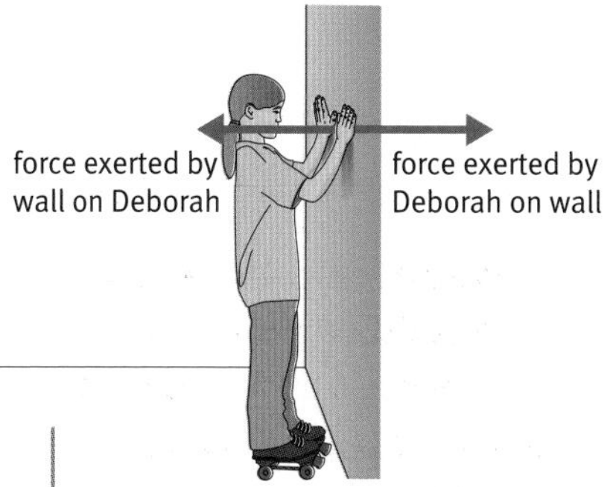

Deborah pushes against the wall. The other force, in the interaction pair, starts her moving away from the wall.

Freefall

So how does a tennis ball actually move, when it is falling? If we look closely, we see that it picks up speed steadily from the moment it is released. The multi-flash photograph on the right shows this. The time interval between flashes is constant. As the ball falls, it moves further between one flash and the next. Its speed increases. It accelerates.

If we drop an object that is light relative to its size, like a cupcake case, we notice that something different happens. It speeds up at first, then falls at a steady speed. Why the difference? The reason is that **air resistance** has a bigger effect on it. Air resistance is a force that arises when anything moves through the air. The faster an object moves, the bigger the air resistance force on it becomes. A falling light object quickly reaches the speed at which the air resistance force on it balances the force of gravity. It then continues to fall at a steady speed.

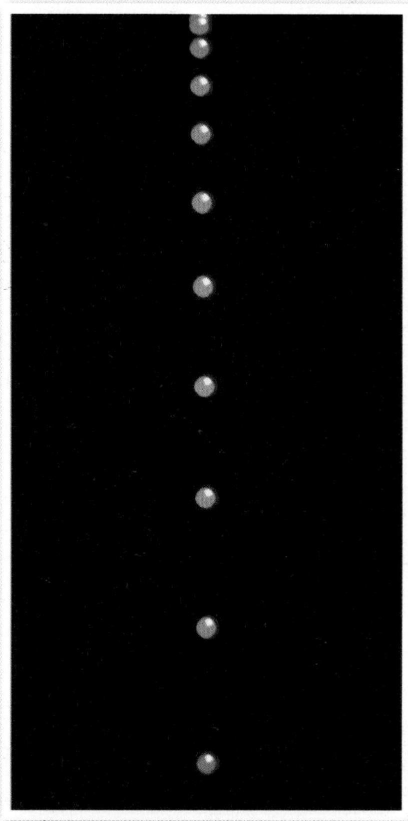

Freefall – a falling ball gets steadily faster as it falls.

Questions

1. What happens to a 'hard' surface when something sits on it? Is it really as hard as it seems?
2. Imagine the bag in the diagram opposite hanging from a string. The string must be exerting an upward force on the bag, equal to the downward force of gravity on it. How does the string exert this upwards force? Use the ideas on these pages to suggest an explanation.
3. Draw a diagram to show the forces acting on a cupcake case when it is falling.

Key words

- ✓ reaction (of a surface)
- ✓ air resistance

Describing motion

Find out about

- how to calculate the speed of a moving object
- how to calculate the acceleration of a moving object

In the previous sections you have seen how forces arise in interactions, and that there are some links between forces and motion. To explore these more fully, we need to be able to describe the motion of an object more clearly.

Distance and displacement

One obvious question to ask about the motion of an object is – how far did it move? But this could mean two different things. A group of walkers follow the route shown in the map. When they finish their walk, how far have they gone? The **distance** along the trail is 6 kilometres. But another answer to the question is that they have gone 2 kilometres east. This is called their **displacement**. The displacement is the straight-line distance and direction from the starting point. For walkers, distance is usually the more important quantity to know, as it helps you to judge how long the walk will take. But for sailors displacement is often more useful and important.

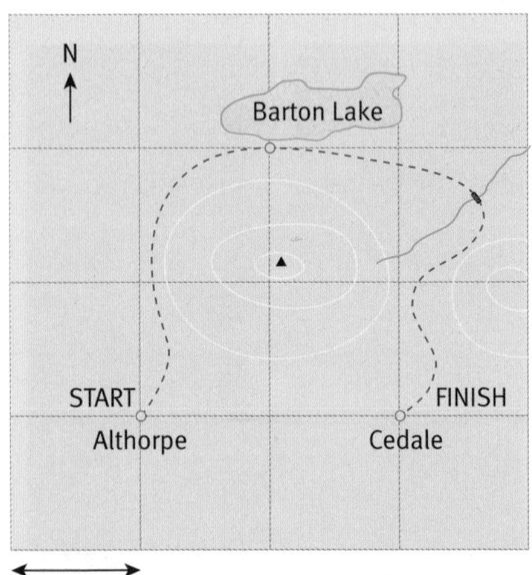

The distance from start to finish is 6 kilometres. But the displacement at the finish is 2 kilometres east of the start.

Speed

To find the speed of an object, you measure the time it takes to travel a known distance. You can then calculate its **average speed**, using the equation:

$$\text{average speed (metres per second, m/s)} = \frac{\text{distance travelled (metres, m)}}{\text{time taken (seconds, s)}}$$

The **instantaneous speed** of an object is the speed at which it is travelling at a particular instant. To estimate the instantaneous speed of an object, we measure its average speed over a very short distance (and hence over a very short time interval). The shorter we make this time interval, the less likely it is that the speed has changed much during it. On the other hand, if we make it very short, it is harder to measure the distance and the time accurately.

Question

1 The walkers use a straight farm track from Cedale back to Althorpe to pick up their cars. When they get back to their cars:
   a what is the total distance they have walked?
   b what is their total displacement?

A car's speedometer measures its average speed over a short time interval, giving a good indication of the instantaneous speed.

Worked example

An athlete runs a 100 metre race. The diagram shows her position at 1 second intervals during the race. She runs 100 metres in 12.5 seconds.

$$\text{average speed} = \frac{100 \text{ m}}{12.5 \text{ s}} = 8 \text{ m/s}$$

But she didn't run at 8 m/s for the whole race, sometimes she ran more quickly, sometimes she ran more slowly; this is why we call this value the *average* speed.

As she crossed the finish line, she ran approximately 10 metres in one second, so her *instantaneous* speed as she crossed the finish line was about 10 m/s.

Velocity and acceleration

People often use the words 'speed' and 'velocity' to mean the same thing. The **velocity** of an object, however, also tells you the direction in which it is moving. So, for example, a cyclist is pedalling at 8 m/s along a road that runs due west. Her instantaneous speed is 8 m/s, but her instantaneous velocity is 8 m/s in a westerly direction.

In everyday language, if the speed of an object is increasing, we say that it is accelerating. Drivers are often interested in the **acceleration** of their car. This might be stated as 0–60 miles per hour in 8 seconds. In situations like this, where the direction of motion does not matter, we can use the equation:

$$\text{acceleration} = \frac{\text{change of speed}}{\text{time for the change to occur}}$$

A more complete definition of acceleration that applies to all situations is:

$$\text{acceleration (metres per second) per second, (m/s}^2\text{)} = \frac{\text{change of velocity (metres per second, m/s)}}{\text{time taken for the change (seconds, s)}}$$

Questions

2 Calculate the average speed of the athlete over the first 20 metres of the race. Explain how this shows that she accelerated during the race.

3 During which time interval do you think the athlete was running fastest? Explain your answer.

4 A high-performance car can accelerate from 0–27 m/s (0–60 mph) in 6 seconds. Calculate the average acceleration of the car in m/s^2.

Key words

- distance
- displacement
- average speed
- instantaneous speed
- velocity
- acceleration

F Picturing motion

Find out about

- ✓ **how graphs can be used to summarise and analyse the motion of an object**

Graphs are very useful for summarising information about the motion of an object over a period of time. It would take many words to describe the motion of an object as clearly and exactly as a graph can do. A graph provides information about the moving object at any instant. From its shape and slope, we can quickly see how the object moved.

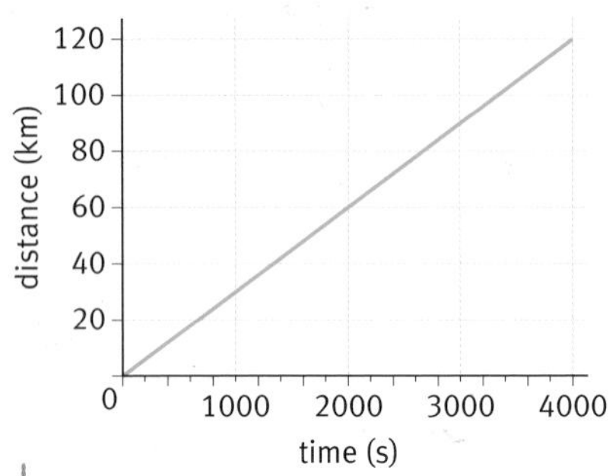

Distance–time graph for a car journey along the motorway.

Distance–time graphs

A **distance–time graph** shows the distance a moving object has travelled at every instant during its motion. The graph on the left is a distance–time graph for a car travelling along a motorway. Use the graph to find the distances the car has travelled after 1000 seconds, 2000 seconds, 3000 seconds, and 4000 seconds. The distance increases steadily. The speed of the car is constant. The constant **slope** of the distance–time graph indicates a steady speed.

The second graph below shows a more complicated journey. This is a cycle ride taken by Vijay during his school holidays. The graph has four sections. In each section the slope of the graph is constant. This means that Vijay's speed is constant during that section of the ride.

- In the first hour, Vijay travels 15 kilometres. His speed is a steady 15 km/h.
- In the second hour he travels only 5 kilometres, because the route goes steeply uphill. The shallower slope indicates a lower steady speed.
- In the third section (from 2.0 to 2.5 hours), his distance travelled does not change. He has stopped. This is what a horizontal section of a distance–time graph means.

distance from home (km): 30, 25, 20, 15, 10, 5

time (h): 0, 0.5, 1.0, 1.5, 2.0, 2.5, 3.0, 3.5

5 km

1 hour

I travel 5 km between these two points

It takes me 1 hour

So my speed is $\frac{\text{distance}}{\text{time}} = \frac{5\text{ km}}{1\text{ hour}}$ = 5 kmph

Vijay's cycle ride.

Questions

1 How far does Vijay travel during the final section of his cycle ride (from 2.5 to 3.5 hours)? So what is his speed during this section?

2 Vijay was travelling faster in this final section of the ride than in the second (uphill) section, but not as fast as in the first section. Explain how you could tell this, just by looking at the graph.

This distance–time graph for Vijay's journey is not very realistic. In a real journey, the speed would change gradually rather than suddenly.

Look at the distance–time graph on the right. It shows a car journey. The gradient tells you the speed of the car. If the gradient gets steeper, the car is speeding up, or accelerating. If the gradient is getting less steep, then the car's speed is decreasing.

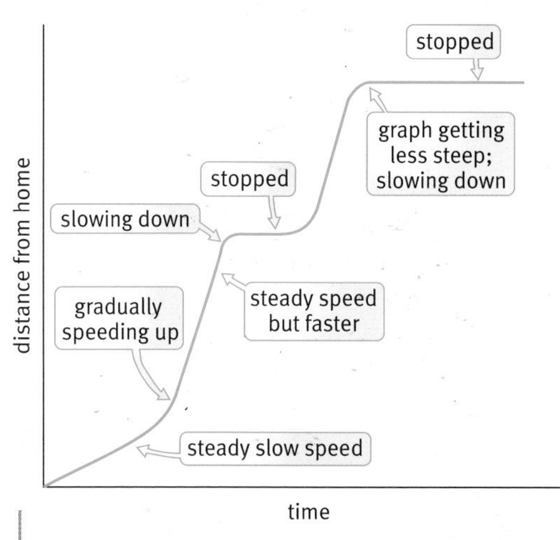

A more realistic distance–time graph.

Displacement–time graphs

If an object moves in a straight line, we can also draw a **displacement–time graph** of its motion. The graph shows how the size of the object's displacement changes with time. If the direction of motion does not change, the displacement–time graph is exactly the same as the distance–time graph. But if the direction of motion changes, the two graphs are different. The graphs below show the motion of a ball thrown vertically upwards from the moment it leaves the thrower's hand until the moment it arrives back again.

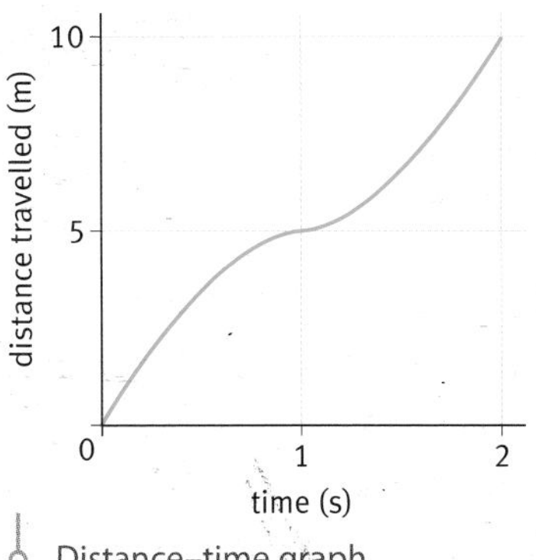

Distance–time graph.

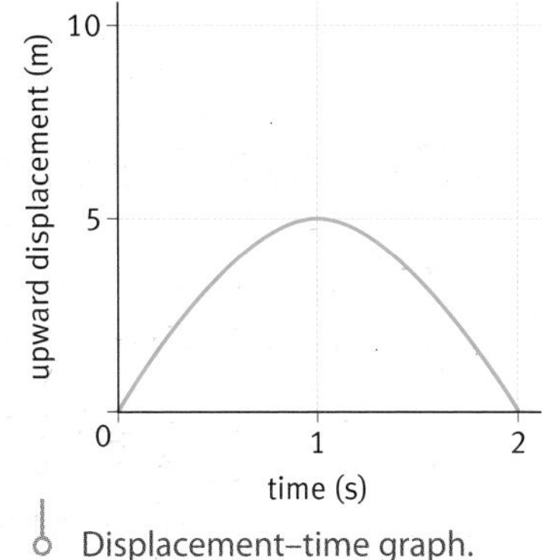

Displacement–time graph.

Two graphs showing the motion of a ball thrown up in the air.

Questions

3 Look at the graphs showing the motion of the ball thrown in the air.

- **a** Describe how the slope of both graphs changes over the first second of the ball's motion. What does this mean is happening to its speed? Is this what you would expect?
- **b** What is the speed of the ball after 1 s of its motion? Explain how you can tell this from the graphs.
- **c** How high does the ball go?
- **d** Explain why the displacement–time graph turns down after 1 s, whereas the distance–time graph continues to go up.

4 Roberta, an athlete, trains by jogging 20 m at a steady speed, then sprinting 20 m at a faster speed. She repeats this five times. Sketch a speed–time graph of her motion during a training session.

5 Think about the motion of a ball thrown upwards. Its speed gets steadily less on the way up and increases steadily again on the way down. Draw its:

- **a** speed–time graph
- **b** velocity–time graph

from the moment it leaves your hand until the moment it lands back in your hand.

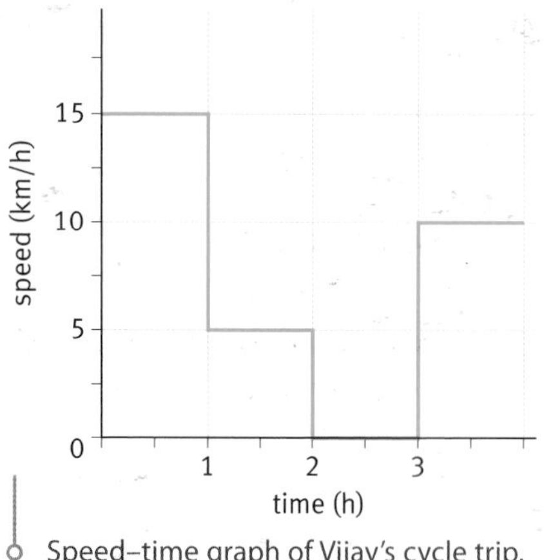

Speed–time graph of Vijay's cycle trip.

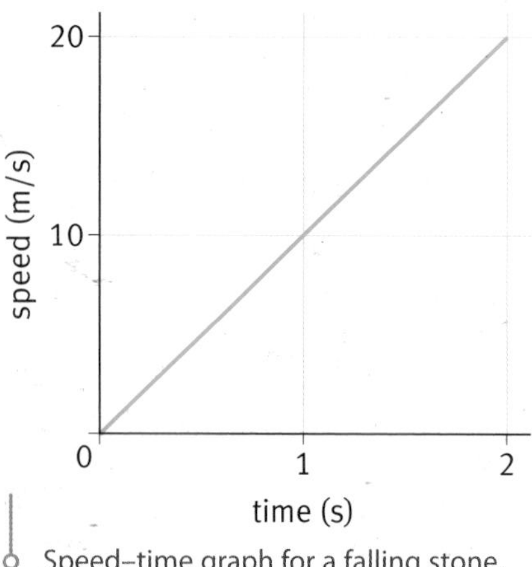

Speed–time graph for a falling stone.

Key words

- distance–time graph
- displacement–time graph
- slope
- velocity–time graph

Speed–time graphs

A speed–time graph shows the speed of a moving object at every instant during its journey. The speed–time graph for Vijay's trip on page 86 would look like the graph on the left. A steady speed is now shown by a straight horizontal line.

Again, the sudden changes of speed shown on this graph are not realistic. It would take time for the speed of a moving object to change. A more realistic speed–time graph would have smoother, more gradual changes from one speed to another.

The second speed–time graph on the left is for a stone being dropped from a high bridge into a river. It has no horizontal sections, so the speed of the stone is changing all the time. The constant slope of the speed–time graph shows that the speed is changing at a steady rate. It has steady (or uniform) acceleration downwards.

From this speed–time graph, we can calculate the acceleration of the stone. Its change of speed is 20 m/s (from 0 to 20 m/s) in 2 seconds. So its acceleration is:

$$\frac{\text{change of speed}}{\text{time taken for the change}} = \frac{20\ \text{m/s}}{2\ \text{s}} = 10\ \text{m/s}^2$$

Velocity–time graphs

Velocity (as explained earlier) has a direction as well as a size. It is not possible to show both size and direction on a graph. However, for an object that is moving in a straight line, we can draw a graph of *the size of its velocity* against time. This is called a **velocity–time graph.**

Look at the graph below. It shows the motion of Karl's skateboard up and then down a slope. At first Karl is travelling at 10 m/s. He gradually slows down until his velocity is zero. But then the line of the graph keeps on going down. The velocity becomes negative. This may seem strange, but it is used to show that the skateboard's direction of motion has changed. Karl is now travelling in the opposite direction.

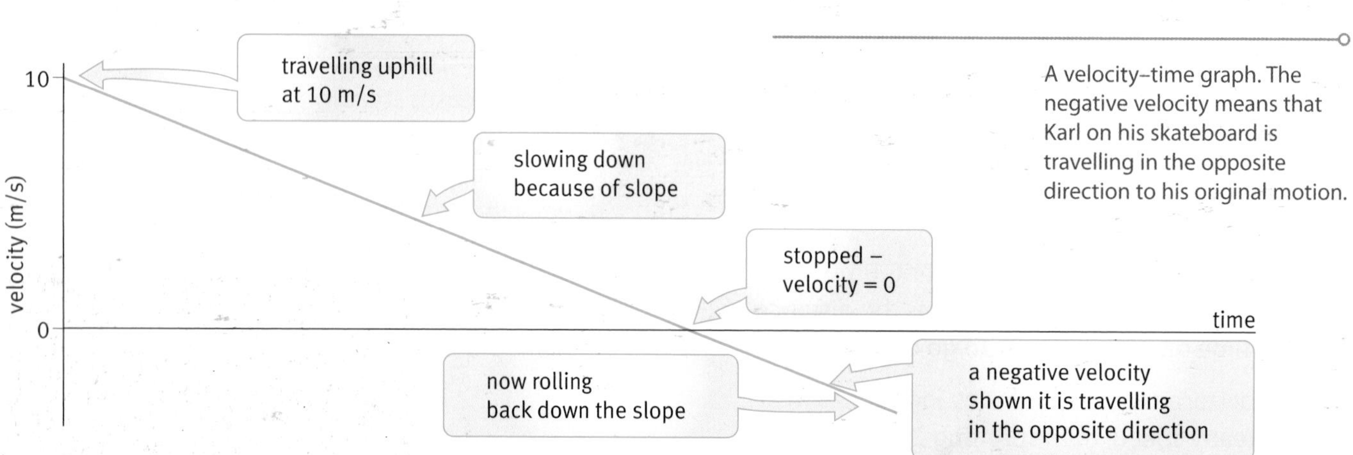

A velocity–time graph. The negative velocity means that Karl on his skateboard is travelling in the opposite direction to his original motion.

Forces and motion

You can now *describe* motion clearly; let's move on to *explain* motion. The key idea for explaining motion is force. If you know all the forces acting on an object, you can explain its motion.

Find out about

- momentum
- the link between change of momentum, force, and time

Forces change motion

A force changes the motion of an object. Imagine a small toy cart, with smooth, well-oiled wheels, sitting on a level table top. Someone (or something) exerts a constant force on it towards the right. The cart will immediately start to move – and its speed will keep on increasing as long as the force continues.

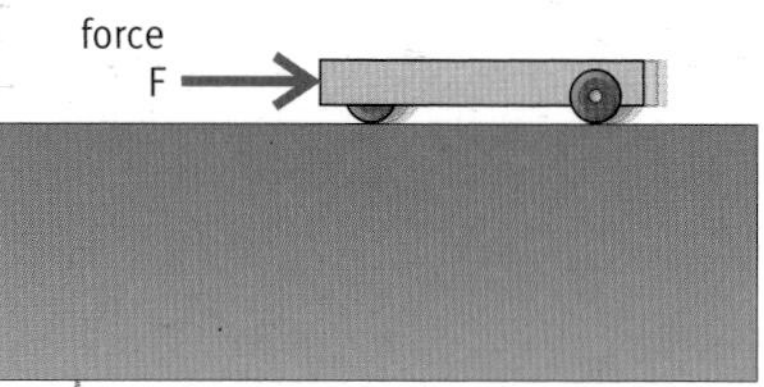

The speed of the cart keeps increasing as long as the force acts on it.

To study the effect of a constant force in more detail, you can use the arrangement shown in the second diagram. The force of gravity on the hanging weight is constant, so the string exerts a steady force on the cart. You could then measure the speed of the cart at different time intervals after this pulling force is applied. The speed–time graph is a straight line. The speed of the cart increases steadily with time. A constant force gives the cart a constant acceleration.

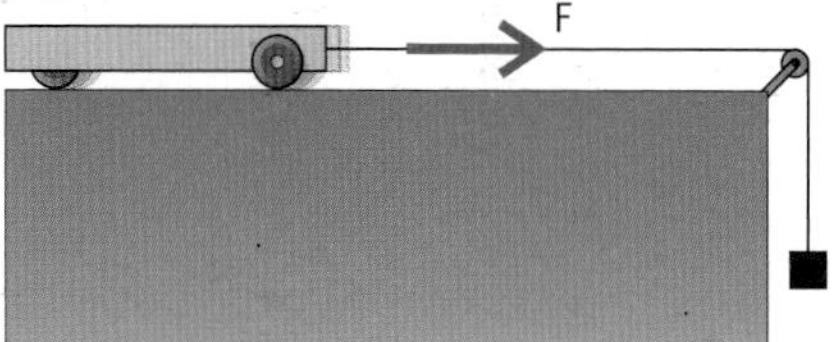

The string exerts a steady force on the cart.

Freefall is another example of motion with constant acceleration. In the flash photograph of a falling ball in Section D the gaps between the positions of the ball keep increasing as it falls. It gets steadily faster. The speed–time graph of a falling stone in Section F summarises the motion of a freely falling stone. The motion of a falling object is due to the gravitational force exerted on it by the Earth. This is a constant force on the object, so it causes its speed to increase steadily, a constant acceleration.

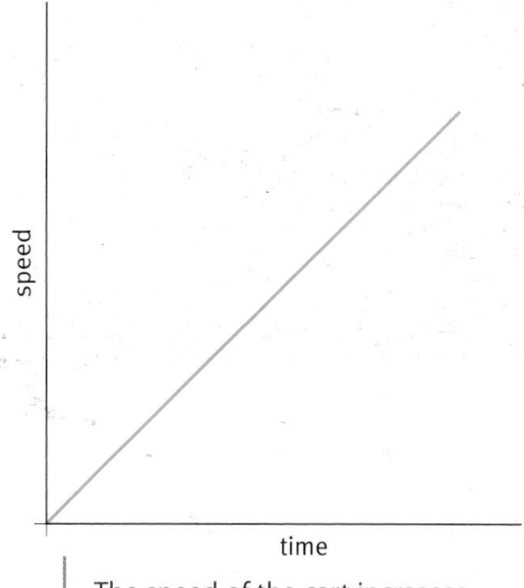

The speed of the cart increases steadily with time.

Momentum

A force acting on an object causes a *change* in its motion. To explore the connection between force and change of motion, let's consider a situation where two objects experience forces of exactly the same size. This happens in an interaction, for example, when two stationary objects spring apart.

In the diagrams on the next page, a spring-loaded cart is released, making the two objects move apart.

Key words

- momentum
- change of momentum

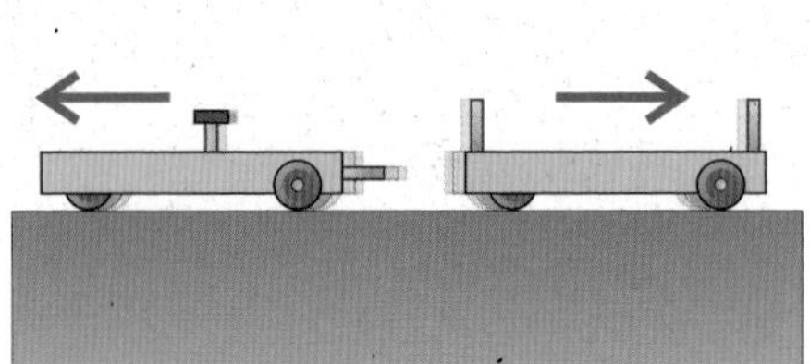

1 The left hand cart is spring loaded. If you put a second cart in front of it and release the spring, both move, in opposite directions. The interaction causes two forces, one on each cart. If the cart are identical, they both move with the same speed.

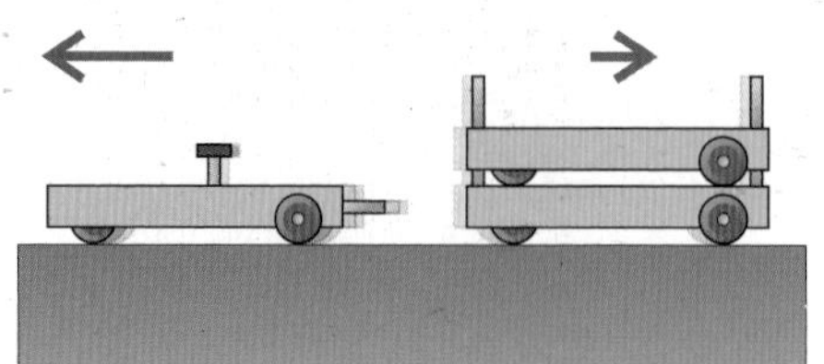

2 Here one cart is twice as heavy as the other. When the spring is released, both move. But the heavier one has only half the speed of the lighter one.

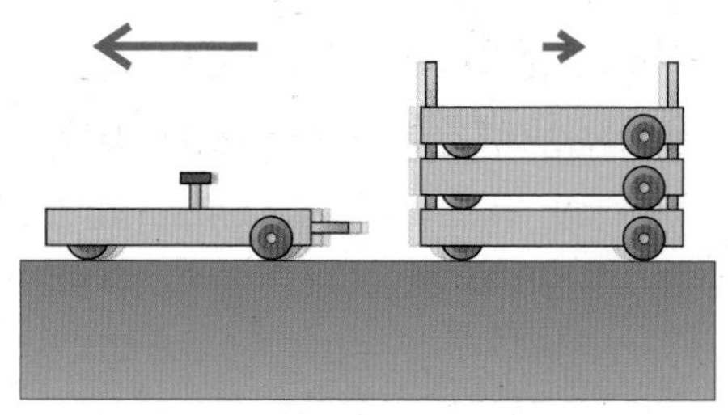

3 If we repeat this with a stack of three carts, we again find that both objects move. The speed of the stack of three carts is now one-third of the speed of the single cart.

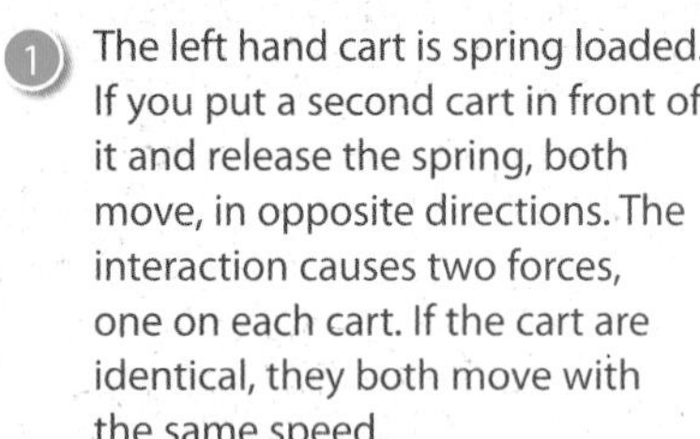

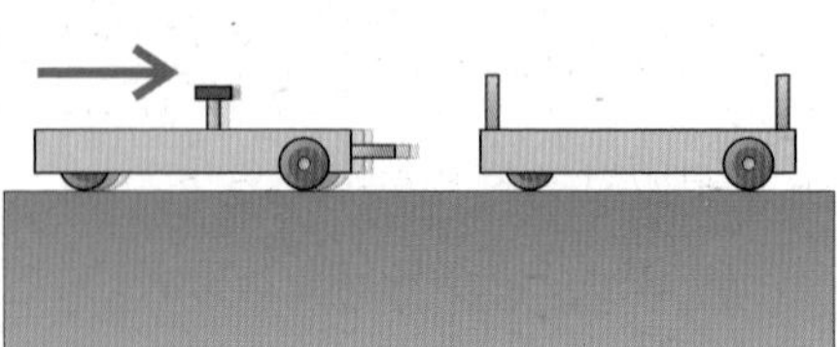

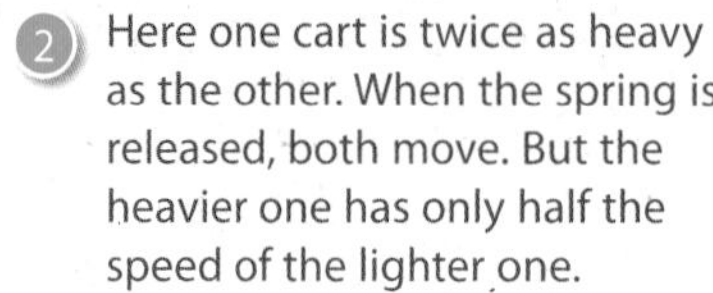

The momentum of the moving cart before the collision is the same as the momentum of the two carts after the collision. In all collisions, the total momentum is the same before and after – another indication that momentum is an important quantity for understanding motion.

In all of the interactions above, the size of the quantity (mass × velocity) is the same for both of the objects involved. This suggests that (mass × velocity) is a useful and important quantity. It is called the **momentum** of the moving object.

$$\underset{(\text{kg m/s})}{\text{momentum}} = \underset{(\text{kg})}{\text{mass}} \times \underset{(\text{m/s})}{\text{velocity}}$$

The faster an object is moving, the more momentum it has. A heavy object has more momentum than a lighter object moving at the same speed.

Force and change of momentum

In an interaction (like the three in the diagram above), the **change of momentum** is the same size for both objects. Two other things are also the same for both objects:

- the resultant force acting on the object (because the two forces in an interaction are always equal in size)
- the time for which it acts (the duration of the interaction, which has to be the same for both).

This suggests that there is a link between the force acting on an object, the time for which it acts, and the change of momentum it produces:

$$\underset{(\text{kg m/s})}{\text{change of momentum}} = \underset{(\text{N})}{\text{resultant force}} \times \underset{(\text{s})}{\text{time for which it acts}}$$

Questions

1 What is the momentum of:
  a a skier of mass 50 kg moving at 5 m/s?
  b a netball of mass 0.5 kg moving at 3 m/s?
  c a whale of mass 5000 kg swimming at 2 m/s?

2 When a force makes an object move, which two factors determine the change of momentum of the object?

3 Calculate the change of momentum for:
  a a force of 40 N acting for 3 s
  b a force of 200 N acting for 0.5 s
  c a force of 3 N acting for 50 s.

4 Use ideas about force and momentum to explain how the burning gases make a rocket or jet engine move.

Using the change of momentum equation

We can use the change of momentum equation to calculate the force applied.

Kicking a football

When a footballer takes a free kick, there is an interaction between his foot and the ball. His foot exerts a force on the ball.

This force lasts for only a very short time, the time for which the foot and the ball are actually in contact. After that, the player's foot can no longer affect the motion of the ball. The kick has not given the ball some 'force' but it has given it some momentum.

Taking a free kick. The interaction between the footballer's foot and the ball causes a change of momentum.

Worked example

How big is the force?

A football has a mass of around 1 kg. A free kick gives it a speed of 20 m/s. What is its momentum?

The football's momentum is given by the equation

$$\text{momentum} = \text{mass} \times \text{velocity}$$
$$= 1\ \text{kg} \times 20\ \text{m/s}$$
$$= 20\ \text{kg m/s}$$

It started with speed zero so the ball's change of momentum during the kick is 20 kg m/s.

The contact time when a football is kicked is around 0.05 s.

Estimate the force on the ball during the kick.

Use the equation

$$\text{change of momentum} = \text{force} \times \text{time it acts}$$
$$20\ \text{kg m/s} = \text{force} \times 0.05\ \text{s}$$

Dividing both sides by 0.05 s, you get

$$\frac{20\ \text{kg m/s}}{0.05\ \text{s}} = \text{force} = 400\ \text{N}$$

So the average force on the ball during the kick is 400 N. The force will not be constant during the time the ball is in contact. It will rise to a maximum (more than 400 N) and fall again to zero in 0.05 s.

Newton's Laws

This link between force, time, and change of momentum was proposed by Isaac Newton in 1687 in his famous book, the *Principia* (the *Mathematical Principles of Natural Science*). Newton did not use the word 'momentum' but wrote instead about an object's 'amount of motion' and how a force acting for a time could change it. The way he uses the word makes it clear that he is talking about the quantity we now call momentum.

Newton's hypothesis has been supported by many observations and measurements since then. There are all kinds of examples of motion, from everyday objects to the motion of the planets and stars. We now accept it as a reliable rule for explaining and predicting the motion of anything, except the very small (atomic scale) and the very fast (approaching the speed of light).

Car safety

Find out about

- ✔ safety features in cars

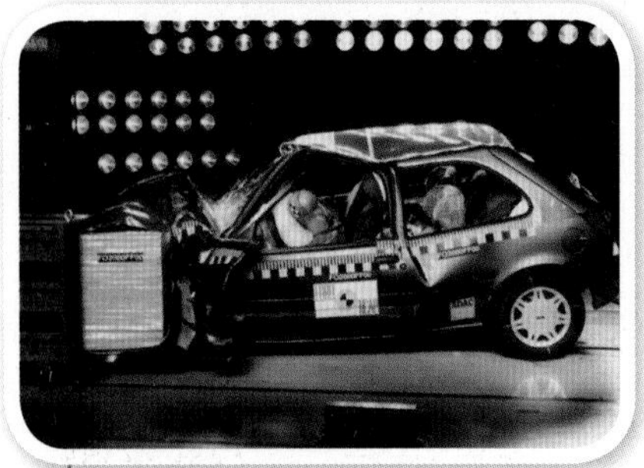

This driver, Hybrid 111, has experienced many crashes. It is packed with sensing equipment to record forces on different areas, like the head, chest, and neck. Each dummy costs more than £100,000 to build.

Key word

- ✔ risk

Questions

1 When you jump down from a wall or ledge, it is almost automatic to bend your knees as you land. Use the ideas on this page to explain why this reduces the risk of injury.

2 In railway stations, there are buffers at the end of the track. These are a safety measure, designed to stop the train if the brakes fail. Use the ideas on this page to explain how buffers would reduce the forces acting on the train and on the passengers.

Cars today are much safer to travel in than cars 10 or 20 years ago. As a result of crash tests like the one shown on the left, designs have changed and are still changing.

If a car is travelling at 100 km/h, the driver and passengers are also travelling at that speed. If the car comes to a very sudden stop, due to a collision, the occupants will experience a very sudden change in their momentum. This could cause serious injury.

Crumple zones

Look at the diagram below. Which car would be safer in a collision? The answer may not be so obvious. You need to think about the change of momentum during the collision and the time the collision lasts.

Would you be safer in car **a** or car **b**?

The momentum of a moving car depends on:

- its mass
- its velocity.

In a collision, the car is suddenly brought to a stop. Its momentum is then zero. The size of the average force exerted on the car during the collision depends on the time the collision lasts:

change of momentum = average force × time for which it acts

The bigger the time, the smaller the average force – for the same change of momentum. This is why cars are fitted with front and rear crumple zones, with a rigid box in the middle. They are designed to crumple gradually in a collision. This makes the duration of the collision (the time it lasts) longer. This then makes the average force exerted on the car less.

The passengers inside the car also experience a sudden change of momentum. They were moving at the same speed as the car and are suddenly brought to a stop. A force exerted on their bodies (by whatever they come into contact with) causes this change of momentum. The longer time it takes to change a passenger's speed to zero, the smaller the average force experienced.

Seat belts and air bags

Some people think that seat belts work by stopping you moving in a crash. In fact, to work, a seat belt actually has to stretch. Seat belts work on the same principle as crumple zones. They make the change of momentum take longer. So the average force that causes the change is less. A crash helmet works in the same way. As the helmet deforms, it increases the time it takes for your head to stop, so the force is less.

With a seat belt, the top half of your body will still move forward, and you may hit yourself against parts of the car. Air bags can help to cushion the impact. Again, they reduce your momentum over a longer time interval so that the average force you experience is less.

Could you save yourself?

Some people think they could survive a car accident without a seat belt, especially if they are travelling in the back seats. A car is travelling at 50 km/h (or roughly 14 m/s). Without a seat belt, the driver would hit the steering wheel and windscreen about 0.07 seconds after the impact. Back-seat passengers would hit the back of the front seats at roughly the same time. As your reaction time is typically about 0.14 seconds, this would all happen before you even had time to react. Even if you could react in time, the force needed to change your speed from 14 m/s to zero in 0.1 s is larger than your arms or legs could possibly exert.

Safety and risk

Car safety features, such as seat belts, are designed to reduce the **risk** of injury in a car accident. Travel can never be made completely safe. We cannot reduce the risk to zero. But many studies, in different countries, have shown that wearing seat belts greatly reduces the risk of serious injury. Many countries have regulations requiring drivers and passengers to use seat belts.

Questions

3 Using the diagrams on the right, estimate how long it takes for the seat belt to bring the driver's body to a stop. If his mass is 70 kg, what is the average force that the belt has to exert to do this?

4 A recent survey found that 95% of front-seat passengers wear seat belts. But only 69% of adult back-seat passengers wear a seat belt, though 96% of child back-seat passengers do.
 a Suggest why a few front-seat passengers do not wear a seat belt, despite the evidence that it reduces the risk of death.
 b Suggest why a higher proportion of children wear seatbelts.

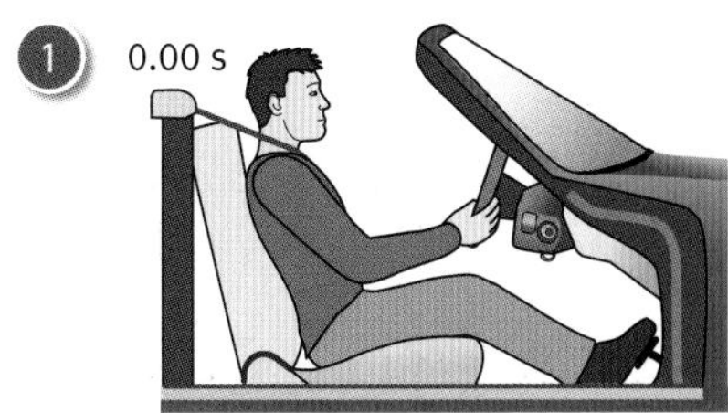

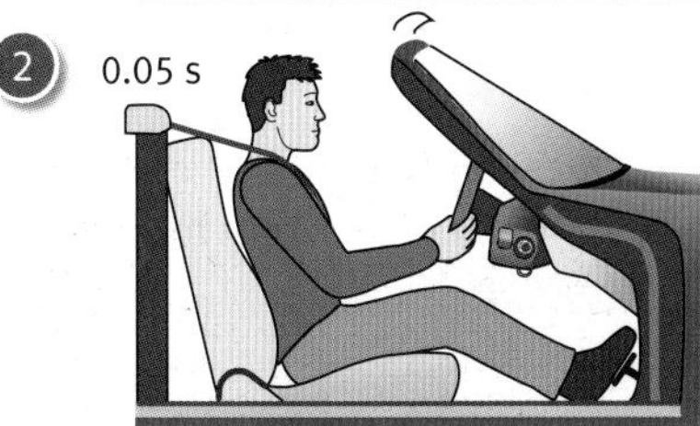

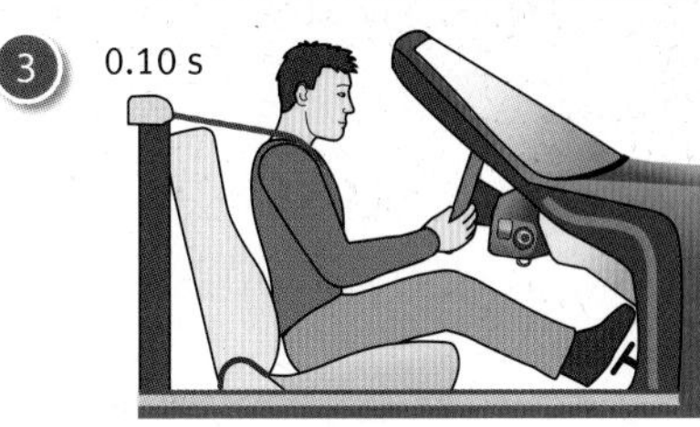

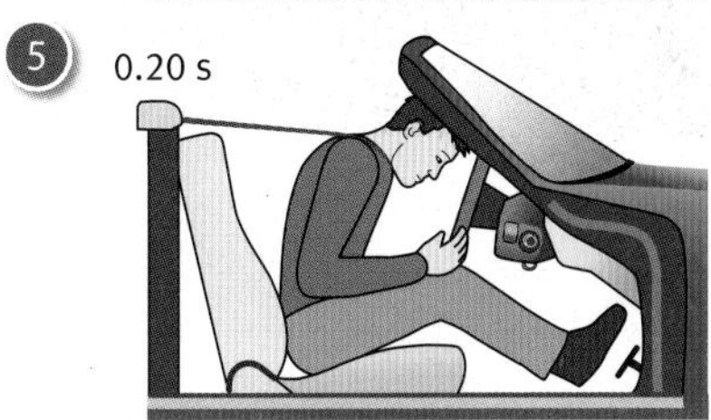

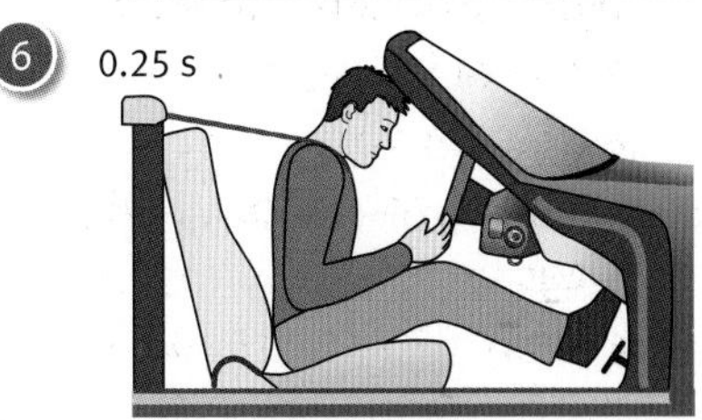

How seatbelts work. Notice how the seatbelt stretches during the collision. This 'spreads' the change of the driver's momentum over a longer period, making the force he experiences smaller.

Find out about

- ✔ **the laws (or rules) that apply to every example of motion**
- ✔ **how a resultant force is needed to change an object's motion**

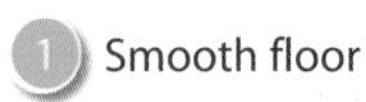
Smooth floor

Imagine pushing this curling stone across a smooth floor. It will keep going after it leaves your hand, because of the momentum you have given it during the interaction with your hand. But it immediately begins to slow down because of friction, and soon it will stop.

2 Ice

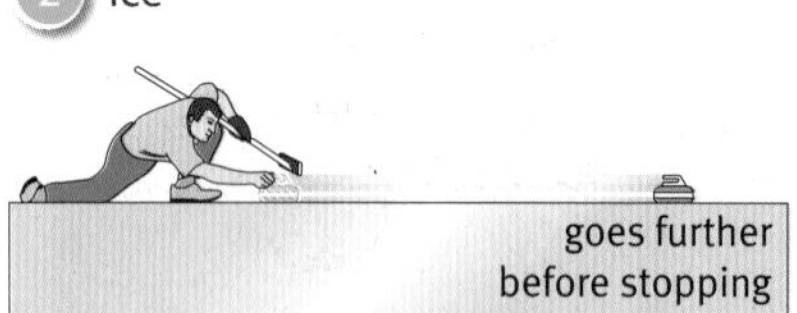

Now think what would happen if you gave the same stone exactly the same push, but this time on ice. It would not slow down as quickly. But it would slow down eventually, because there is still some friction. Eventually it would stop.

3 'Perfect' ice

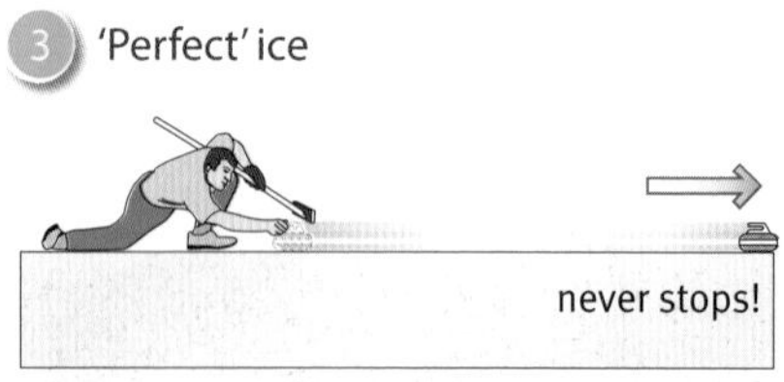

Now imagine 'perfect' ice, so slippery that there is no friction force between it and the stone. If you were able to give the stone the same push as before, it would not slow down after it left your hand. There is no friction force, so it just keeps on going, at the same speed, for ever.

Steady speed

Already in this chapter, we've seen two 'rules' about forces and motion:

- if an object is stationary, the resultant force on it is zero
- if there is a resultant force acting on an object, it causes a change in the momentum of the object. If the object was initially at rest (stopped), a resultant force makes it accelerate (move with increasing speed).

What about an object moving at a steady speed? These rules do not say anything about this situation. What forces are involved? To answer this, think about the situation shown in the three diagrams.

So motion at a steady speed does not need a force to maintain it. If the resultant force on an object is zero, its motion will remain unchanged. If it happens to be stationary, it will stay stationary. If it happens to be moving, it will carry on moving at the same speed in the same direction.

In the real world

In the real world there is no 'perfect' ice as in the third diagram. So you can never have a moving object with no forces acting on it. But it is possible to have a situation where there is no *resultant* force on a moving object. This happens when the driving force that is causing the motion exactly balances the counter-forces (of friction and air resistance). When this occurs, the moving object does not slow down or stop. It keeps moving at whatever speed it had when the forces became equal.

To see how this works, think about the forces involved in riding a bicycle.

When you press on the pedals of a bike, the chain makes the back wheel turn. The tyre pushes back along the ground. The other force in this interaction pair is the force exerted by the ground on the tyre. This pushes the bike forward. It is therefore called the **driving force**. As you move, air resistance and friction at the axles cause a **counter-force**. This is in the opposite direction to your motion.

Slowing down

What happens if the cyclist stops pedalling? There is now no driving force, but there is still a counter-force. So the resultant force is now in the opposite direction to her direction of motion. It causes a change of momentum in the direction of the resultant force. This is a negative change, meaning that the cyclist's momentum gets smaller rather than bigger.

Some people find it hard to believe that an object can move in one direction, whilst the resultant force acting on it is in the other direction. But this happens all the time, in any situation where something has been set in motion and is now slowing down. The key is to remember that a force does not cause motion; it causes a *change* of motion.

Laws of motion

The laws of motion (worked out by Newton) can be stated as follows:

Law 1: If the resultant force acting on an object is zero, the momentum of the object does not change.

Law 2: If there is a resultant force acting on an object, the momentum of the object will change. The size of the change of momentum is equal to the resultant force × the time for which it acts. The change is in the same direction as the resultant force.

Law 3: When two objects interact, each experiences a force. The two forces are equal in size, but opposite in direction.

These laws apply to all objects and to every situation (apart from objects at the subatomic level or moving at speeds near the speed of light). They enable us to explain and predict the motion of objects.

When you start off, the counter-force is small. Your driving force is bigger. You move forward, and your speed increases.

As you go faster, the air resistance force on you gets bigger. The counter-force increases. You still get faster, but not as quickly as before.

Eventually you reach a speed where the counter-force exactly balances your driving force. Your speed stops increasing. You are travelling, at a steady speed.

Questions

1 List three examples from everyday life of a situation where the resultant force on an object is zero. Explain how these are in agreement with the first law of motion.

2 List three examples from everyday life of a situation where there is a resultant force acting on an object. Explain how these are in agreement with the second law of motion.

3 Draw a fourth diagram in the series on this page to show the forces acting when the cyclist stops pedalling and freewheels. Write a caption, like those for the first three diagrams, to explain the motion.

Key words

- ✓ **driving force**
- ✓ **counter-force**

J Work and energy

Find out about

- how to calculate the work done by a force
- the link between work done on an object and the energy transferred
- how to use energy ideas to predict the motion of objects

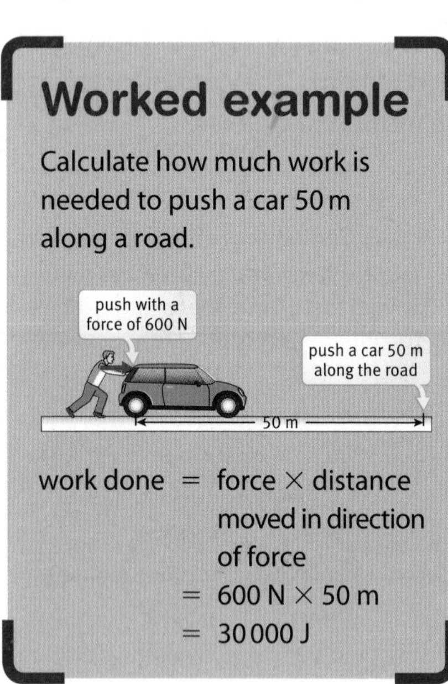

Worked example

Calculate how much work is needed to push a car 50 m along a road.

work done = force × distance moved in direction of force
= 600 N × 50 m
= 30 000 J

Transferring energy

When you push something and start it moving, the force you exert gives the object momentum. But your push also transfers energy to the object. This comes from the energy stored in your body (which gets less) to the energy of the moving object (which gets bigger). The energy of a moving object is called **kinetic energy**.

In general, when a force makes something move, it transfers energy to the moving object or to its surroundings. We say that the force 'does **work**'. The amount of work it does depends on:

- the size of the force
- the distance the object moves in the direction of the force.

The equation for calculating the amount of work done by a force is:

work done by a force	=	force	×	distance moved in the direction of the force
(joules, J)		(newtons, N)		(metres, m)

This also tells us the amount of energy transferred, because:

amount of energy transferred = amount of work done by the force

Energy and work are both measured in joules (J). A force of 1 newton applied over a distance of 1 metre does 1 J of work, and transfers 1 joule of energy.

Work

In physics, the word 'work' has a meaning that is slightly different from its everyday meaning.

Imagine that you are out in your car and you break down. Luckily, there is a garage just down the road. You ask your passenger to steer the car while you push it to the garage. If the garage is a long way down the road, you are going to have to do more work than if it is nearby.

However, if you push a really heavy object and cannot get it to move, you will feel that you are doing hard work that is having no result! From the physics point of view, you are not doing work because the force you are applying is not moving anything. The same is true if you are holding a heavy object. It feels like hard work, but the force you are applying to it is not making it move, so you are not doing any work.

Lifting things: gravitational potential energy

Lifting an object also involves doing work. Imagine lifting luggage into the car boot. You are transferring energy from your body's store of chemical energy. The **gravitational potential energy** of the luggage increases. The increase is equal to the amount of work you have done.

Suppose you have a suitcase that weighs 300 N. To lift it up, you have to exert an upward force of 300 N. If you lift it 1 metre into the boot of the car, then

$$\begin{aligned}\text{work done} &= \text{force} \times \text{distance moved in the direction of the force}\\ &= 300\ \text{N} \times 1\ \text{m}\\ &= 300\ \text{J}\end{aligned}$$

The suitcase now has 300 J more gravitational potential energy. In general, when anything is lifted up, you can calculate the change in gravitational potential energy from the equation

$$\underset{\text{(joules, J)}}{\text{gravitational potential energy}} = \underset{\text{(newtons, N)}}{\text{weight}} \times \underset{\text{(metres, m)}}{\text{vertical height difference}}$$

Notice that it is only the vertical height difference that matters. If you slide a suitcase up a ramp, the gain in gravitational potential energy is the same as if you lift it vertically.

Doing work by lifting: increasing gravitational potential energy.

Questions

1. If your mass is 40 kg, then your weight is roughly 400 N. How much work do you have to do each time you go upstairs – a vertical height gain of 2.5 m?
2. A mother is pushing a child along in a buggy. She is doing work. So the amount of energy stored in her muscles is getting less. Where is this energy being transferred to? (Careful! The buggy is going at a steady speed.)

Calculating kinetic energy

The equation for calculating the **kinetic energy** of a moving object is:

$$\underset{\text{(joules, J)}}{\text{kinetic energy}} = \frac{1}{2} \times \underset{\text{(kilogram, kg)}}{\text{mass}} \times \underset{\text{(metres per second, m/s)}^2}{(\text{velocity})^2}$$

The kinetic energy is proportional to the mass of the moving object. But notice that the amount of kinetic energy depends on the velocity squared. So small changes in velocity mean quite big changes in kinetic energy.

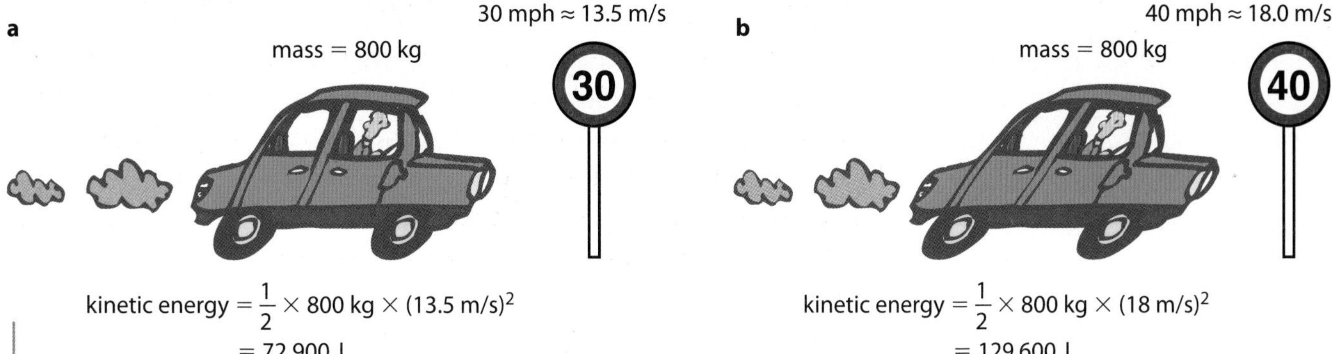

kinetic energy $= \frac{1}{2} \times 800\ \text{kg} \times (13.5\ \text{m/s})^2 = 72\,900\ \text{J}$

kinetic energy $= \frac{1}{2} \times 800\ \text{kg} \times (18\ \text{m/s})^2 = 129\,600\ \text{J}$

A car travelling at 40 mph (**b**) has nearly twice as much kinetic energy as the same car at 30 mph (**a**). It uses a lot more fuel to get to this speed.

Worked example

Calculate the change in kinetic energy of a trolley pushed with a force of 6 N over a distance of 5 m (assume there are no frictional forces acting).

change in kinetic energy of trolley = work done by pushing force
= force × distance
= 6 N × 5 m
= 30 J

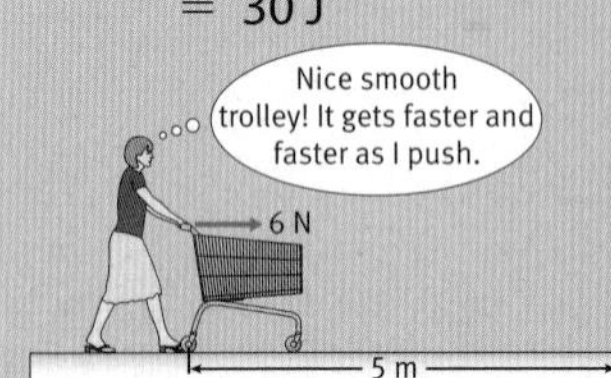

On a down slope, gravitational potential energy decreases and kinetic energy increases. On an up slope, the change is in the opposite direction.

Key words

- work
- gravitational potential energy
- kinetic energy
- conservation of energy

Making things speed up: changing their kinetic energy

Imagine pushing a well-oiled supermarket trolley along a level floor. As you push, you are doing work. The trolley keeps speeding up. Its speed increases as long as you keep pushing. You are transferring energy from your body's store of chemical energy to the trolley, where it is stored as kinetic energy. If the trolley is absolutely smooth running, the amount of work you do pushing it is equal to the change in the trolley's kinetic energy.

A real trolley will always have some friction, so its change in kinetic energy will be less than this. Some work is wasted in causing unwanted heating.

Conservation of energy

Energy ideas are very useful for working out what will happen when something falls through a certain height difference, or slides down a smooth ramp. If friction is small enough to be ignored, the amount of gravitational potential energy stored in the system gets less, and the kinetic energy of the moving object gets bigger, by the same amount.

decrease in gravitational potential energy = increase in kinetic energy

The opposite happens if a moving object rises vertically, or up a smooth slope. Its kinetic energy decreases. The gravitational potential energy stored in the system gets bigger, by the same amount.

The really useful thing about this is that it doesn't depend on the shape of the path the object follows, as long as it is smooth enough to allow us to ignore friction. A fairground roller coaster is a good example of these ideas in action. As the roller coaster goes round the track, gravitational potential energy is changing to kinetic energy (on the down slopes) and vice versa (on the up slopes).

If the slope has a complicated shape, it would be very difficult, maybe even impossible, to work out how fast the roller coaster is going at the bottom of the slope using the ideas of force and momentum. It is easier to use the principle of **conservation of energy**.

To see how this works out in practice, look at the following example:

Worked example

Calculate the speed of the roller coaster shown below at the bottom of the ride (assuming there are no friction forces).

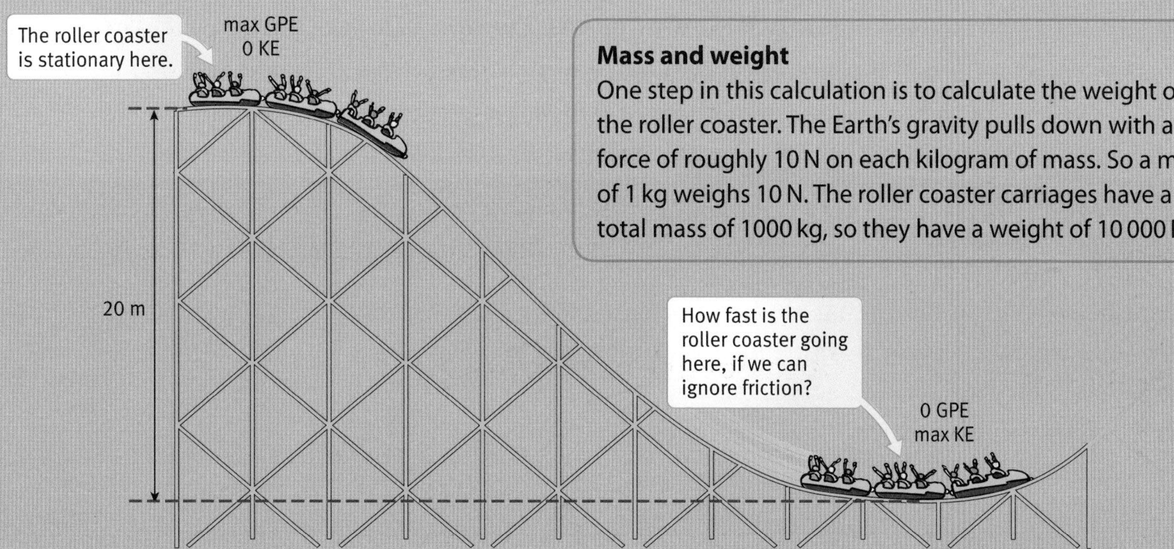

Mass and weight

One step in this calculation is to calculate the weight of the roller coaster. The Earth's gravity pulls down with a force of roughly 10 N on each kilogram of mass. So a mass of 1 kg weighs 10 N. The roller coaster carriages have a total mass of 1000 kg, so they have a weight of 10 000 N.

The simplest way to solve this problem is to use energy ideas. If friction is small enough to ignore, then the gravitational potential energy that the roller coaster loses as it goes down the slope is equal to the kinetic energy it gains:

$$\text{loss of gravitational potential energy} = \text{weight} \times \text{vertical height change}$$
$$= 10\,000\text{ N} \times 20\text{ m}$$
$$= 200\,000\text{ J}$$

$$\text{loss of gravitational potential energy} = \text{gain in kinetic energy}$$

So $\text{gain in kinetic energy} = 200\,000\text{ J}$

But $\text{gain in kinetic energy} = \frac{1}{2} \times \text{mass} \times (\text{velocity})^2$

So $\frac{1}{2} \times \text{mass} \times (\text{velocity})^2 = 200\,000\text{ J}$

Multiply both sides by 2: $\text{mass} \times (\text{velocity})^2 = 400\,000\text{ J}$

or $1000\text{ kg} \times (\text{velocity})^2 = 400\,000\text{ J}$

Divide both sides by 1000 kg: $(\text{velocity})^2 = \dfrac{400\,000\text{ J}}{1000\text{ kg}}$

$$= 400\ (\text{m/s})^2$$

Take the square root of both sides: $\text{velocity} = 20\text{ m/s}$

The shape of the slope does not matter at all. Only the vertical height difference is important in working out the change in gravitational potential energy – and hence the increase in kinetic energy.

Questions

3 A ten-pin bowling ball has a mass of 4 kg. It is moving at 8 m/s. How much kinetic energy does it have?

4 Which of the following has more kinetic energy:
  a a car of mass 500 kg travelling at 20 m/s?
  b a car of mass 1000 kg travelling at 10 m/s?

5 Repeat the calculation above for a roller coaster that is only half as heavy (weight 5000 N). What do you notice about its speed at the bottom? How would you explain this?

Science Explanations

Forces and motion form the basis of our understanding of how the world works. Every example of motion we observe can be explained by a few simple rules (or laws) that are so exact and precise that they can be used to predict the motion of an object very accurately.

You should know:

- about the interaction pair of forces that always arise when two objects interact
- that vehicles, and people, move by pushing back on something and this interaction causes a forward force to act on them
- about the friction interaction between two objects that are sliding (or tending to slide) past each other
- that the resultant force on an object is the sum of all the individual forces acting on it, taking their directions into account
- about the reaction force on an object that arises because it pushes down on a surface
- how to calculate the average speed of a moving object
- about instantaneous speed and how this is different to the average speed
- what is meant by distance, displacement, speed, velocity, and acceleration
- how to draw and interpret distance–time, speed–time, displacement–time, and velocity–time graphs
- how to calculate the acceleration of an object
- how to find acceleration from the gradient of a velocity–time graph
- that when a resultant force acts on an object it causes a change in momentum
- how to calculate momentum and the change in momentum due to a force
- that many vehicle safety features increase the duration of an event (such as a collision), so that the average force is less, for the same change of momentum
- that if there is no resultant force on an object, its momentum does not change – it either remains stationary or keeps moving at a steady speed in a straight line
- about the work done when a force moves an object and how to calculate the work done, which is equal to the energy transferred
- that when work is done on an object, energy is transferred to the object and when work is done by an object, energy is transferred from the object to something else
- about gravitational potential energy and how to calculate the change in gravitational potential energy as an object is raised or lowered
- about kinetic energy, how to calculate it, and that doing work on an object can increase its kinetic energy by making it move faster
- that when an object falls, if friction and air resistance can be ignored, the decrease in gravitational potential energy is equal to the increase in kinetic energy and this can be used to work out the speed of the object.

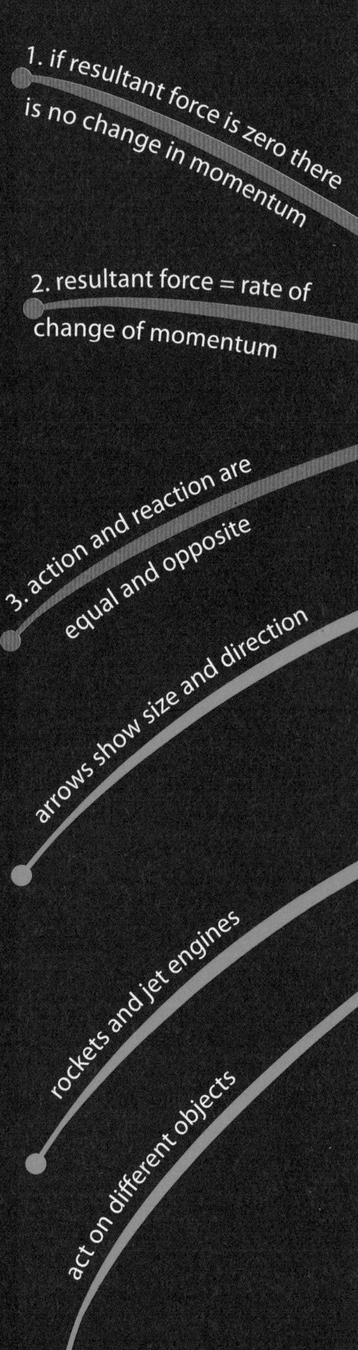

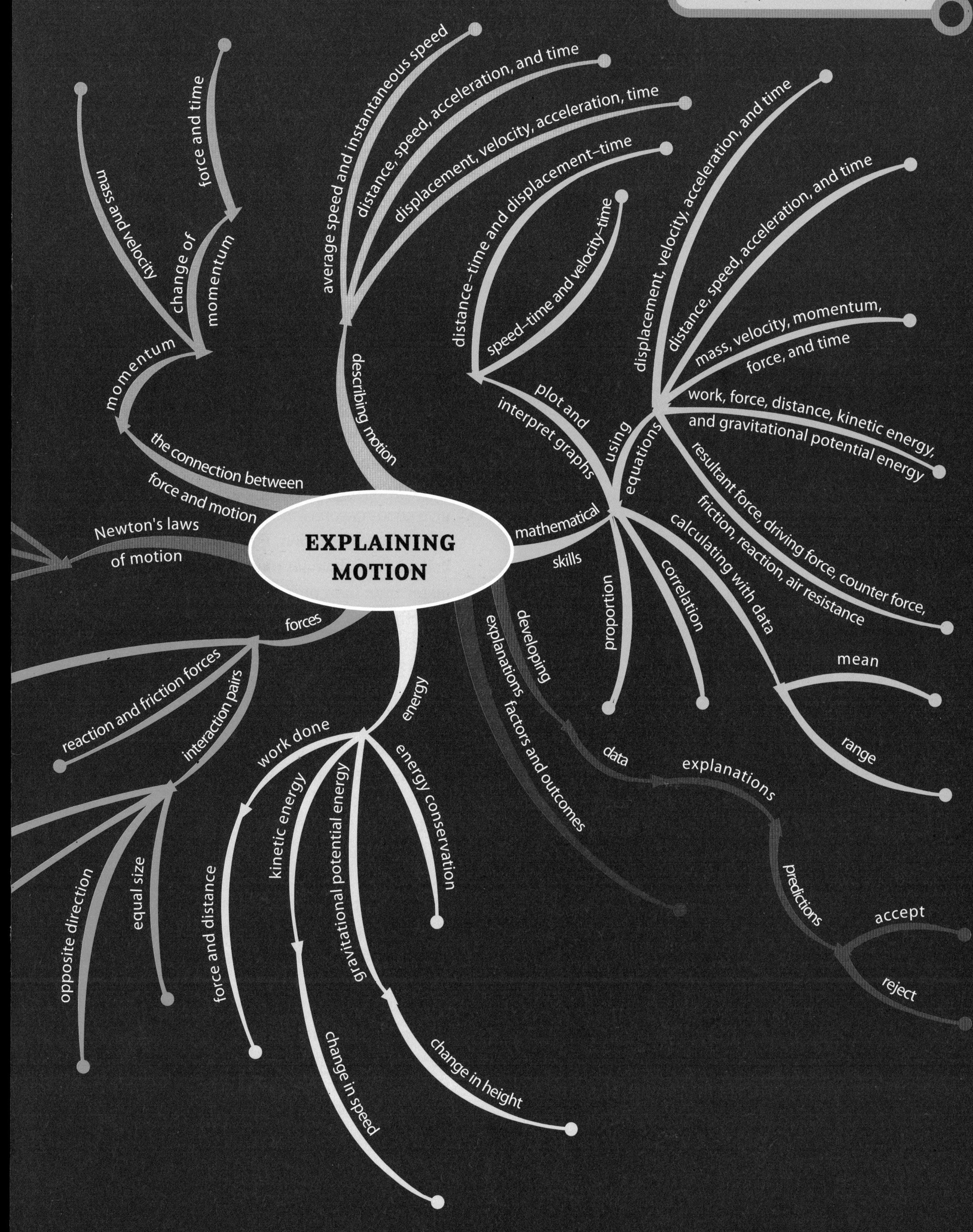
EXPLAINING MOTION
describing motion
average speed and instantaneous speed
distance, speed, acceleration, and time
displacement, velocity, acceleration, time
the connection between force and motion
momentum
mass and velocity
change of momentum
force and time
Newton's laws of motion
forces
reaction and friction forces
interaction pairs
opposite direction
equal size
energy
work done
force and distance
kinetic energy
change in speed
gravitational potential energy
change in height
energy conservation
mathematical skills
plot and interpret graphs
distance–time and displacement–time
speed–time and velocity–time
using equations
displacement, velocity, acceleration, and time
distance, speed, acceleration, and time
mass, velocity, momentum, force, and time
work, force, distance, kinetic energy, and gravitational potential energy
resultant force, driving force, counter force, friction, reaction, air resistance
calculating with data
mean
range
correlation
proportion
developing explanations
factors and outcomes
data
explanations
predictions
accept
reject

Ideas about Science

In addition to understanding forces and motion, you need to understand how scientific explanations are developed. The list below links these ideas about science with some examples from the module.

In developing scientific explanations you should be able to:

- identify statements that are data and statements that are explanations. For example, the statement 'The acceleration of a falling object has a constant value' is data; the statement 'When a constant force acts on an object it causes a constant acceleration' is an explanation.
- recognise that an explanation may be incorrect, even if the data is correct. For example, some measurements show that pushing with a constant force on a toy car causes it to travel at a constant speed, but the explanation 'A constant force causes a car to travel at a constant speed' is incorrect. The pushing force is balancing the friction force – there is no resultant force on the car.
- identify where creative thinking is involved in the development of an explanation. For example, the idea of an interaction pair of forces means that when you push on a wall the wall pushes back – this is not an obvious idea, but Isaac Newton suggested this as his third law of motion: action and reaction are equal and opposite.
- recognise data or observations that are accounted for, or conflict with, an explanation. For example, if data shows that the force on an object is proportional to its acceleration, this agrees with Newton's second law: the force on an object is equal to its rate of change of momentum.
- give good reasons for accepting or rejecting a scientific explanation. For example, Aristotle believed that heavier objects always fall faster than light objects. A good reason for rejecting this would be a slow-motion film of objects falling in a vacuum – they fall at the same speed.
- decide, giving reasons, which of two scientific explanations is better. Galileo said that light and heavy objects fall at the same speed. This matches observations better, when there is no air resistance.
- understand that when a prediction agrees with an observation this increases confidence in the explanation on which the prediction is based, but does not prove it is correct. For example, Newton's laws of motion have correctly predicted the behaviour of many moving objects.
- understand that when a prediction disagrees with an observation this indicates that one or the other is wrong and decreases confidence in the explanation on which the prediction is based. For example, observations do not support Aristotle's ideas about motion.

Review Questions

1 Think about the following situations:

- **i** Amjad on his skateboard, throwing a heavy ball to his friend (main objects to consider: Amjad, the skateboard, and the ball)
- **ii** a furniture remover trying to pull a piano across the floor, but it will not move (main objects to consider: the furniture remover, the piano, and the floor)
- **iii** a hanging basket of flowers outside a café (main objects to consider: the basket and the chain it is hanging from).

For each situation:

- **a** Sketch a diagram (looking at it from the side).
- **b** Sketch separate diagrams of the main objects in the situation (these are listed for each).
- **c** On these separate diagrams, draw arrows to show the forces acting on that object. Use the length of the arrow to show how big each force is.
- **d** Write a label beside each arrow to show what the force is.

2 A tin of beans on a kitchen shelf is not falling, even though gravity is still acting on it. The shelf exerts an upward force, which balances the force of gravity. Explain in a short paragraph how it is possible for a shelf to exert a force. Draw a sketch diagram if it helps your explanation.

3

- **a** The winner of a 50 m swimming event completes the distance in 80 s. What is his average speed?
- **b** How far could Leonie cycle in 10 minutes if her average speed is 8 m/s?
- **c** The average speed of a bus in city traffic is 5 m/s. How much time should the timetable allow for the bus to cover a 6-km route?

4

- **a** A bus leaves a bus stop and reaches a speed of 15 m/s in 10 s. Calculate its acceleration.
- **b** A car accelerates at 3 m/s^2 for 8 s. By how much will its speed have increased?

5 What is the momentum of:

- **a** a hockey ball of mass 0.4 kg moving at 5 m/s?
- **b** a jogger of mass 55 kg, running at 4 m/s?
- **c** a van of mass 10 000 kg, travelling at 15 m/s?
- **d** a car ferry of mass 20 000 000 kg, moving at 0.5 m/s?

6 A weightlifter raises a bar of mass 50 kg until it is above his head – a total height gain of 2.2 m. How much gravitational potential energy has it gained? How much more work must he do to hold it there for 5 s?

B5 Growth and development

Why study growth and development?

How does a human embryo develop? What makes cells with the same genes develop differently? Exploring questions like these is part of the fast-moving world of modern biology.

What you already know

- Genes affect the way organisms including humans develop.
- Clones such as plants grown from cuttings are more similar to each other than organisms with a combination of their parents' genes.
- Organisms are made up of cells, and these divide and develop into the whole organism.

Find out about

- the structure of DNA, and how it controls the proteins a cell makes
- how cells divide to make sex cells and to make new body cells
- how cells become specialised
- the differences between plant and animal growth.

The Science

Sex cells carry the genetic information to make a new individual. Cells become specialised because of the different proteins they make. DNA is the chemical that genes are made of, and its unique structure determines the proteins a cell makes.

Ideas about Science

Data and explanations of data are different. Creative thinking is needed to develop explanations from data.

A Growing and changing

Find out about

- different cells, tissues, and organs
- growing up

You began life as a single cell. By the time you were born you were made of millions of cells. You probably weighed about 3 or 4 kilograms. Now you probably weigh over 50 kilograms. Not only have you grown, but you have changed in many ways. In other words, you have developed.

Since you were born, your **development** has been gradual. In some plants and animals, development involves big changes. For example, the young and the adults in these photographs look very different.

A

B

C

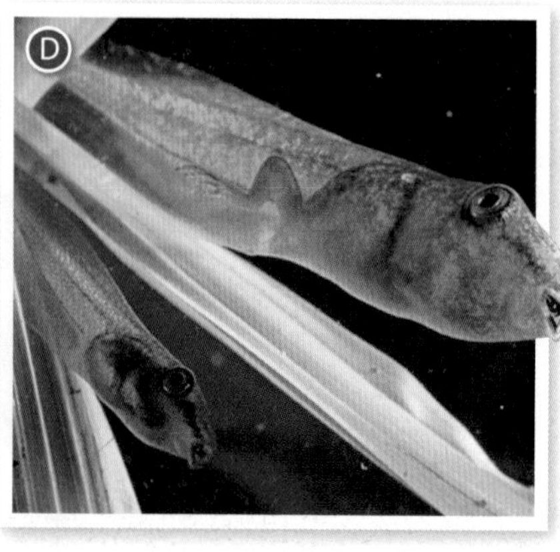
D

E

F

Questions

1 Match the young and the adults in pictures A to F.

2 Which animal in the pictures has a life cycle most like a fly?

Building blocks

Like you, the plants and animals in the pictures above are multicellular. Not every cell is the same. Your body has more than 300 different kinds of cell. Each kind of cell is **specialised** to do a particular job. Muscle cells can contract to cause movement, nerve cells are specially shaped to carry impulses, and red blood cells lack nuclei, giving extra room to carry oxygen.

Tissues and organs

All newly formed human cells look much the same. Then they develop into groups of specialised cells called **tissues**. Our body has muscle tissue, nervous tissue, and fatty tissue. Bone, cartilage, and blood are also tissues.

Plant cells are different from animal cells, but they are specialised too. Plant cells have cell walls outside the cell membrane, and some have spaces called vacuoles.

As an animal embryo or plant grows, groups of tissues arrange themselves into **organs**, for example, the heart and brain in humans, and roots, leaves, and flowers in plants. The heart is made up of muscle tissue, nervous tissue, connective tissue, and some fatty tissue. The diagram below shows the different plant tissues found in a leaf, which is a plant organ.

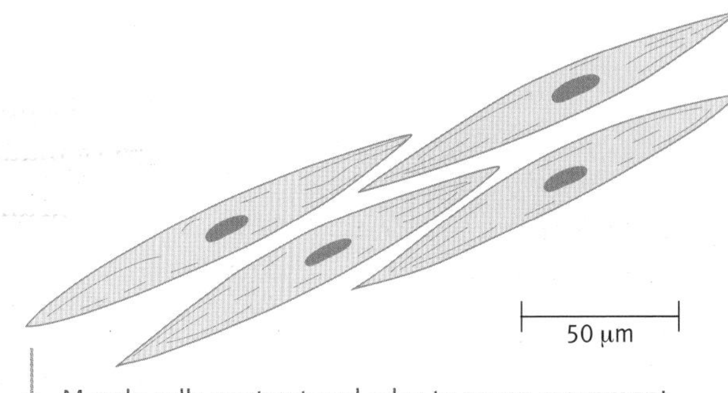

Muscle cells contract and relax to cause movement.

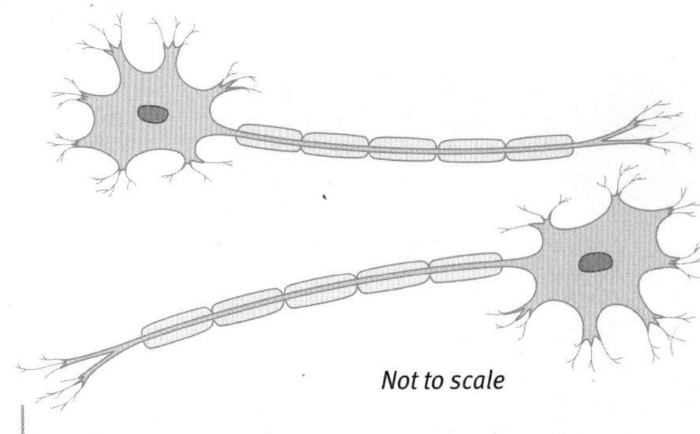

Nerve cells carry nerve impulses.

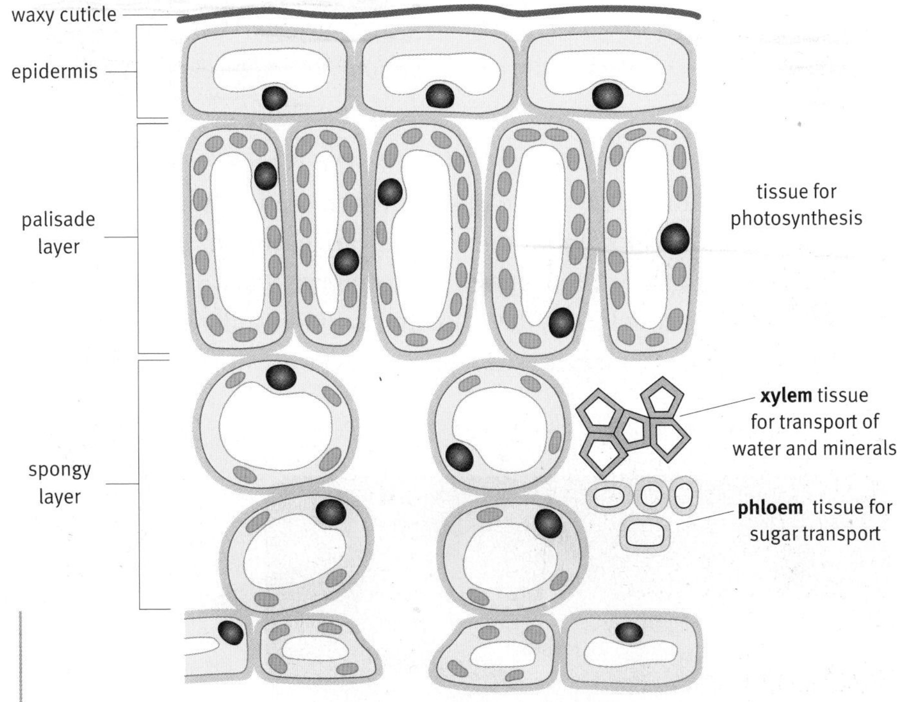

waxy cuticle
epidermis
palisade layer
tissue for photosynthesis
spongy layer
xylem tissue for transport of water and minerals
phloem tissue for sugar transport

Different specialised tissues in the leaf develop from unspecialised cells.

Question

3 Explain the difference between a tissue and an organ.

Key words

- ✓ development
- ✓ specialised
- ✓ tissues
- ✓ organs
- ✓ xylem
- ✓ phloem

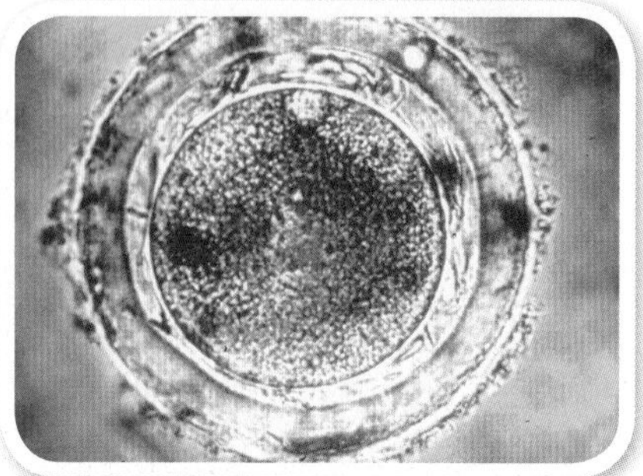

A fertilised human egg cell.

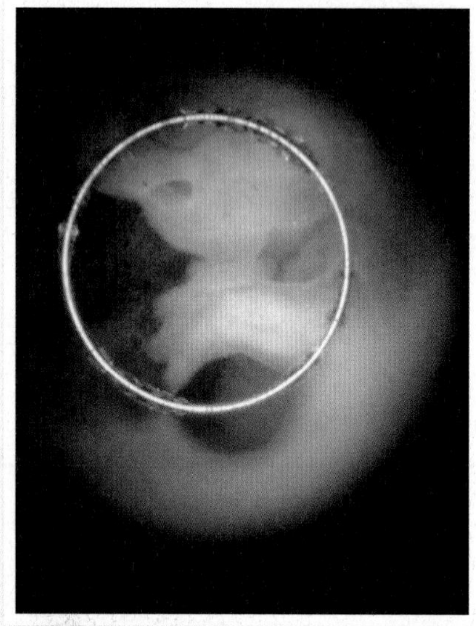

A human fetus at eight weeks. This photograph was taken from inside the mother's uterus. The fetus is about 2.5 cm long.

From single cell to adult

All the cells in your body come from just one original cell – a fertilised egg cell or **zygote**.

So the zygote must contain instructions for making all the different types of cells in your body, for example, muscle cells, bone cells, and blood cells. It also has the information to make sure that each type of cell develops in the right place and at the right time. This information is in your DNA – the chemical that your genes are made of.

In humans:

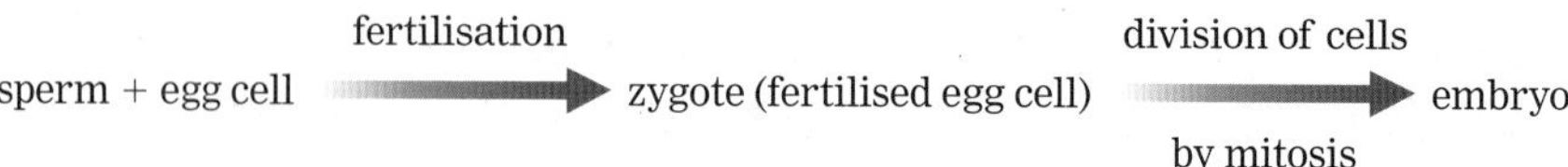

The growing baby

During the first week of growth, the zygote divides by mitosis to form a ball of about 100 cells. The nucleus of each cell contains an exact copy of the original DNA. As the embryo grows, some of the new cells become specialised and form tissues. After about two months, the main organs have formed and the developing baby is called a **fetus**. A six-day embryo is made of about 50 cells. Adults contain about 10^{14} cells, each with the DNA exactly copied.

When the embryo is a ball of eight cells or fewer, it occasionally splits into two. A separate embryo develops from each section. When this happens, identical twins are produced. They are **clones** of each other. This shows that there are cells in the early embryo that are identical and unspecialised. These cells can develop into complete individuals, and are called **embryonic stem cells**. You can read more about these cells in Section J.

Key words

- zygote
- clone
- embryonic stem cells
- meristem cells
- fetus

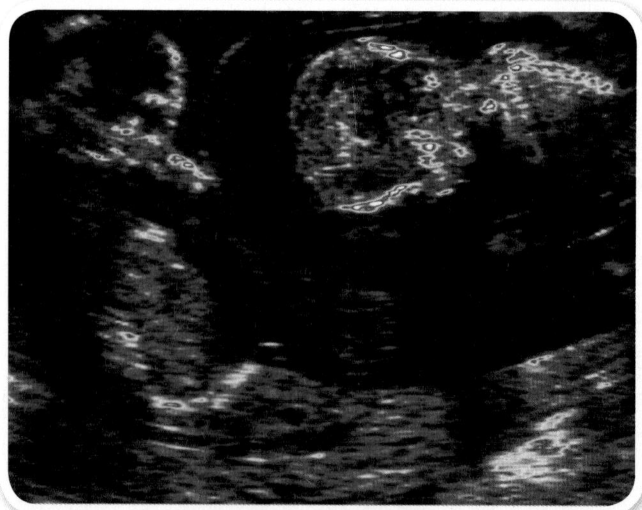

This ultrasound scan shows that twins are expected. Both of the babies' heads can be seen.

Questions

4 What is a zygote?

5 When does a human embryo become a fetus?

Growth patterns

For living things to grow bigger, some of their cells must divide to make new cells. You will probably stop growing taller by the time you are about 18–20 years old.

Flowering plants continue to grow throughout their lives.

- Their stems grow taller.
- Their roots grow longer.
- To hold themselves upright, most stems increase in girth or have some other means of support.

Plants increase in length by making new cells at the tips of both shoots and roots. They also have rings of dividing cells in their stems and roots to increase their girth. These dividing cells are called **meristem cells**.

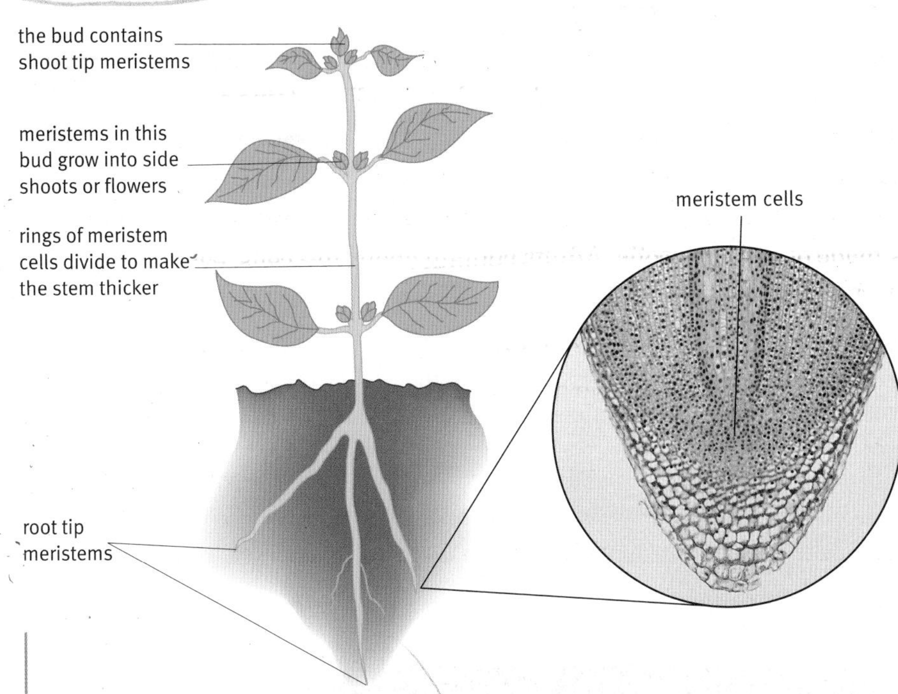

Meristem cells divide to make stems and roots longer and make the stem thicker. On the right is a root tip meristem (photographed through a microscope).

This giant sequoia tree is over 2000 years old, 83 m tall, and 26 m in girth (circumference). It is known as 'General Sherman' and is officially the largest giant sequoia tree and the largest living thing on Earth.

Honeysuckle twines its stem around other plants for support.

Questions

6 Why is it important that all living things have cells that can divide?

7 Name the type of cell in a plant that can divide to make new tissues.

8 Explain how plants:
 a grow taller
 b grow longer roots
 c grow thicker in girth.

Find out about

- ✔ why plants are so good at repairing damage

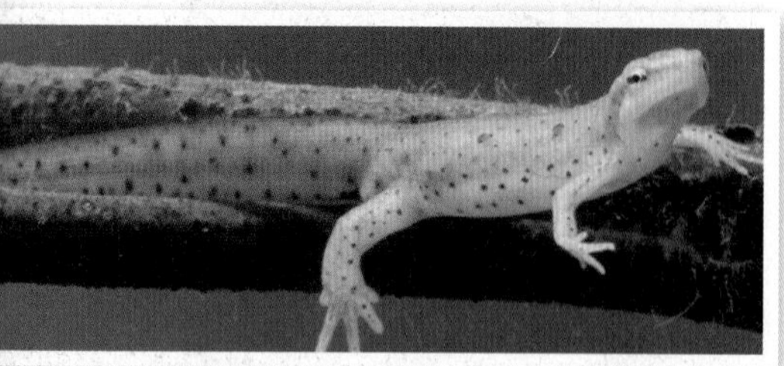

Cells in your body divide when you are growing. If you cut yourself, cells divide to repair your body, but only for small repairs. Many plants and some animals can replace whole organs.

Why can plants grow back?

Plant meristem cells are **unspecialised**. Plants keep some meristem cells all through their lives. These are spare back-up cells that divide to make any kind of cell the body needs. So plants can make new xylem or phloem tissues, or can regrow whole organs, such as leaves, if they are damaged. Scientists who develop genetically modified crops can put this to use. They can regenerate whole plants from small pieces of unspecialised plant tissue that they have modified.

How do newts grow?

Animals also have spare back-up cells called **stem cells**. These cells divide, grow, and develop into specialised cells that the body needs.

Animal growth

Newts' stem cells stay unspecialised throughout their lives. So newts can grow new legs if they need to – or even an eye.

The stem cells in adult humans are not as useful, because they have already started to specialise. For example, the stem cells in your skin can only develop into skin cells.

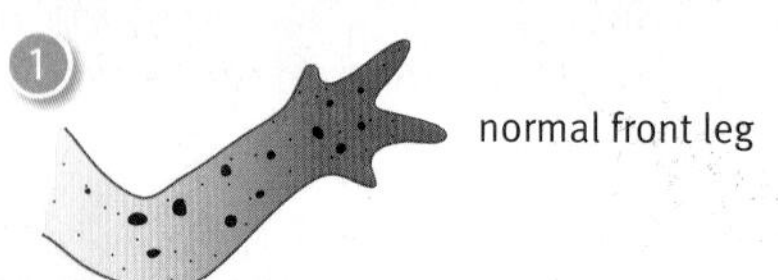

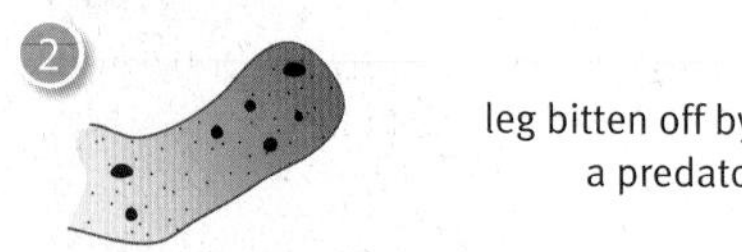

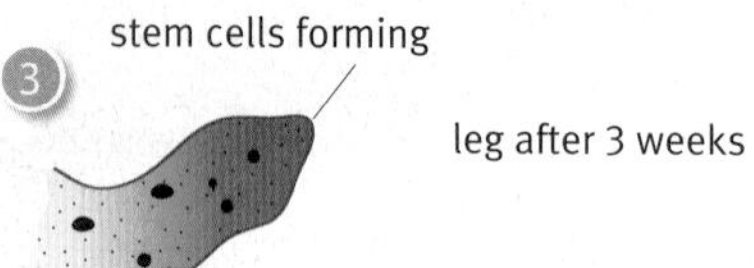

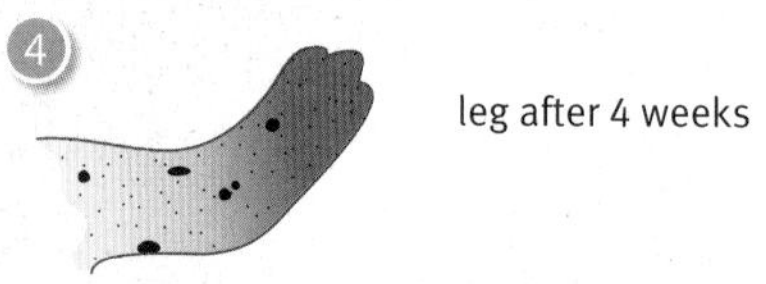

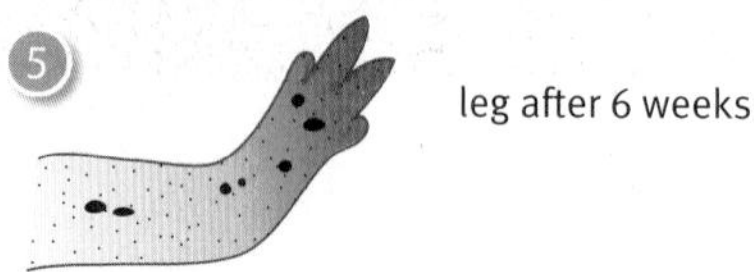

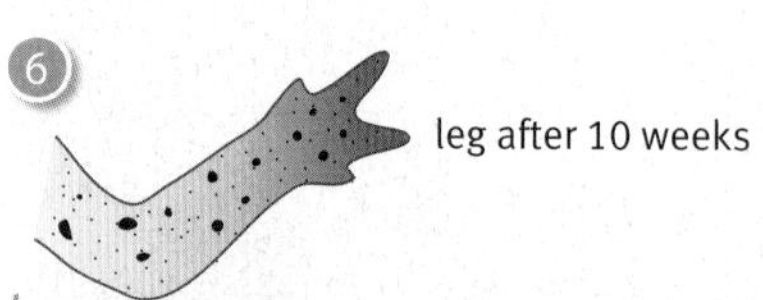

If a newt's limb is bitten off by a predator, it can grow a replacement. Most animals can only make small repairs to their body.

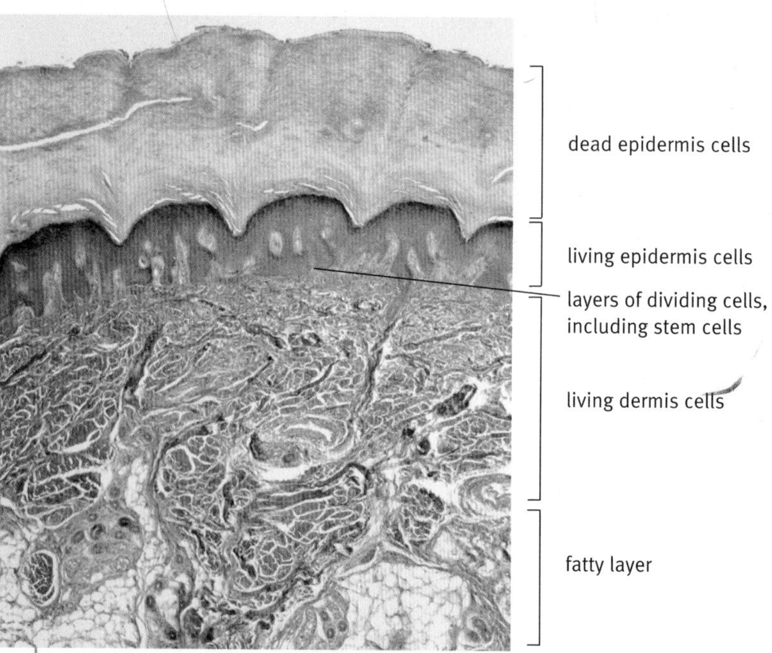

A cross section through human skin. Some of the stem cells continue to grow and divide. Others replace skin cells at a wound or those that wear off at the surface. (Magnification ×36.)

You replace millions of skin cells every day. Most of the dust in your home is worn-off skin cells. Other tissues in your body need a constant supply of new cells. For example, bone marrow contains stem cells to make new blood cells.

Using meristems to make more plants

In module B1, you saw that gardeners use meristems when they grow new plants by taking **cuttings**. Cuttings are just shoots or leaves cut from a plant. In the right conditions they develop roots and grow into new plants.

Some cuttings grow new roots when you put them in water or compost. Others grow better when you dip the cut ends in **rooting powder** before you plant them. Rooting powder contains plant hormones called **auxins**. Auxins cause the new cells produced by the base of the shoot meristem to develop into roots.

By taking cuttings, gardeners can produce lots of new plants quickly and cheaply. But this is not the only reason that they do it. All the cuttings taken from one plant have identical DNA. As they are genetically identical, they are called clones. So taking cuttings is a good way of reproducing a plant with exactly the features that you want. When flowering plants produce seeds, they are reproducing sexually. So new plants grown from seeds vary. They are not identical.

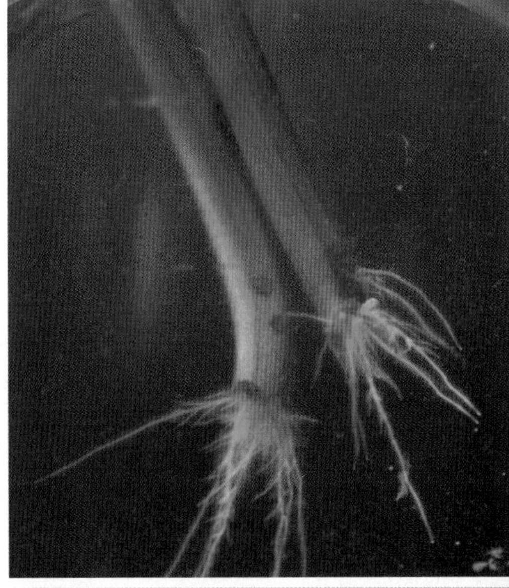

This willow cutting needed only water to grow.

Rooting powder.

Questions

1 For each of these types of cell, say whether they are fully unspecialised or not:
   a meristem cells
   b adult human stem cells.

2 Explain why a newt can regrow a leg but a human cannot.

3 Give two reasons for growing plants from cuttings.

4 Explain how rooting powder helps a plant cutting to grow.

5 A gardener wants to grow dahlias with a variety of colours and sizes. Should she grow them from cuttings or seeds? Explain your answer.

Key words

- ✓ **unspecialised**
- ✓ **stem cells**
- ✓ **cuttings**
- ✓ **auxins**
- ✓ **rooting powder**

C Phototropism

Find out about

- ✔ why plants grow towards light

Plants rooted in soil cannot move from place to place – not even the 'walking palm' tree in the picture below. Plant seeds can land anywhere. To survive, the growing plant will need to get enough light for photosynthesis. The plant needs to sense and respond to the environment. In nature, plants continually compete with each other in an environment that is constantly changing.

This houseplant has grown towards the window to increase the amount of light falling on its leaves.

The walking palm tree, *Socratea durissima*, in Costa Rica. New roots grow towards a sunny patch and pull the stem and leaves towards the light. Hormones control the direction of root growth. Older roots in the shade die.

You may have noticed that plants on windowsills seem to bend towards the light. They are not moving, but growing. The direction the light comes from affects the direction of plant growth. This is called **phototropism**. Phototropism increases a plant's chances of survival.

Questions

1. Write a definition for phototropism.
2. How does the walking palm tree grow towards the light?
3. Explain why a plant benefits from bending towards the light.

Darwin's phototropism experiments

Charles Darwin experimented with phototropism. He showed that the young shoots of grasses:

- normally grew towards light
- remained straight when he covered their tips.

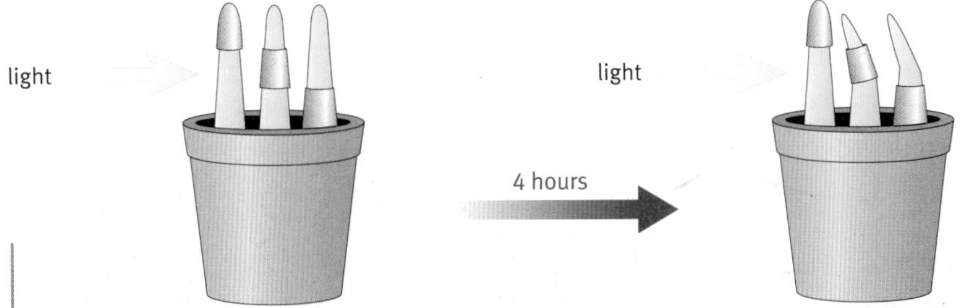

Foil covers different parts of the barley shoots.

In the experiment, shown in the picture above, covering the lower parts of the shoot did not stop bending towards the light. This shows that only the tip is sensitive to light. The shoot bends below the tip – where cells are no longer dividing but are increasing in length.

Darwin did not know how bending towards light happened but his results allowed him to explain which parts of the plant sense and respond to light. Now scientists have found that higher concentrations of plant hormones called auxins cause shoot cells to expand. The diagrams on the right explain how this causes phototropism.

Questions

4 Look at the diagram showing phototropism experiments in barley. Suggest where:
  a the shoot detects light
  b the cells are growing very quickly to cause bending.

5 Explain why Darwin could not develop a full explanation from the data he collected on phototropism.

6 Distinguish between hypothesis, theory, explanation, and data in the experiment on phototropism.

1

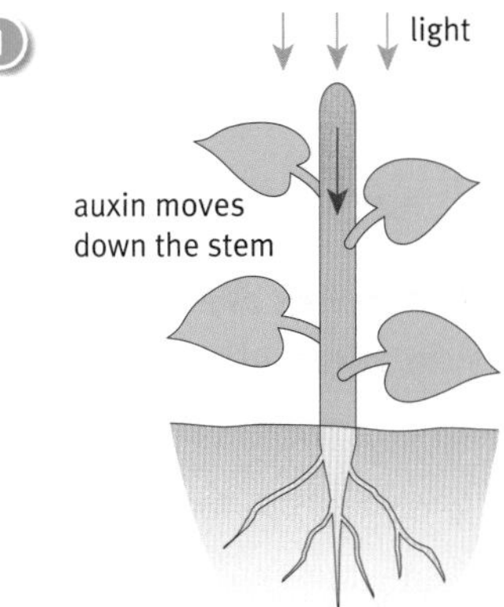

When a growing shoot gets light from above, the auxins spread out evenly and the shoot grows straight up.

2

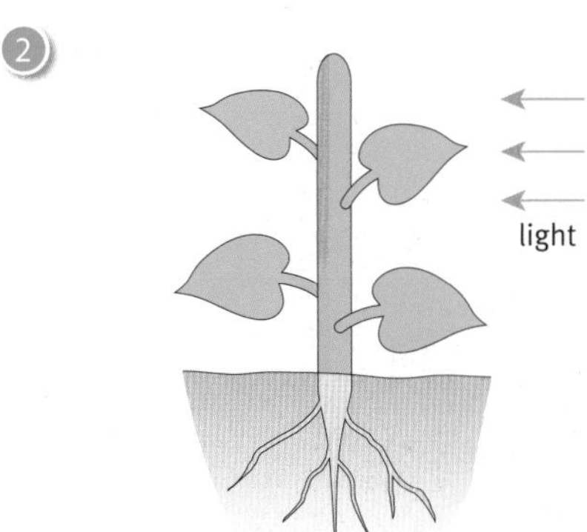

When light comes from one side, the auxins move over to the shaded side.

3

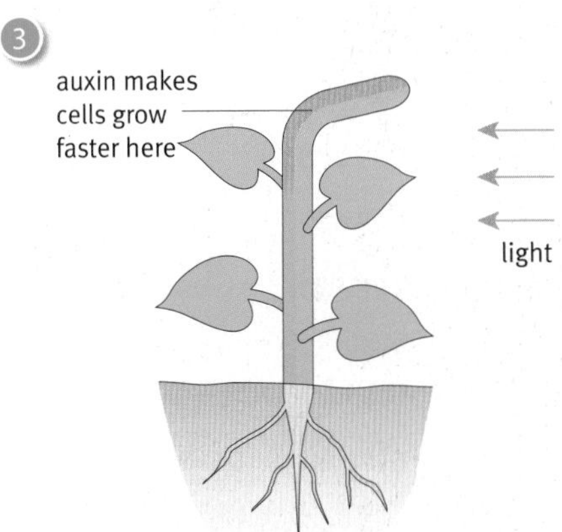

The shoot grows faster on the shaded side, making the shoot bend towards the light.

How auxins explain phototropism.

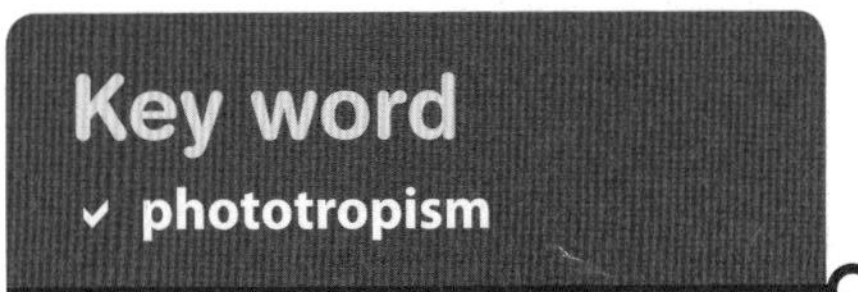

Key word

- phototropism

A look inside the nucleus

Find out about

- ✔ where genes are kept inside your cells

All cells start their lives with a nucleus. A few specialised cells lose their nuclei when they finish growing. Human red blood cells are one example. Their job is to carry oxygen attached to haemoglobin molecules.

Red blood cells develop from stem cells in your bone marrow. As they develop, they make more and more haemoglobin. By the time they leave the bone marrow, they are full of haemoglobin and their nuclei have broken down.

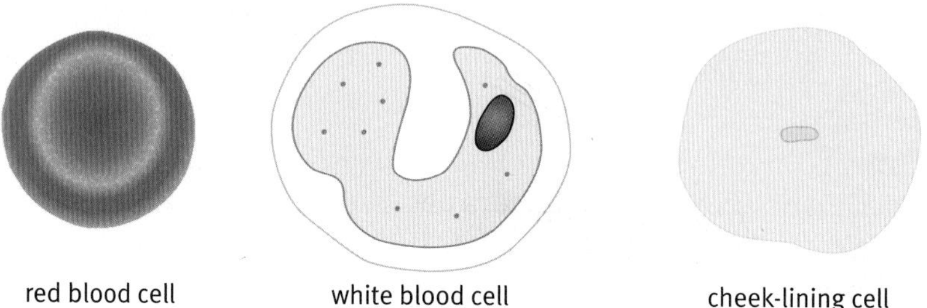

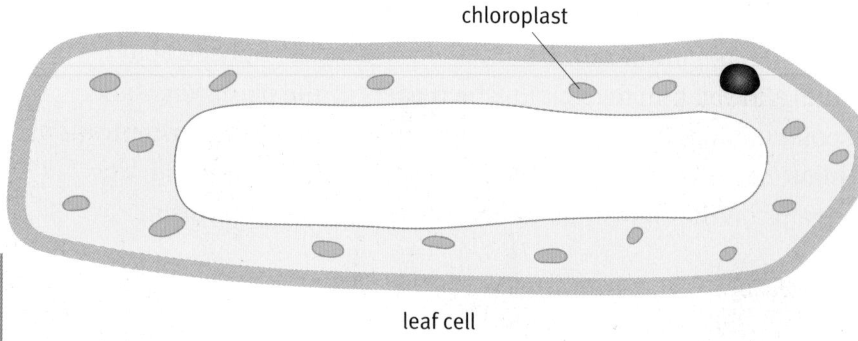

Cells vary in size and shape.

Organism	Estimated gene number	Chromosome number
human	~30 000	46
mouse	~30 000	40
fruit fly	13 600	8
Arabidopsis thaliana (plant)	25 500	5
roundworm	19 100	6
yeast	6 300	16
Escherichia coli (bacterium)	3 200	1

Chromosomes

A **chromosome** is a long molecule of DNA wound around a protein framework. You have about a metre of DNA in each of your nuclei. This is made up of about 30 000 **genes**. Each gene codes for a protein or part of a protein.

Different species have different numbers of chromosomes and different numbers of genes (see the table on the left).

You have 23 pairs of chromosomes in your nuclei. You got one set of 23 from your mother's egg cell nucleus and the other set from the nucleus of your father's sperm.

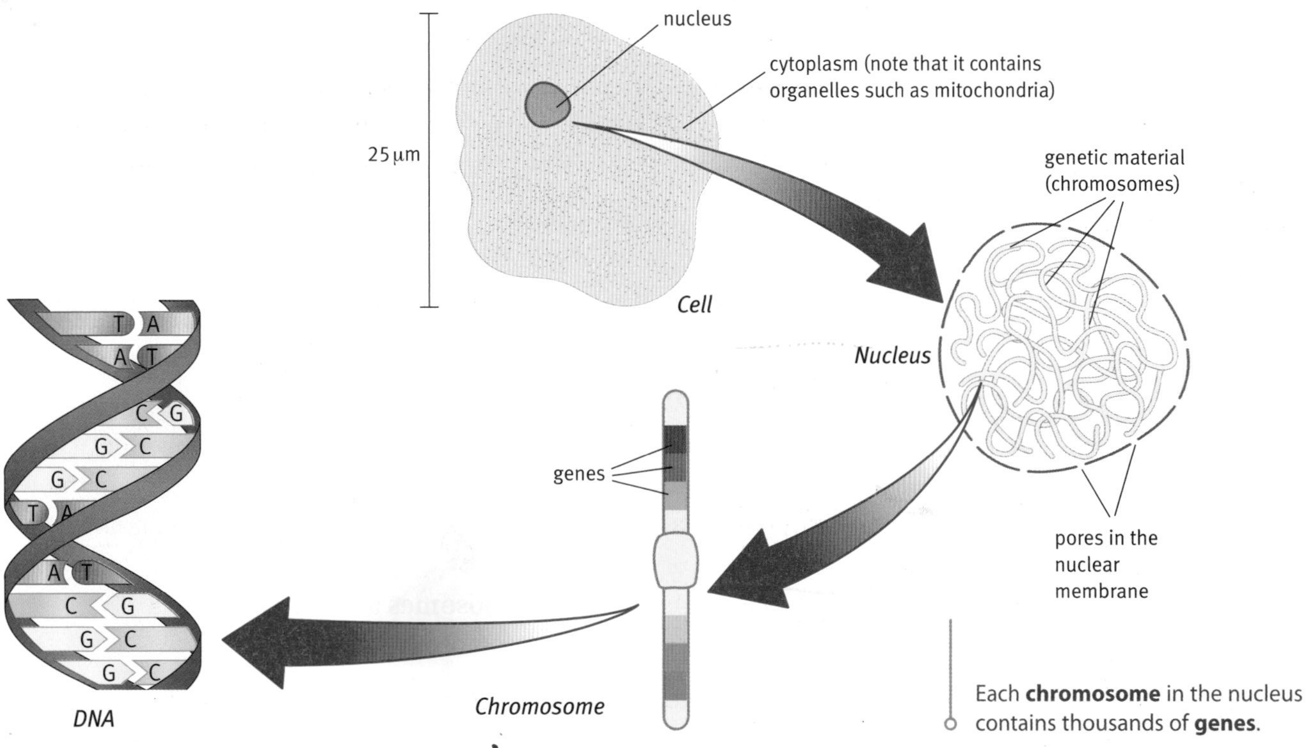

Each **chromosome** in the nucleus contains thousands of **genes**.

These people are more alike than it appears. 99.9 % of their genes are the same.

Key words

- chromosomes
- genes

What is special about DNA?

You will find out later in the module that the molecule of DNA has a particular structure that allows it to:

- make exact copies of itself
- provide instructions so that the cell can make the right proteins at the right time.

Questions

1. Name two ways in which your red blood cells are different from the other cells in your body.
2. Suggest why red blood cells wear out and have to be replaced every 2–3 months.
3. How many different types of protein are there likely to be in yeast?
4. Which organism has half as many genes as yeast?
5. What two properties does DNA have that make it work as genetic material?

E Making new cells

Find out about

- ✓ how your cells divide for growth and to repair your body

Imagine a space probe bringing back objects like this from Mars. Scientists would need to find out whether they were alive or not. It might be a living organism able to colonise Earth!

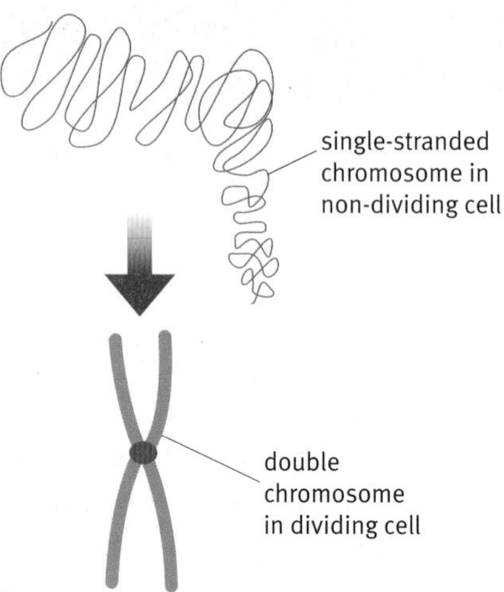

You can see chromosomes only in dividing cells.

Key words

- ✓ organelles
- ✓ mitosis
- ✓ ribosomes
- ✓ mitochondria

It is not always easy to tell whether something is a living organism or not.

You could ask yourself:

- can it grow and reproduce?
- is it made of cells?

Life cannot exist without the growth, repair, and reproduction of cells.

Cell division

When new body cells are made, they contain the same number of chromosomes as each other and the parent cell. They also contain the same cell parts, called **organelles**. So, before a cell divides, it must grow and make copies of:

- other **organelles** – such as **ribosomes** and **mitochondria**
- its nucleus, including the chromosomes.

Only then does the cell divide. This part of the process is called **mitosis**.

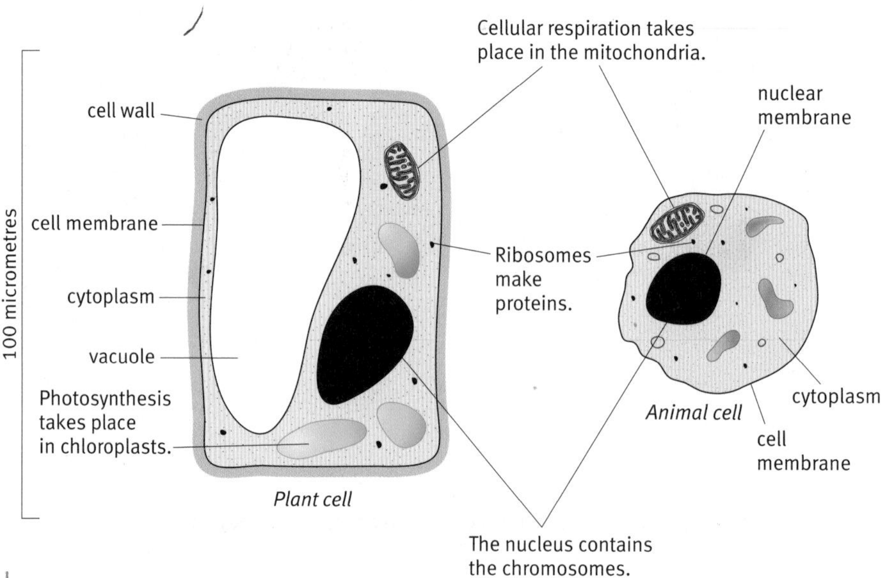

Cell organelles, such as mitochondria, are copied before a cell divides.

Copying chromosomes

You can see the chromosomes in dividing cells by using a light microscope. The DNA is too spread out in other cells to be visible. After the chromosomes are copied, the DNA strands become shorter and fatter. You can read more about how DNA is copied in Section G.

Mitosis

During mitosis, copies of chromosomes separate and the whole cell divides.

First, a complete set of chromosomes goes to each end of the dividing cell and two new nuclei are formed. A complete set of organelles also goes to each end. Then the cytoplasm divides to form two identical cells. In plant cells, a new cell wall forms too.

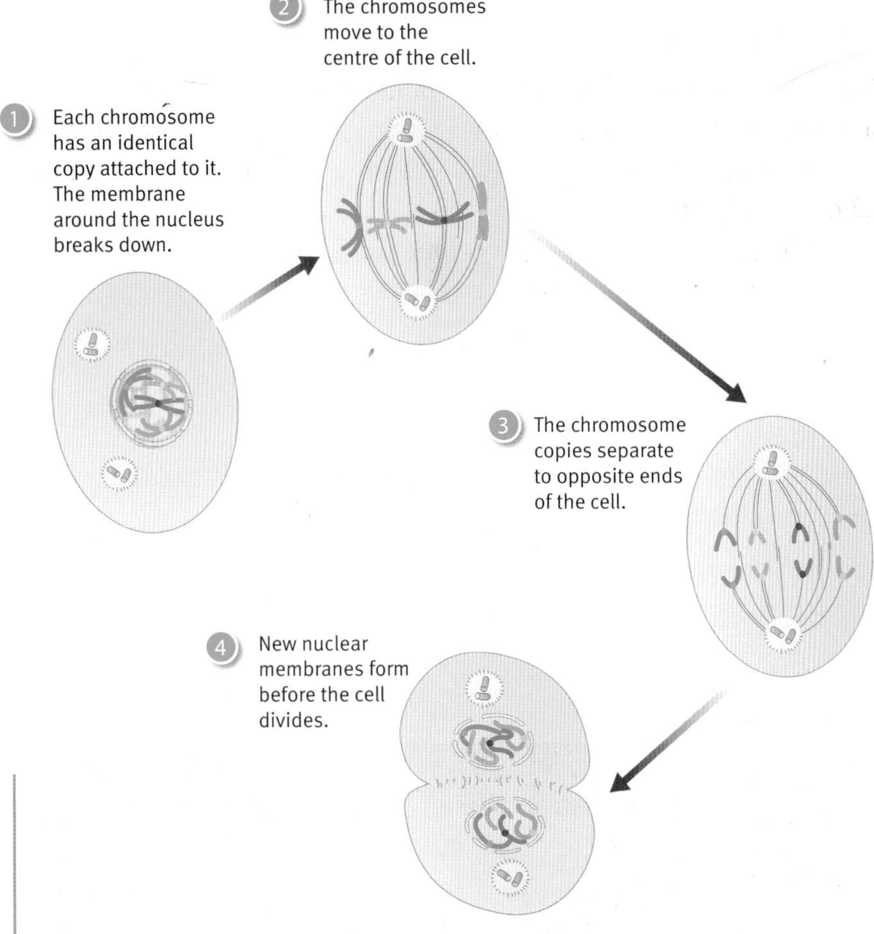

The pictures show what happens in mitosis in an animal cell. They show only four chromosomes.

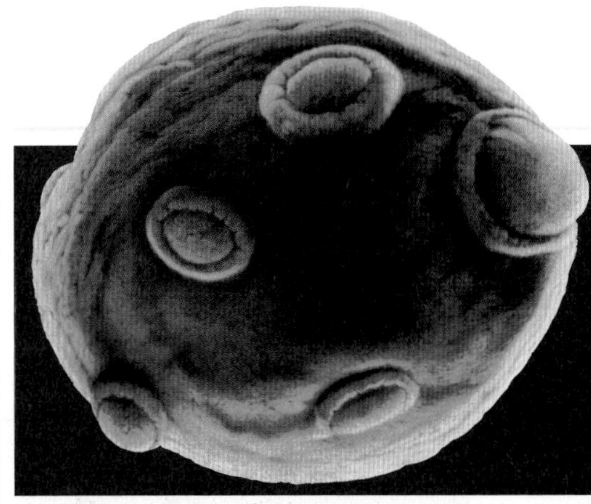

In yeast, new cells grow as buds from the parent.

Daffodil bulbs divide to form new ones.

Mitosis and asexual reproduction

Some plants and animals reproduce asexually. They use mitosis to produce cells for a new individual.

This means that each of the individuals produced in asexual reproduction is genetically identical to the parent, so it is a clone of its parent.

Questions

1 a How many cells are made by mitosis?
  b What are these new cells like compared with their parent cell?

2 What might scientists do to find out whether the objects from Mars in the cartoon opposite are living things or not?

3 Before a cell divides, it grows. What two steps happen during cell growth?

4 What two main steps happen during cell division by mitosis?

F Sexual reproduction

Find out about

- cell division to make gametes

Most plants and animals reproduce sexually. Males and females make sex cells or **gametes**, which join up at fertilisation. The fertilised egg, or zygote, develops into the new life.

It is not always obvious which plant or animal is male and which is female. Some plants and animals are both.

Only the male peacock has magnificent tail feathers.

The berries show that the holly on the right is female. You cannot be sure about the one on the left.

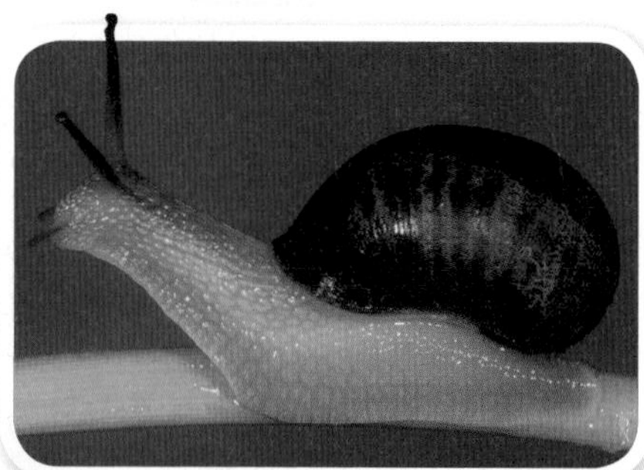

A snail has both male and female sex organs.

The only way to be sure about the sex of an organism is to look at its gametes. Males have small gametes that move. Females have large gametes that stay in one place.

Human males produce sperm in their testes. Females produce egg cells in their ovaries.

250 micrometres

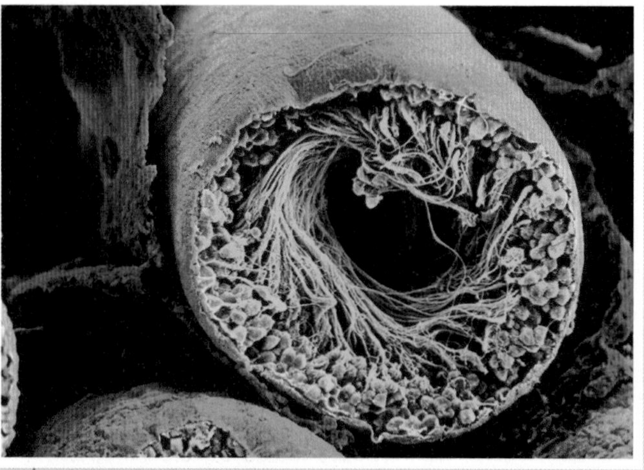

Sperm develop in tubules in the testes.

600 micrometres

Pollen contains the male gametes of a flowering plant.

Male gametes are usually made in very large numbers. They move to the female gamete by swimming or being carried by the wind or an insect.

What is special about gametes?

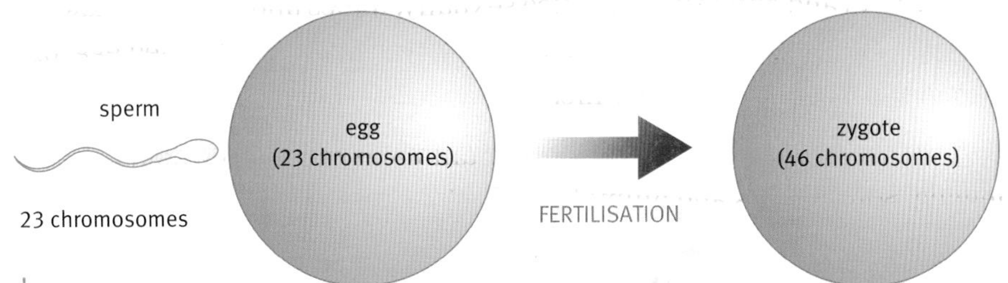

Meiosis halves the number of chromosomes in gametes. Fertilisation restores the number in the zygote.

Human body cells have 23 pairs of chromosomes – 46 in total. Gametes have only 23 single chromosomes. This is important because when a sperm cell fertilises an egg cell, their nuclei join up. The fertilised egg cell (zygote) gets the correct number of chromosomes: 23 pairs – 46 in total. Half your chromosomes come from your mother and half come from your father. Gametes are made by a special kind of division called **meiosis**.

In humans, meiosis makes gametes that:

- have 23 single chromosomes (one from each pair)
- are all different – no two gametes have exactly the same genetic information.

Offspring from sexual reproduction are different from each other and from their parents. We say that they show **genetic variation**.

Meiosis

Meiosis starts with normal body cells. It only happens in sex organs.

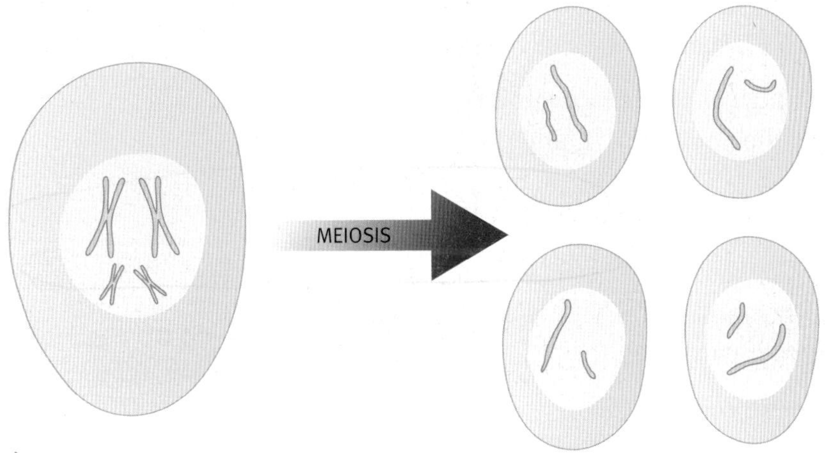

The chromosomes in this diagram have been copied. The parent cell divides twice, producing four cells. There are four cells after meiosis. They have half the number of chromosomes as the parent cell.

Key words

- gametes
- meiosis
- genetic variation

Questions

1 Look again at the photos of holly at the top of the opposite page. You can be sure that the holly on the right is female. Why can you not be sure about the sex of the holly on the left?

2 Why is it important that gametes have only one set of chromosomes?

3 Why are male gametes made in such large numbers?

The mystery of inheritance

Find out about

- the structure of DNA
- how DNA is copied for cell division

In 1865 Gregor Mendel published his work on pea plants. You learnt about this in module B3. Mendel's data showed how features could be passed on from parents to their young. The explanation for how information is passed on did not simply emerge from Mendel's data. It took the work and creative thinking of many scientists to explain how it happens.

1859	A chemical was extracted from nuclei and named 'nuclein'.
1944	'Nuclein' was recognised as genetic material.
Late 1940s	Erwin Chargaff discovered a pattern in the number of bases in DNA.
1951	Linus Pauling and Robert Corey showed that proteins have a helix structure.
1952–3	Rosalind Franklin and Maurice Wilkins produced X-ray diffraction pictures of DNA. They showed that the molecule had a regular, repeating structure.

Some of the discoveries that led up to the discovery of DNA structure. These discoveries provided explanations that accounted for Mendel's data. Predictions based on the explanations were then tested by further experiments.

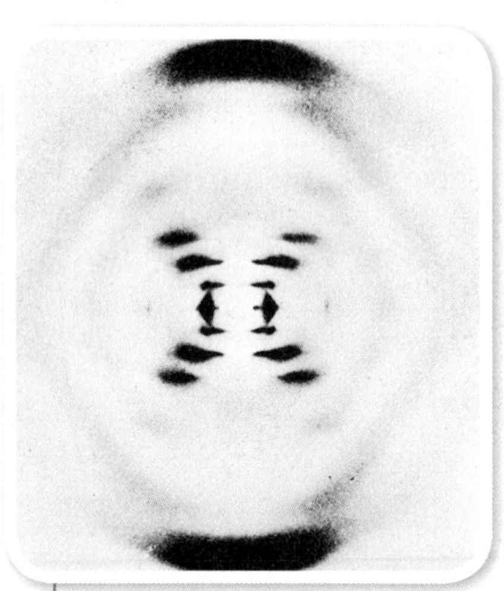

X-ray diffraction pictures of DNA, like this, show a repeating pattern.

Solving the mystery

In 1953, Francis Crick and James Watson published their now famous paper 'A Structure for Deoxyribose Nucleic Acid', in the scientific journal *Nature*. Their paper brought together all the work done on DNA. They used it to work out the **double-helix** structure of DNA.

Watson (left) and Crick reveal their model of DNA.

Base pairing

There are four special molecules called 'bases' in DNA: adenine (A), thymine (T), guanine (G), and cytosine (C).

Erwin Chargaff had discovered that the amount of A is always the same as the amount of T, and the amount of G is the same as the amount of C. This is true no matter what organism the DNA comes from.

Crick and Watson concluded from this evidence that:

- A always pairs with T
- G always pairs with C.

This is **base pairing**.

The double helix

Watson made cardboard models of the bases. He found that A+T was the same size as G+C. Suddenly, he realised what that meant. In a molecular model, he and Crick fitted the pairs of bases between two chains of the other chemicals in DNA, a sugar called deoxyribose and phosphates. It worked. The shape turned out to be a double helix – a bit like a twisted ladder. Better still, it matched the X-ray evidence.

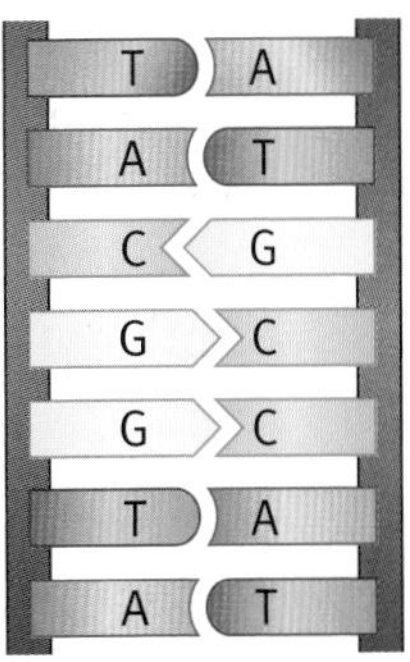

The rungs of the ladder are the pairs of bases held together by weak chemical bonds.

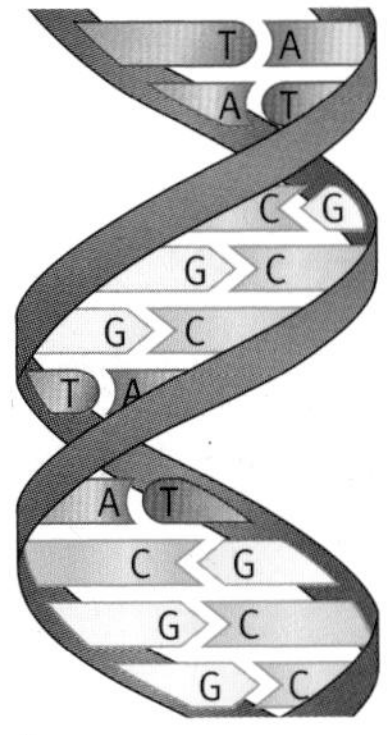

There are ten pairs of bases for each twist in the helix.

How DNA passes on information

Watson and Crick's structure not only accounted for what was known about DNA so far, it also let them predict how DNA might copy itself exactly. Base pairing means that it is possible to make exact copies of DNA:

- Weak bonds between the bases split, unzipping the DNA from one end to form two strands.
- Immediately, new strands start to form from free bases in the cell.
- As A always pairs with T, and G always pairs with C, the two new chains are identical to the original.

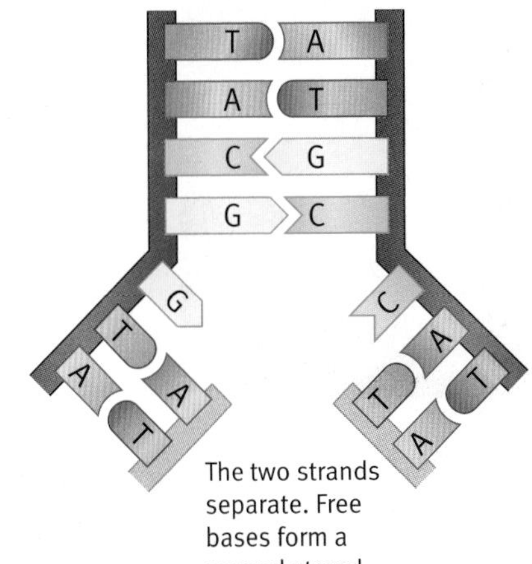

The two strands separate. Free bases form a second strand.

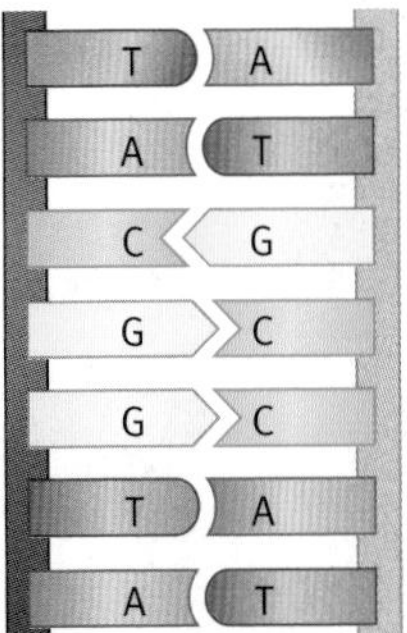

Each DNA molecule is made of half old DNA (black) and half new DNA (grey).

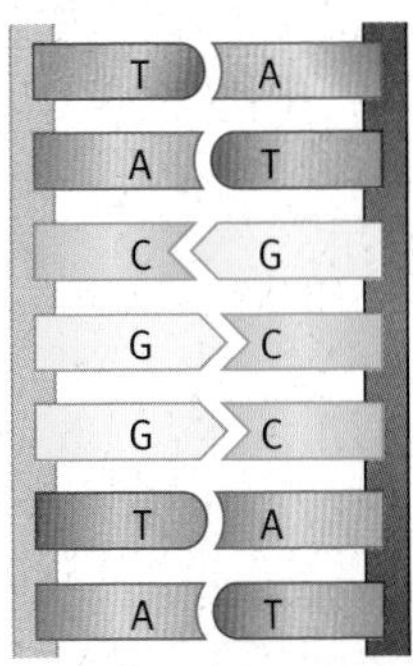

Each half of the split DNA molecule is complete, making two identical DNA molecules.

The structure of DNA (schematic).

Questions

1. The shape of a DNA molecule is sometimes described as a twisted ladder.
   a. What makes the sides of the ladder?
   b. What part of the ladder are the bases?
2. The bases in a DNA molecule always pair up the same way. Which base pairs with:
   a. A?
   b. C?
   c. G?
   d. T?
3. Describe what happens when a DNA molecule is copied.
4. Which observations made by other scientists did Watson and Crick's model account for?
5. Which aspect of Watson and Crick's work could be described as 'creative thinking' rather than making simple deductions from the data?

Key words

- ✓ **double helix**
- ✓ **base pairing**

Find out about

- how DNA controls which protein a cell makes

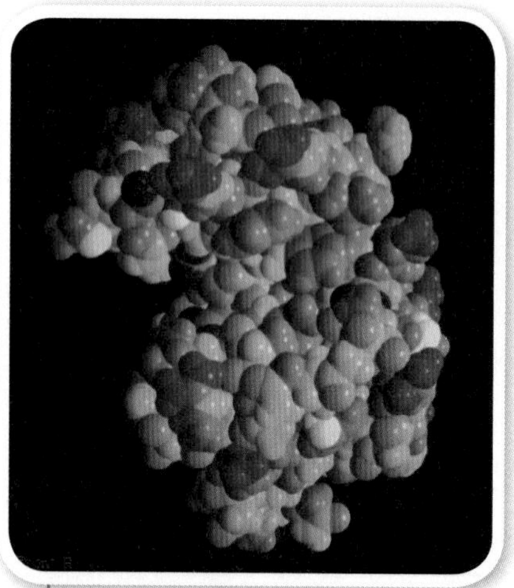

Enzymes work because of the shape of their active sites.

The great number of jobs carried out by proteins means that they are very different from one another. The exact shape of a protein can be very important to how it works. Cells make proteins from about 20 different **amino acids**. Chains of between 50 and many thousands of amino acids are formed. In each protein, the amino acids are joined in a particular order, but there are thousands of possibilities. In the same way we can make thousands of different words using only 26 letters in our alphabet. The order of the amino acids fixes the way the chains of amino acids fold to form the three-dimensional shape of the protein.

The genetic code

In 1961, Crick worked out that three bases on a DNA strand code for each amino acid. This is called a **triplet code**. Different combinations of the four bases (A, T, G, and C) produce 64 triplet codes. So there is more than one code for each amino acid. Crick guessed this because any fewer bases would not give enough codes for every amino acid. There are also codes for start and stop. They mark the beginning and end of a gene. Subsequent experiments have demonstrated that Watson and Crick's explanations were correct. Their work forms the foundation of the science of genetic engineering.

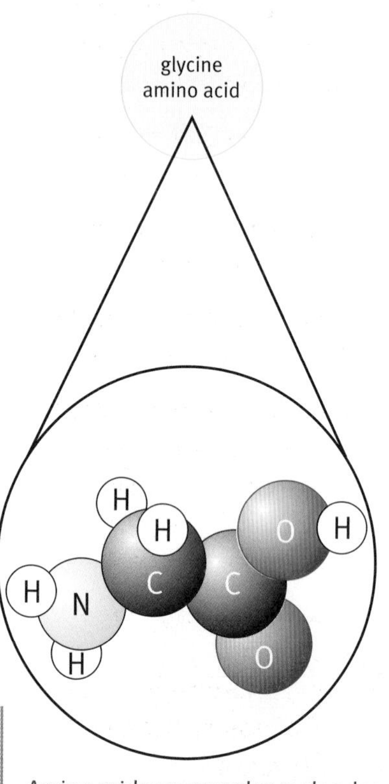

Amino acids are complex molecules. The diagrams on this page and the next represent the molecule simply.

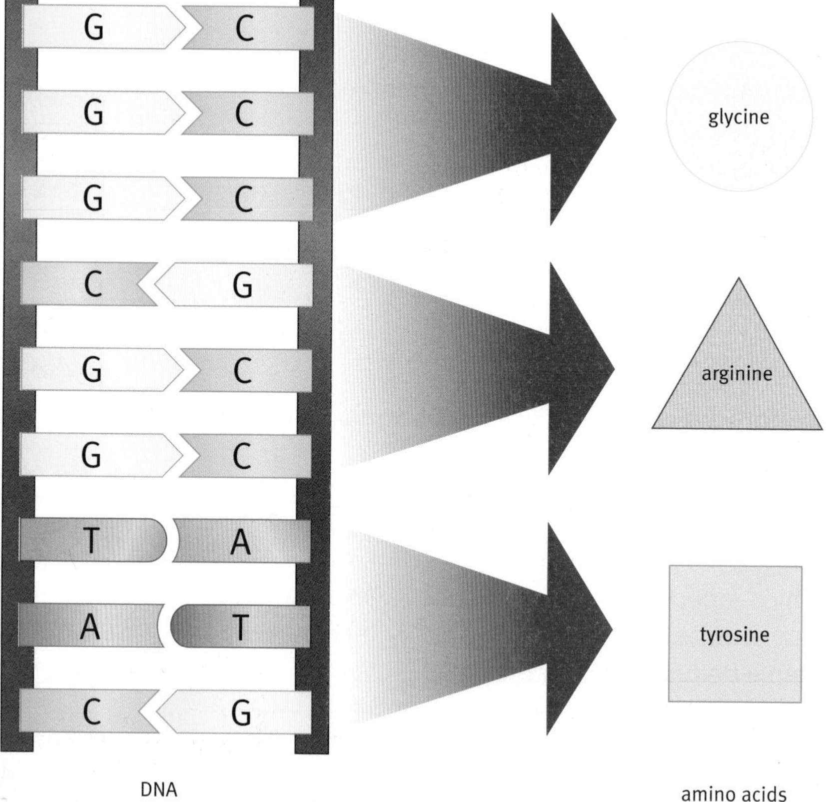

Three bases on the DNA code for each amino acid. For example, TTT codes for lysine.

Which part of a cell makes proteins?

DNA in the nucleus contains the genetic code for making the proteins. The proteins are made on tiny organelles in the cytoplasm, called ribosomes. Genes cannot leave the nucleus. So how do ribosomes get the instruction for making a protein? A molecule small enough to get through the pores of the nuclear membrane transfers the genetic code to the ribosomes. This smaller molecule is called messenger RNA (mRNA).

The differences between DNA and mRNA are that mRNA has:

- only one strand
- the base uracil (U) in place of the thymine (T) in DNA.

The diagram below shows how a protein is made.

1. The gene unzips, and mRNA bases pair with DNA bases to form a strand of mRNA. U matches up with A instead of T.

2. The RNA moves out of the nucleus to one of the many ribosomes in the cytoplasm.

3. The ribosome attaches to one end of the mRNA. As it moves along the mRNA, the ribosome reads the genetic code so that it can join the amino acids together in the correct order. When it has finished, the ribosome releases the protein into the cytoplasm and starts to make another one.

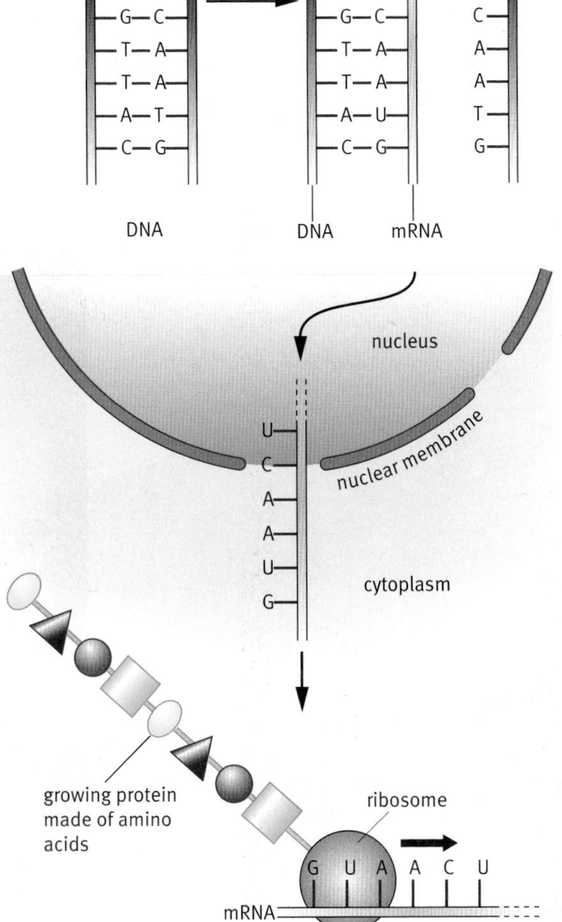

DNA makes protein with a ribosome's help (schematic).

Key word

- amino acids

Questions

1. Why are instructions for making proteins copied onto mRNA?
2. How many DNA bases code for each amino acid?
3. What is the DNA triplet code for tyrosine?
4. Which amino acid has the DNA code GCC?
5. Make bullet-point notes to explain how a protein is made. Your first bullet point should be:
   - the gene unzips.

   Your last bullet point should be:
   - the ribosome releases the protein into the cytoplasm.

Specialised cells – special proteins

Find out about

- some different proteins in your body
- why cells become specialised

An oak tree has about 30 different types of cell. Your body has more than 300 types of cell. Each cell type has its own set of proteins.

Some proteins make up the framework of cells and tissues. These are **structural proteins**. If we take away all the water in an animal cell, 90% of the rest is proteins.

Protein	Found in ...	Property
keratin	hair, nails, skin	strong and insoluble
elastin	skin	springy
collagen	skin, bone, tendons, ligaments	tough and not very stretchy

Different structural proteins have different properties.

The flesh of meat and fish is the animals' muscles. Muscles are mainly protein and provide the protein in many people's diets. Plant seeds such as soya and other beans are rich in proteins and are the basis of many vegetarian meals.

Other proteins are essential for the chemical reactions that keep our bodies working. For example, **enzymes** speed up the chemical reactions in a cell. **Antibodies** are the proteins that help to defend us against disease.

An oak tree has about 30 types of cell.

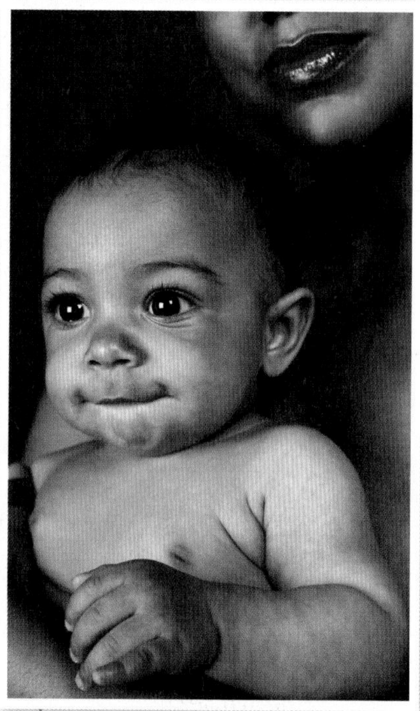

Humans have over 300 different types of cell.

Questions

1 Name three types of protein in your body.

2 Name one structural protein and say how it is suited to do its job.

What is the link between genes and proteins?

All the cells in your body come from just one original cell, the zygote. This divides to form a ball of cells. Soon, cells start to specialise. They make the proteins needed to become a particular type of cell.

DNA is a cell's genetic code. Each gene is the instruction for a cell to make a different protein. By controlling what proteins a cell makes, genes control how a cell develops.

Each of your cells has a copy of all your genes. Something must make some cells turn into nerve cells and others into heart cells and all the other cell types. There must be **genetic switches**, but how they work is still being researched by scientists.

Key words

- ✓ enzymes
- ✓ genetic switches
- ✓ structural proteins
- ✓ antibodies

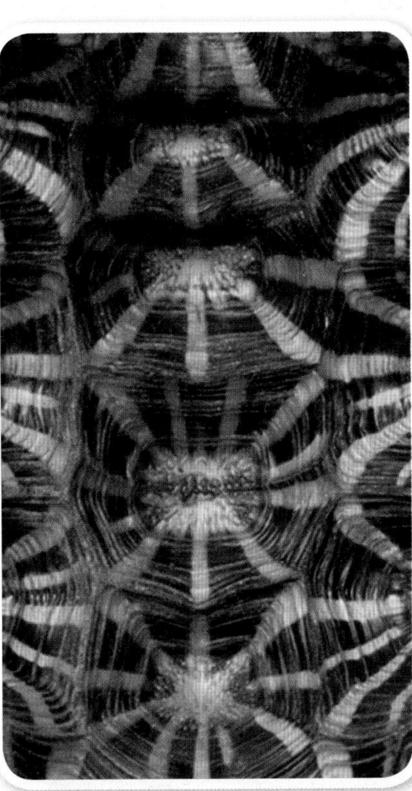
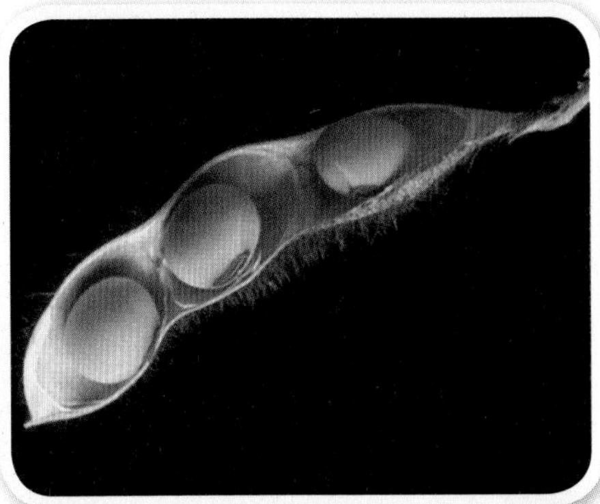

Rhino horn, tortoise shell, soya beans, steak … lots of different proteins.

Questions

3 DNA contains a cell's genetic code. What does DNA do in a cell?

4 How do genes control how a cell develops?

5 Where would you find cells that make these proteins:
- a collagen?
- b amylase?
- c haemoglobin?

This is one chromosome from a salivary gland of a midge. The green areas are the active genes where DNA unravels whilst the protein is being made. One of these is instructing the cell to make a lot of amylase.

Up to the eight-cell stage, each cell of a human embryo can develop into any kind of cell – or even into a whole organism.

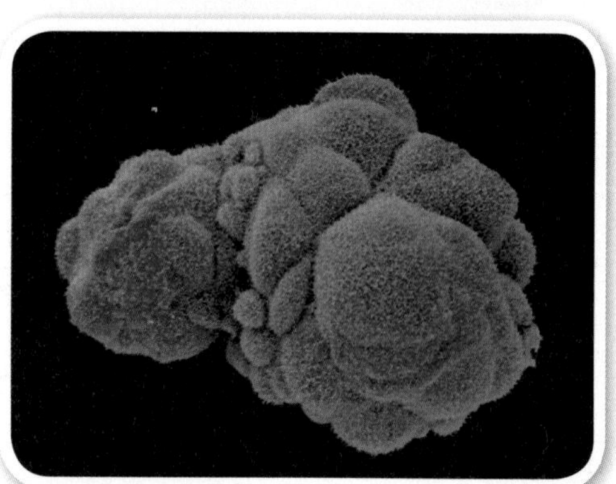

At six days and about 50 cells, the cells in this human embryo are already specialised and will form different types of tissue. The different cells can no longer develop into a whole organism.

Gene switching

Each gene controls the manufacture of one type of protein. So, in any organism, there are as many genes as there are different types of protein. In humans there are 20 000 to 25 000 genes.

Not all these genes are active in every cell. As cells grow and specialise, some genes switch off.

In a hair cell, the genes for the enzymes that make keratin will be switched on:

hair cell genes switched on → enzymes for making keratin → hair grows

But the genes for those that make amylase will be switched off.

In a salivary gland cell:

salivary gland cell genes switched on → amylase secreted → starch digested

Gene switching in embryos

An early embryo is made entirely of embryonic stem cells. These cells are unspecialised up to the eight-cell stage. All the genes in these cells are switched on. As the embryo develops, cells specialise. Different genes switch off in different cells.

unspecialised cell

All the genes are switched on in this chromosome.

hair cell

salivary gland cell

Different genes are switched off in specialised cells.

Key

gene switched on

gene switched off

Some proteins are found in each type of cell, for example, the enzymes needed for respiration. All cells respire, so the genes needed for respiration are switched on in all cells.

In adults, there are stem cells in parts of the body where there is regular replacement of worn out cells. These can only develop into cells of the particular organ where they are positioned. So some of their genes must be switched off.

Right cell, right place

Compare the fingers on your right hand with the same fingers on your left hand. They are probably almost mirror images of each other. As we grow, the position and type of cells must be controlled, so each tissue and organ develops in the right place.

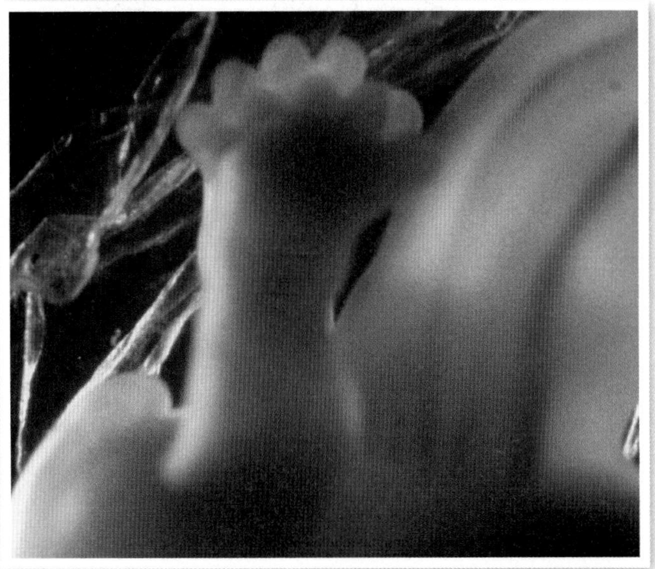

Cells near the end of a limb will make fingers. Cells nearer the body will make the arm. This happens because of the difference in the concentrations of chemical signals in each region of the embryo.

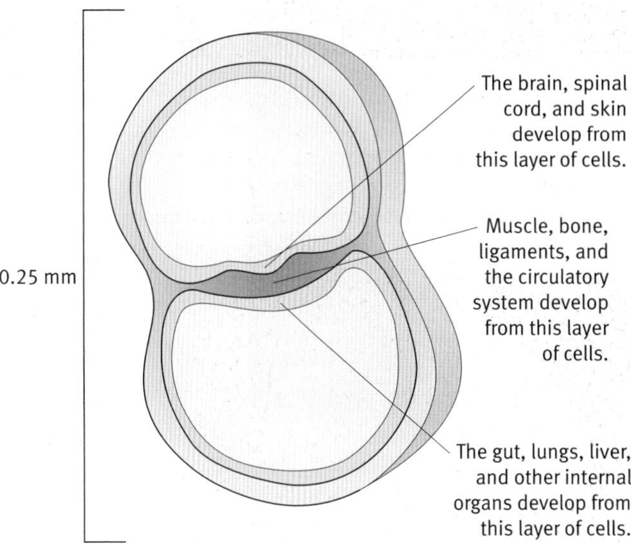

It is possible to map specialised parts of the body onto a diagram of a 14-day-old human embryo. Here you can see which groups of cells in the embryo will develop into future tissues and organs.

Questions

6 Suggest a function, other than respiration, that all cells carry out.

7 At the eight-cell stage of any embryo, how many genes are switched on?

8 What is the evidence that some genes are switched off at the 50–100-cell stage?

Stem cells

Find out about

- ✓ scientific research to use stem cells for treating some diseases

A lot of research into stem cells is currently going on. This is because many scientists see the possibility of using them for:

- the treatment of some diseases
- the replacement of damaged tissue.

Imagine if scientists could produce …	They might use them to treat …
nerve cells	Parkinson's disease and spinal cord injuries
heart muscle cells	damage caused by a heart attack
insulin-secreting cells	diabetes
skin cells	burns and ulcers
retina cells	some kinds of blindness

Scientists grew this skin from skin stem cells in sterile conditions. Doctors use it for skin grafts.

The problem is to find stem cells of the correct type and then to grow them to produce enough cells. Stem cells can come from early embryos, umbilical cord blood, and adults. Embryonic stem cells are the most useful because the cells are not yet specialised. All their genes are still switched on up to the eight-cell stage.

There are problems with this new technology, which is called **therapeutic cloning**. For example, tissues from the embryonic stem cells do not have the same genes as the person getting the transpant. Transplanted tissue is rejected if your body recognises that the cells are not from your body.

Key word

- ✓ therapeutic cloning

Cloning from your own cells

Another possibility is to remove the nucleus of a zygote and replace it with the nucleus from a patient's own body cell. The new embryo would have the same genes as the patient. So the embryonic stem cells produced match those of the patient.

Therapeutic cloning.

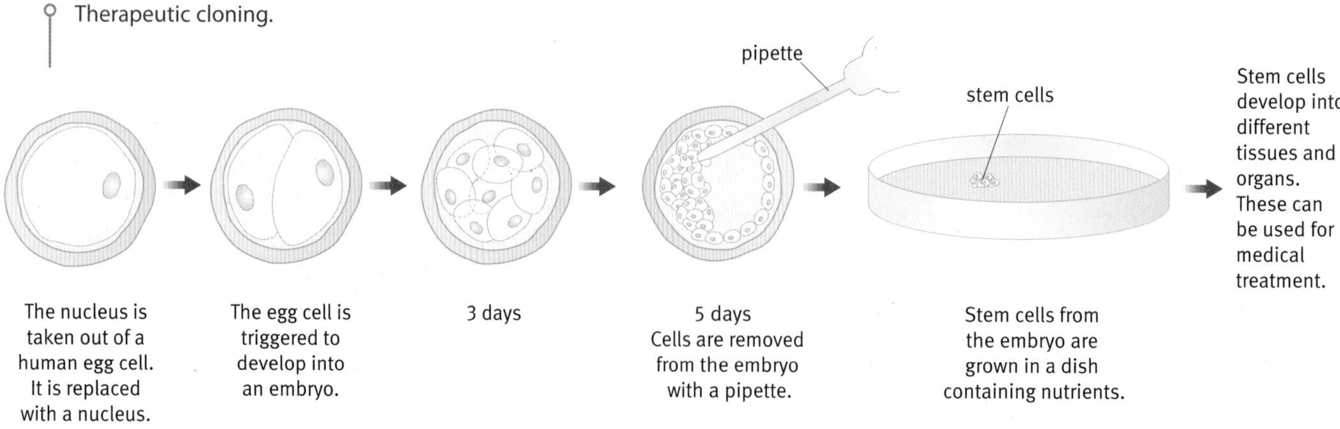

Using adult stem cells

A way to avoid using embryos would be use to use the 'adult' stem cells already in your body. Adult bone marrow stem cells from a donor are routinely used to treat leukaemia (cancer of the blood). The stem cells in the transplant restore healthy blood-cell production in the recipient's own bone marrow.

More useful 'adult' stem cells can be harvested from the placenta once a child is born. Some companies will store your milk teeth in a special freezer so that the stem cells they contain could be used to treat you later. These stem cells can grow into a range of specialised cells.

Most adult human stem cells grow into only a few cell types, because many genes are switched off. Scientists have recently found ways to switch common adult stem cell genes back on. It might be possible to use a person's own stem cells to replace any of their damaged cells. There wouldn't be a risk of rejection and the cells might restore organs that could not otherwise heal themselves.

Some companies will store your baby teeth. The stem cells that they contain might help to treat you years later.

Some successes

The national regulatory bodies have not approved any treatments yet for widespread use, but trials and experiments are underway. Doctors need to be sure that the cells will behave in the right way when they are implanted into the body.

Claudia Castillo received a new windpipe made with cartilage produced by her own stem cells. Her lungs had been badly damaged by illness.

- A donor windpipe was treated with enzymes so that all the cells were removed.
- Stem cells were harvested from the patients' bone marrow.
- In the lab the stem cells developed into cartilage cells.
- The 'scaffolding' of connective tissue was coated with the cells.
- Surgeons implanted the new windpipe.
- The patient recovered – there was no rejection.

This patient received a new windpipe. The cartilage had been made using her own stem cells.

Questions

1. Why is therapeutic cloning using embryonic stem cells controversial?
2. Describe how stem cells from a donor could treat leukaemia.
3. Name two potential sources of adult stem cells.
4. What are the advantages of using stem cells from your own body to treat an illness?
5. Independent regulators oversee the use of gametes and embryos in fertility treatment and research in the UK. Why is it important that regulators are independent from government and the scientists carrying out research?

Science Explanations

Genetic technologies are at the cutting edge of modern science. The study of proteins made in cells, stem cell technology, and cellular control are key areas of research. These fields promise to provide huge benefits to present and future generations.

You should know:

- that in multicellular organisms, cells are often specialised to do a particular job
- that up to the eight-cell stage, all the cells in a human embryo are identical (embryonic stem cells) and can produce any type of human cell
- why some cells (adult stem cells) remain unspecialised, becoming specialised at a later stage
- that in plants, only cells within special regions called meristems undergo cell division – these can develop into any kind of plant cell and are used to produce clones
- about the range of tissues that unspecialised plant cells can become
- that plant hormones (auxins) cause cut stems from a plant to develop roots and grow into a complete plant, which is a clone of the parent
- that the environment affects growth and development of plants, for example, plants grow towards light (phototropism), and how this happens
- how genetic information is stored within cells
- why cell division by mitosis produces two new cells identical to each other and to the parent cell
- what happens during the main processes of the cell cycle (cell growth and mitosis)
- why a special type of cell division called meiosis is needed to produce gametes
- that instructions for making proteins are coded for by genes in the nucleus, but proteins are produced in the cell cytoplasm
- that four different bases make up both strands of the DNA molecule and that these always pair up in the same way, A with T and C with G
- that the order of bases in a gene is the genetic code to produce amino acids to make a protein
- why body cells in an organism contain the same genes, but they only produce the specific proteins they need because those genes that are not needed are switched off
- how any gene can be switched on during development in embryonic stem cells to produce any type of specialised cell
- the applications that adult stem cells and embryonic stem cells can potentially be used for.

GROWTH AND DEVELOPMENT
growth
multicellular
specialisation
plant
animal
phototropism
auxins
xylem
phloem
stem cells
embryonic
adult
meristems
cloning
decisions
ethics
DNA
genetic code
A
T
C
G
bases
genes
switched on
switched off
developing scientific explanations
cell division
mitosis
cell cycle
organelles
chomosomes replicated
meiosis
gametes
female 1/2 chromosomes
male 1/2 chromosomes
zygote

Ideas about Science

In this module you learn more about growth and development. You should become more aware that scientists need to work out explanations of their observations and measurements.

- Watson and Crick discovered the structure of DNA in 1953. They looked at all the evidence gathered in the previous 50 years and produced a model that satisfied the evidence. It showed that DNA is a double helix.

You need to distinguish between data and the explanations that explain the data. You also need to suggest whether the data and the explanation agree or not. You need to recognise when creative thought is involved in producing an explanation.

- Watson and Crick's DNA structure accounted for what was known about DNA so far. It also let them predict how DNA might copy itself exactly.

Scientific explanations allow predictions to be made. If the prediction proves to be correct, this increases the confidence in the explanation. When a prediction proves to be wrong, either the prediction or the explanation may be incorrect.

The use of stem cells has ethical implications.

- The potential of stem cells to treat diseases and mend damaged tissue is enormous. Some stem cells come from human embryos, but there is a high risk of rejection if these stem cells do not have the same genes as the person getting the transplant. There are also ethical issues about using embryos in this way.
- One solution is to replace the nucleus of a zygote with a nucleus from a body cell of the recipient. This would then develop into an embryo – a clone of the person. Stem cells from the cloned embryo could be used to treat diseases without the fear of rejection. Some people think that producing clones of people in this way is wrong. Scientists are now studying how to switch genes back on in body cells to form cells of all tissue types.
- All of this work is subject to government regulation.

It is sometimes difficult to decide what is right and what is wrong. Some people think that the right decision is one that leads to the best outcome for the greatest number of people involved.

Other people think that certain actions are either right or wrong whatever the consequences. You need to identify arguments that are based on these two different ideas.

Review Questions

1 This question is about genes.

a What is the job of genes?

i store glucose
ii describe how to make proteins
iii release energy by respiration
iv transport materials around a cell

b Complete the following statement by choosing from the table below.
Genes are sections of very ….

… long DNA molecules that make up chromosomes
… short DNA molecules that make up chromosomes
… short chromosomes that make up DNA molecules
… long chromosomes that make up DNA molecules

2 The growth and development of each cell is controlled by its DNA.
What are the features of DNA?
Make and complete a table like the one below.

DNA feature	Answer
number of strands	
number of different bases	
arrangement of bases between strands	single pairs triplets fours
shape of molecule	

3 Meiosis occurs during the formation of gametes.
Four people try to explain the link between meiosis and fertilisation.

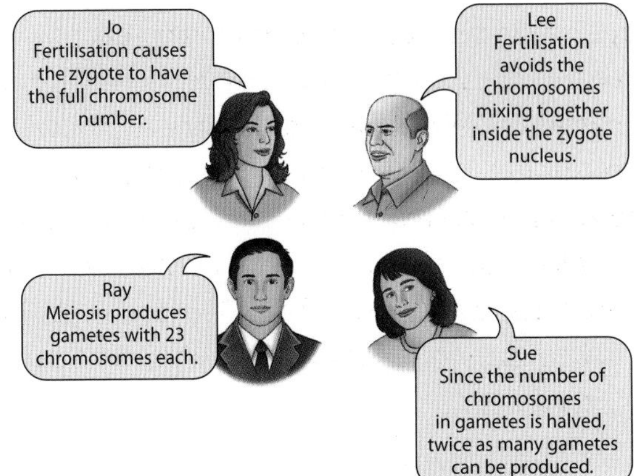

Which two people's ideas together give the best explanation of the link between meiosis and fertilisation?

4 Cells in an embryo become specialised.
What are the results of this change?
Decide whether the following statements are true or false.

The cells no longer contain the same genes.

Some of the genes are no longer active.

Each cell produces only the specific proteins it needs.

The cells form different types of tissues.

5 The order of bases in a gene is the code for building up amino acids in the correct order to make a particular protein.
Explain how the code works.

C5 Chemicals of the natural environment

Why study chemicals of the natural environment?

Conditions on Earth are special. The temperature is just right for most water to be liquid. The atmosphere has enough oxygen for living things to breathe, but not so much that everything catches fire. Rocks are the source of many of the chemicals that meet our daily needs.

What you already know

- The atmosphere is a protective blanket of gases around the Earth, which supports life.
- Bonds between atoms in molecules are strong, while forces between molecules are very weak.
- An atom consists of a central nucleus surrounded by electrons.
- Ionic compounds have a lattice structure.
- Ionic compounds conduct electricity when molten or in solution, because the ions are free to move.
- During a chemical reaction, atoms and mass are both conserved.
- A chemical change caused by the flow of an electric current is called electrolysis.
- Making, using, and disposing of products can affect the environment.

Find out about

- the chemicals in spheres of the Earth
- theories of structure and bonding
- tests for analysing water quality
- methods used to extract metals from their ores
- how to calculate chemical quantities.

The Science

Theories of structure and bonding explain how atoms are arranged and held together in all the chemicals that make up our Earth. There are three types of strong bonding that give rise to useful materials. There are weaker forces of attraction that allow molecules to stick to each other enough to make liquids and solids.

Ideas about Science

The properties of metals make them very useful and important in our lives. But mining, mineral processing, and metal extraction can all have a serious impact on the environment. Scientists are applying their understanding of chemistry in the environment to work out how we can use natural resources in a sustainable way.

A Chemicals in spheres

Find out about

- naturally occurring elements and compounds
- abundances of elements in different spheres
- cycling of elements and compounds between the spheres

People who study the Earth often think of it as being made up of spheres (see the diagram below). Starting from the middle, first comes the core, then the mantle, then the crust.

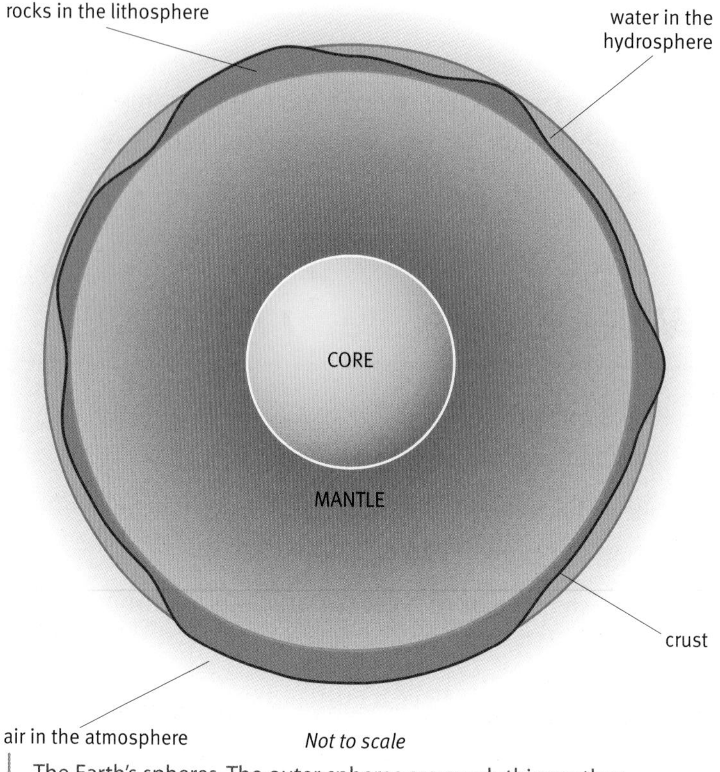

The Earth's spheres. The outer spheres are much thinner than this relative to the core and mantle.

This module looks at the spheres that make up the Earth's surface and provide us with natural resources: the lithosphere, hydrosphere, and atmosphere.

The **lithosphere** is broken into giant plates that fit around the globe like puzzle pieces. These are the tectonic plates, which move a little bit each year as they slide on top of the upper part of the mantle. The **crust** and upper **mantle** make up the lithosphere. The lithosphere is about 100 km thick.

The oceans and rivers make up the **hydrosphere**. This is not a complete sphere, but the oceans cover two-thirds of the globe, so it almost is.

Finally, wrapped like a big fluffy duvet around the Earth, keeping it warm, is the layer of air we call the **atmosphere**.

There are living things in parts of all three outer spheres.

Key words

- lithosphere
- crust
- mantle
- hydrosphere
- atmosphere

Elements in the spheres

The spheres vary greatly in their chemical composition. The atmosphere consists mainly of two uncombined elements: nitrogen and oxygen. The hydrosphere is almost entirely the compound water.

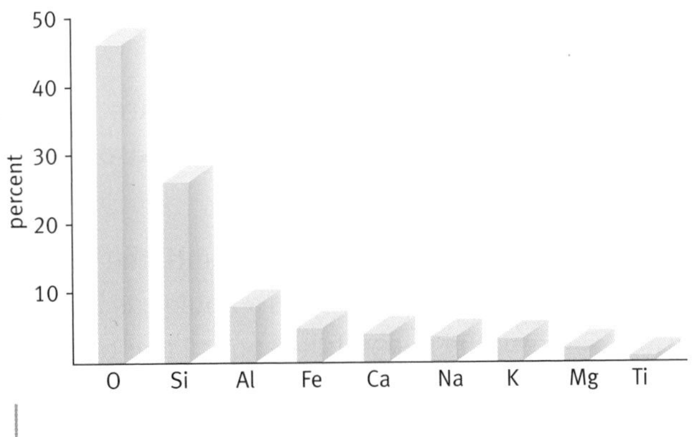

The elements in the lithosphere.

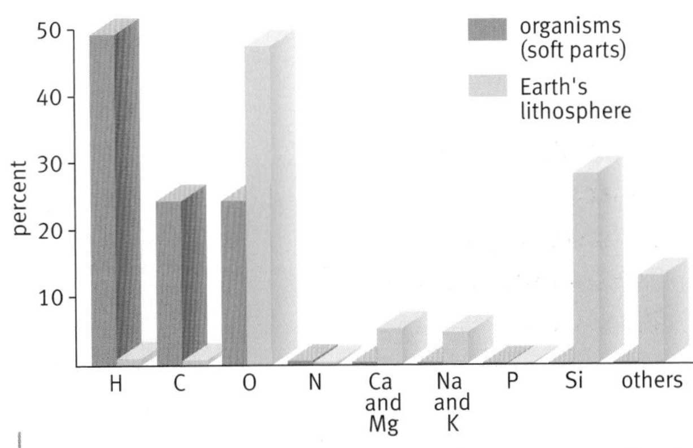

The percentage of different elements in living tissue and in the lithosphere compared.

The rocks of the lithosphere are made up mainly of silicates. These are compounds of silicon combined with oxygen and much smaller quantities of other elements.

The elements exist as different chemical species in the three spheres. In the lithosphere, carbon is combined with hydrogen to make the hydrocarbons in crude oil. Carbon is also hidden in the calcium carbonate of chalk and limestone. In the atmosphere, carbon is present as the gas carbon dioxide. Carbon is a vitally important element for life: most of the chemicals that make up living things are compounds of carbon combined with hydrogen, oxygen, and a few other elements.

Flowing between the spheres

Chemicals do not always stay in one sphere. They are constantly on the move between the spheres. Think of a carbon atom: it may start in the atmosphere; be taken into a plant; be washed into water in the hydrosphere; then buried in sediment of the lithosphere.

Water flows freely between the spheres. The obvious place for water is the hydrosphere. But think of clouds in the atmosphere; then rain sinking into the lithosphere, reappearing as a spring, from where it flows into a river, and then into the sea.

Questions

1 Look at the bar charts on this page.
   a What are the three most common elements in living tissue?
   b What are the three most common elements in the lithosphere?
   Comment on your answers.

2 'Living things inhabit the atmosphere, hydrosphere, and lithosphere.' Give some examples to illustrate this statement.

3 Make a list of four different things you use in a typical day that come from the lithosphere.

B Chemicals of the atmosphere

Find out about

- gases in the air
- weak attractions between molecules
- strong covalent bonding

The Earth is just the right size for its gravity to hold onto its atmosphere. It is also just the right distance from the Sun to have the right temperature for water to exist as a liquid. This water, together with the carbon dioxide and oxygen in the atmosphere, allows the Earth to support a great variety of plant and animal life.

The relative proportions of the main gases in the atmosphere are about 78% N_2, 21% O_2, 1% Ar, and 0.04% CO_2. The atmosphere also contains water vapour and small amounts of other gases.

The presence of argon (Ar) in the air was not discovered until the 1890s – over 100 years after the discovery of oxygen. Argon is a noble gas but it is not a rare gas. Every time you breathe in you take about 5 cm^3 of argon into your lungs.

The average temperature of the atmosphere at the Earth's surface is 15 °C. At a height of 80 km the temperature drops to –90 °C, its lowest point. Chemicals in the atmosphere have melting and boiling points low enough for them to exist as gases at temperatures between these extremes.

All the chemicals in the atmosphere are either non-metallic elements (eg, O_2, N_2, Ar) or compounds made from non-metallic elements (eg, CO_2).

nitrogen	N N
oxygen*	O O
argon**	Ar
carbon dioxide	O C O

The main of atmospheric gases.
* The element oxygen is unusual as it can exist as normal oxygen, O_2, or as ozone, O_3.
** The element argon is a noble gas. All the noble gases are made up of single atoms, so argon is represented by Ar.

Atoms and molecules in the air

Most of the chemicals in the atmosphere are made of **small molecules**. Only the noble gases exist as single atoms.

All molecules have a slight tendency to stick together. For example, there is an attraction between one O_2 and another O_2. But these **attractive forces** are very weak. This is why the chemicals that make up the atmosphere are gases with low melting and boiling points.

One way to picture this is that the molecules are moving so quickly that, when two O_2 molecules come close to each other, the attractive force between them is not strong enough to hold them together.

Key words

- small molecules
- attractive forces
- molecular models
- electrostatic attraction
- covalent bonding

Strong bonds in molecules

The forces inside molecules that hold the atoms together are very strong, many times stronger than the weak attractions between molecules. Small molecules such as O_2 or H_2 do not split up into atoms except at very, very high temperatures.

Chemists often use a single line to represent a single bond between atoms in a molecule. For example, the simple molecules in hydrogen, oxygen, and carbon dioxide can be represented by the molecular formulae H_2, O_2, and CO_2. But if you want to show their bonds, they can be represented by:

H—H   O=O   O=C=O

Some **molecular models** use the same idea. A coloured ball represents each atom, and a stick or a spring is used for each bond.

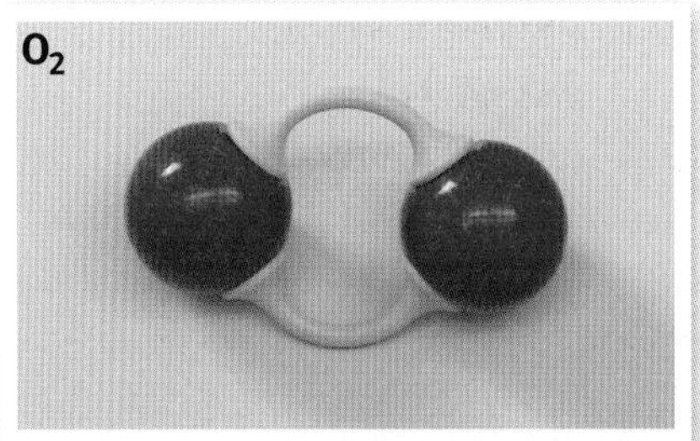

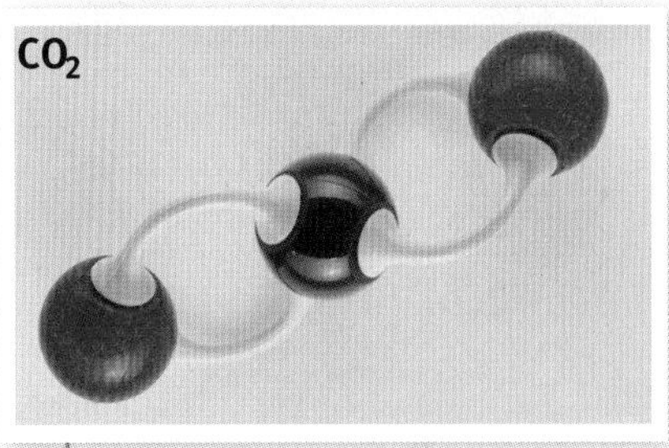

Ball-and-stick models of H_2, O_2, H_2O, and CO_2. Notice that the molecules have a definite shape. There are fixed angles between the bonds in molecules.

Electrons and bonding

A knowledge of atomic structure can help to explain the bonding in molecules (see C4, K: Ionic theory and chemical structure). When non-metal atoms combine to form molecules, they do so by sharing electrons in their outer shells.

The atoms are held strongly together by the attraction of their nuclei for the pair of electrons they share.

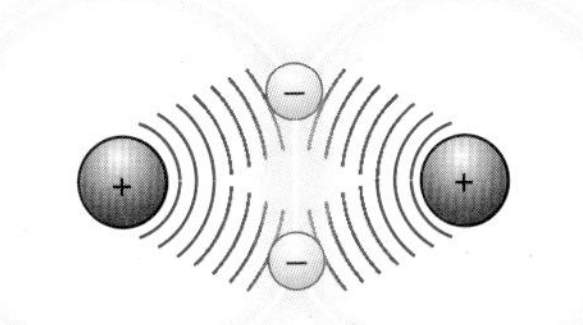

The atoms cannot move any closer together because the repulsion between the two positively charged nuclei will push them apart again.

The formation of a covalent bond between two hydrogen atoms.

The molecule of hydrogen is held together by the **electrostatic attraction** between the two nuclei and the shared pair of electrons. This is a single bond.

This type of strong bonding is called **covalent bonding**. 'Co' means 'together' or 'joint', while the Latin word *valentia* means strength. So we have strength by sharing.

The number of bonds that an element can form depends on the electrons in the outer shell of its atoms. The table gives the number of covalent bonds normally formed by atoms of some non-metal elements.

Atom	Usual number of covalent bonds
H, hydrogen	1
Cl, chlorine	1
O, oxygen	2
N, nitrogen	3
C, carbon	4
Si, silicon	4

Questions

1 Draw diagrams to show the covalent bonding in these molecules:
   - a hydrogen chloride, HCl
   - b ammonia, NH_3
   - c methane, CH_4
   - d ethene, C_2H_4

2 Estimate the volume of argon in the room in which you are sitting.

Find out about

- unusual properties of water
- bonding within and between water molecules
- ions in solution

On Earth, water exists mostly in the liquid state (the oceans and the clouds), with some in the solid state (ice) and a smaller amount as gas (water vapour).

Key words

- salts
- dissolve

Properties of water

We see water so often that we take its special chemical and physical properties for granted.

One of these special properties is that it is a liquid at room temperature. The H_2O molecule has a smaller mass than molecules of O_2, N_2, or CO_2, which are all gases at room temperature. So you might expect H_2O to be a gas at room temperature too. But H_2O melts at 0 °C and boils at 100 °C. So molecules of H_2O must have a greater tendency to stick to each other than the molecules of a gas like oxygen.

Strange things happen when water cools from room temperature. As it cools to 4 °C, the liquid contracts as expected. However, on cooling further to 0 °C, it starts to expand. This means that ice is less dense than the cold water surrounding it, so it floats. This is very important in nature. It means that winter ice does not sink to the bottom of lakes where it would be far from the warming rays of spring sunshine.

Another special property of water is that it is a good solvent for **salts**. Most solvents do not **dissolve** ionic compounds, but water does.

Pure water does not conduct electricity – in this way it resembles other liquids made up of small molecules. This shows that it does not contain charged particles that are free to move. Even though pure water does not conduct, you should never touch electrical devices with wet hands. This is because the water on your hands is not pure, so it will conduct electricity and increases the risk of you receiving an electric shock.

Water molecules

Knowledge of the structure of water molecules can help to explain its remarkable properties. The three atoms in a water molecule are not arranged in a straight line, but are at an angle.

The electrons are not evenly shared in the covalent bonds between atoms. The oxygen atom has more than its fair share. It also has four unshared electrons in its outer shell. These two effects give each water molecule a small negative charge on its oxygen side, and a small positive charge on its hydrogen side. Overall, each molecule is electrically neutral.

The small charges on opposite sides of the molecules cause slightly stronger attractive forces between the molecules. The small charges also help water dissolve ionic compounds by attracting the ions out of their crystals.

The attractions between water molecules and their angular shape mean that they line up in ice to create a very open structure. As a result, ice is less dense than liquid water.

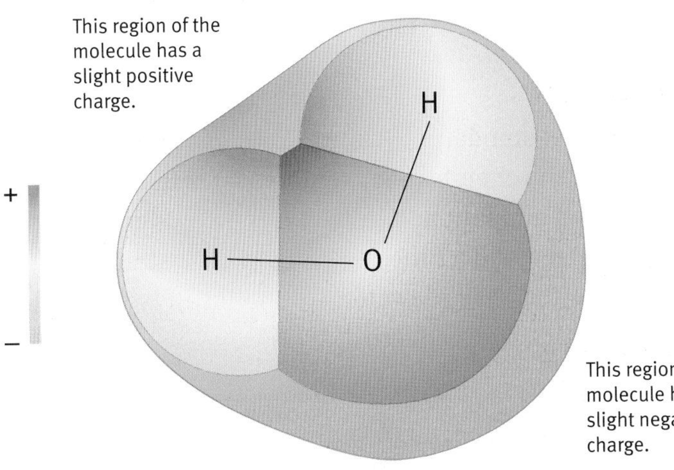

A water molecule, showing that oxygen has a slightly greater share of the pair of electrons in each bond.

water molecule

Ions separate and can move about freely when dissolved in water. In seawater there is a mixture of positive ions and negative ions.

Why is seawater salty?

The diagram below shows how soluble chemicals get carried from rocks to the sea during part of the water cycle.

The main soluble chemical carried into the sea is sodium chloride. River water does not taste salty because the concentration is so low, but the concentration of salt in the sea has built up over millions of years, and so it does taste salty.

Other chemicals that dissolve include potassium chloride, potassium bromide, magnesium chloride, magnesium sulfate, and sodium sulfate. One litre of typical seawater contains about 40 g of dissolved chemicals from rocks. Most of the compounds that are dissolved in seawater are made up of positively charged metal ions and negatively charged non-metal ions. They are salts.

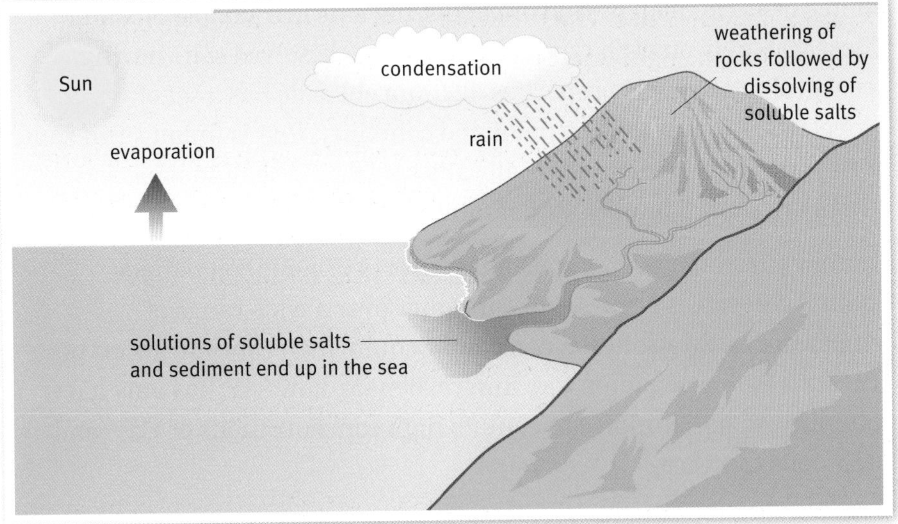

Weathering slowly breaks down rocks. This exposes the inside of the rocks to water. Soluble chemicals in the rocks dissolve in the water and get washed away.

Questions

1. Water has an unusually high boiling point compared with other small molecules. What other unusual properties does water have?
2. Write out the formulae of these salts with the help of the periodic table showing the ion charges in C4 Section K. The sulfate ion is SO_4^{2-}.
   a. Potassium bromide.
   b. Magnesium chloride.
   c. Magnesium sulfate.
   d. Sodium sulfate.
3. How would the possibility of life on Earth have been affected had ice been denser than water?

D Detecting ions in salts

Find out about

- properties of ions
- precipitation reactions
- ionic equations
- tests for ions

Why test water?

Government regulations set standards for the quality of water supplied to our homes, and control the chemicals that industrial companies can discharge into rivers. The Foundation for Environmental Education awards beaches all over the world a Blue Flag if they meet standards that include the quality of the local seawater. We should all be interested in the make-up of the Earth's hydrosphere.

A domestic water supply.

A Blue Flag beach in Devon.

Industrial discharge to a river.

Solubility

Simple tests can help us find out about the ions in a sample of water. These tests rely on each type of ion from the dissolved salts having special, characteristic properties. Sodium chloride has a set of properties, such as melting point and solubility, that is unique. But some of its properties are shared by all salts containing sodium ions, and others are shared by all salts containing chloride ions.

Solubility in water is an important property of ionic compounds. Sodium chloride is quite soluble in water over a wide range of temperatures. So a solution can contain quite high concentrations of Na^+ and Cl^- ions together. Calcium carbonate, however, has only a low solubility, so no solution can contain high concentrations of Ca^{2+} and CO_3^{2-} ions together.

Key words

- precipitate
- ionic equation

If you try to put high concentrations of Ca^{2+} and CO_3^{2-} ions together, you need two solutions: one where you have dissolved lots of a soluble *calcium* salt, and another where you have dissolved lots of a soluble *carbonate* salt. Calcium nitrate and sodium carbonate would be good examples to use. Then you mix the two solutions together. Very quickly, huge numbers of Ca^{2+} and CO_3^{2-} meet each other and cluster together to make solid crystals, which fall to the bottom of the solution as a solid white **precipitate**. The process is exactly the reverse of dissolving. The other two ions, sodium and nitrate, stay dissolved in the solution because sodium nitrate is a soluble salt.

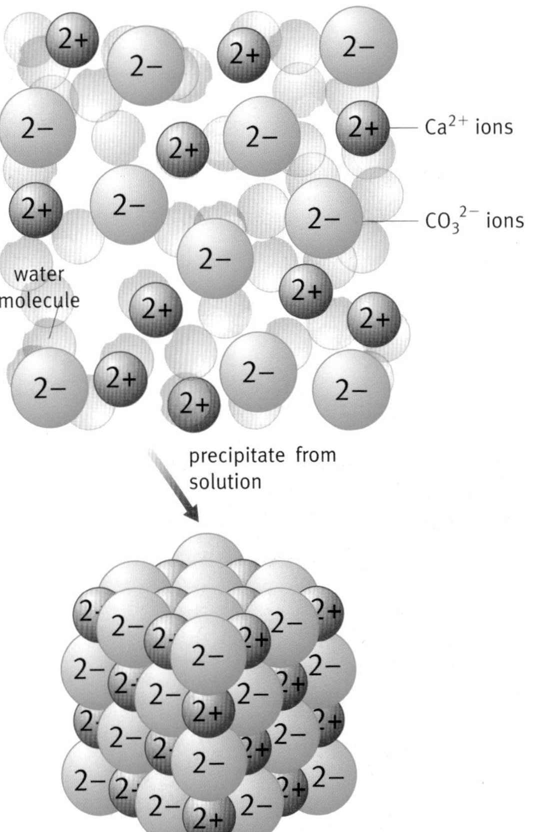

Ions of an insoluble salt cluster together and make a solid precipitate.

Ionic equations

When a solution of calcium nitrate is mixed with a solution of sodium carbonate, only two of the four ions in the mixture make a precipitate. Chemists use **ionic equations** to summarise reactions like this. Ionic equations allow you to focus attention just on the ions involved. For the precipitation of calcium carbonate:

$$Ca^{2+}(aq) + CO_3^{2-}(aq) \longrightarrow CaCO_3(s)$$

It is important to include the state symbols in these equations.

To write ionic equations correctly, use the following steps:

1. Write the correct formula of the precipitate, with the state symbol (s), on the right-hand side of the arrow.
2. Write the ions that make the precipitate separately, with their charges and state symbols (aq), on the left-hand side.
3. If necessary, put numbers in front of the ions on the left to show the ratio in which they stick together. You can work this out from the formula of the precipitate.

Questions

1. What would happen if you mixed solutions of potassium carbonate and calcium nitrate?
2. Write an ionic equation showing the precipitation of magnesium carbonate by mixing solutions of magnesium nitrate and sodium carbonate.

Shellfish use calcium carbonate to make their protective shells. At the surface of the sea, the concentrations of Ca^{2+} and CO_3^{2-} in the seawater are high enough for the calcium carbonate to stay as a solid precipitate and not dissolve. When shellfish die, their remains sink and build up as sedimentary limestone rocks on the seabed.

Salt or hydroxide	Solubility
all nitrates	soluble
all salts of sodium and potassium	soluble
silver chloride	insoluble
silver bromide	insoluble
silver iodide	insoluble
barium sulfate	insoluble
calcium carbonate	insoluble
hydroxides of metals not in Group 1	insoluble

The solubility of some salts and hydroxides.

Predicting when precipitates form

If you know the solubility of salts, you can predict when precipitation reactions will happen. This can help you to detect the ions present in a solution.

Looking for metal ions

Adding a solution of sodium carbonate and observing a white precipitate would be a good test for calcium ions in a solution – but there are other metal ions that would also form insoluble, white carbonate precipitates. So the results would not be conclusive.

Several metal ions form insoluble hydroxide compounds, too. The ions of transition metals are special, though, because they are coloured. This gives their hydroxide precipitates characteristic colours that can be used to identify the metal ion in the original solution. To carry out the test, you add a small volume of a dilute solution of sodium hydroxide to a solution of the substance to be investigated.

The ionic equation for the Fe^{3+} test is:

$$Fe^{3+}(aq) + 3OH^-(aq) \longrightarrow Fe(OH)_3(s)$$

Metal ion tested	Precipitate	Observations
copper, $Cu^{2+}(aq)$	$Cu(OH)_2(s)$	light blue (insoluble in excess NaOH (aq))
iron(II), $Fe^{2+}(aq)$	$Fe(OH)_2(s)$	green (insoluble in excess NaOH (aq))
iron(III), $Fe^{3+}(aq)$	$Fe(OH)_3(s)$	red-brown (insoluble in excess NaOH (aq))
calcium, $Ca^{2+}(aq)$	$Ca(OH)_2(s)$	white (insoluble in excess NaOH (aq))
zinc, $Zn^{2+}(aq)$	$Zn(OH)_2(s)$	white (soluble in excess NaOH (aq))

Tests for positively charged ions.

The hydroxide precipitates $Fe(OH)_2$, $Fe(OH)_3$, and $Cu(OH)_2$.

In the test for Zn^{2+} ions, if you add excess sodium hydroxide the precipitate dissolves, leaving a colourless solution. The other hydroxide precipitates in the table are insoluble in excess sodium hydroxide.

Water companies check the concentrations of transition metal ions in drinking water, and keep them below levels that would be toxic. They also keep the concentrations of Mg^{2+} and Ca^{2+} below levels that would cause too much limescale build-up in water pipes.

Looking for non-metal ions

Some tests for non-metal ions also use precipitation reactions.

The ions chloride (Cl^-), bromide (Br^-), and iodide (I^-) from the halogen group each make insoluble precipitates with silver ions (Ag^+). Each precipitate has a characteristic colour.

The ionic equation for the chloride ion, Cl^-, test is:

$$Ag^+(aq) + Cl^-(aq) \longrightarrow AgCl(s)$$

Sulfate ions (SO_4^{2-}) make a characteristic white precipitate with barium ions (Ba^{2+}).

The ionic equation for the sulfate ion, SO_4^{2-}, test is:

$$Ba^{2+}(aq) + SO_4^{2-}(aq) \longrightarrow BaSO_4(s)$$

Ion tested	Test	Precipitate	Colour of precipitate
$Cl^-(aq)$	acidify with dilute nitric acid, then add silver nitrate solution	$AgCl(s)$	white
$Br^-(aq)$	acidify with dilute nitric acid, then add silver nitrate solution	$AgBr(s)$	cream
$I^-(aq)$	acidify with dilute nitric acid, then add silver nitrate solution	$AgI(s)$	yellow
$SO_4^{2-}(aq)$	acidify, then add barium chloride solution or barium nitrate solution	$BaSO_4(s)$	white

Tests for negatively charged ions.

The precipitates AgCl(s), AgBr(s), and AgI(s).

Water companies test for sulfate ions and keep the concentration below about 1g/litre. Anything higher has a laxative effect.

Carbonate ions make carbon dioxide gas when you add a dilute acid. Bubbles form, creating effervescence and the gas can be tested to see if it makes limewater go cloudy. Carbon dioxide reacts with calcium hydroxide in limewater to make a milky white precipitate of calcium carbonate.

The test for carbonate ions.

Questions

3 What would you expect to see if you added a solution of barium nitrate to an acidified solution of sodium sulfate?

4 Which tests would you carry out to distinguish between solutions of sodium chloride and sodium iodide? What results would you expect?

5 Write ionic equations for the following:
  a Cu^{2+} ions reacting with OH^- ions to make a precipitate of $Cu(OH)_2$
  b Ag^+ ions reacting with acidified I^- ions to make a precipitate of AgI
  c Barium nitrate reacting with potassium sulfate.

6 Solution X gives a green precipitate when sodium hydroxide solution is added, and a white precipitate when acidified barium nitrate solution is added. What salt is dissolved in solution X?

7 Solution Y gives a white precipitate when sodium hydroxide solution is added, which dissolves in excess sodium hydroxide. Solution Y gives a cream precipitate when acidified and a solution of silver nitrate is added. What salt does Y contain?

Chemicals of the lithosphere

Find out about

- the Earth's crust
- ionic bonding
- giant ionic structures

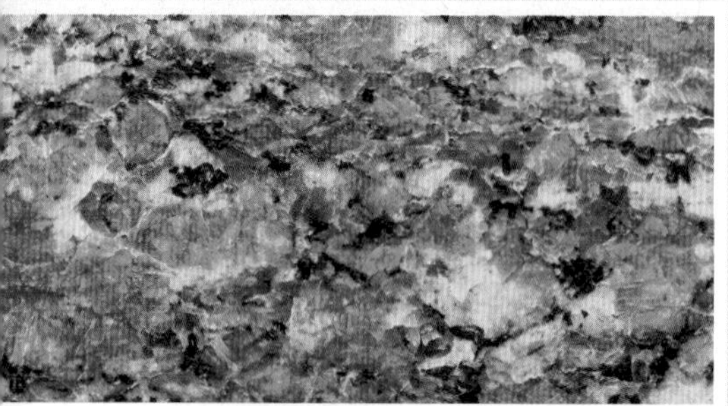

Granite is a mixture of minerals: glassy grains of quartz, black crystals of mica, and large, pink or white crystals of feldspar.

Key words

- rock
- mineral
- ore
- abundant
- ionic compound
- ionic bonding
- giant ionic lattice

Questions

1 Write a sentence that makes clear the differences between the words *rock*, *mineral*, and *lithosphere*.

2 Draw up a table to compare the properties of a molecular compound such as water and an ionic compound such as sodium chloride.

Rocks and minerals

The lithosphere is made of **rocks**. Rocks can be big (mountains), medium-sized (boulders), or small (stones and pebbles).

Rocks may contain one or more **minerals**. Minerals are naturally occurring chemicals; they may be elements or compounds.

Limestone rock is made mainly from the mineral calcite, with the chemical formula $CaCO_3$. Sandstone rock is made mainly from the mineral quartz, with the chemical formula SiO_2. Both calcite and quartz are compounds. Granite rock is a mixture of several minerals; quartz, feldspar, and mica, all of which are compounds.

Gold is found naturally on its own in the lithosphere. So a gold nugget is a rock made from one mineral that is an element.

A metal **ore** (see Section G) is a rock that contains minerals from which a metal can be extracted economically.

Haematite, Fe_3O_4. Crystals of this mineral range from metallic black to dull red.

Calcite, $CaCO_3$.

The non-metals oxygen and silicon are the two most **abundant,** or common, elements in the lithosphere. They form silica and silicate minerals, which make up 95% of the Earth's continental crust. The feldspars are silicates, which also contain ions of metals including aluminium, the third most abundant element in the lithosphere.

Quartz is a form of silica, with silicon and oxygen atoms joined by covalent bonds. Haematite and calcite are examples of minerals made up of ions.

Evaporite minerals

Seawater contains abundant dissolved chemicals. When the water evaporates, **ionic compounds** crystallise. Sodium chloride, NaCl, or rock salt, is a common example. The mineral is halite.

Minerals that are formed in this way are called evaporites. Roughly 200 million years ago, in the Triassic era, vast salt deposits were laid down as seawater evaporated off the northwest coast of England. The salt was later covered with other sediments and is now under the county of Cheshire in England. People have been extracting and trading this salt since before Roman times.

The Great Basin: a hot, dry area in Western USA. Salty water evaporates quickly, leaving salt flats.

The structure and properties of salts

Crystals of sodium chloride are cubes. They are made up of sodium and chloride ions.

In every crystal of sodium chloride, the ions are arranged in the same regular pattern. All crystals of the compound have the same cubic shape.

As a sodium chloride crystal forms, millions of Na^+ ions and millions of Cl^- ions pack closely together. The ions are held together very strongly by the attraction between their opposite charges. This is called **ionic bonding**, and the structure is called a **giant ionic lattice**. Unlike compounds such as water, which is made up of individual molecules of H_2O, there is *not* an individual NaCl molecule.

Because of the very strong attractive forces, it takes a lot of energy to break down the regular arrangement of ions. So NaCl has to be heated to 801°C before it melts, and to 1413°C before it boils. Compare this with melting ice (melting point 0°C) and boiling water (boiling point 100°C), where you only need to supply sufficient energy to overcome the relatively weak attractive forces *between* the H_2O molecules.

Sodium and chloride ions are charged particles; however, solid ionic compounds do not conduct electricity because the ions are not free to move. When an ionic compound is molten or dissolved in water, it can conduct electricity because its ions separate and can move around.

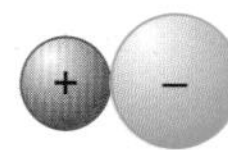

1 The Na^+ attracts other Cl^- ions that are close to it. The Cl^- attracts other Na^+ ions that are close to it.

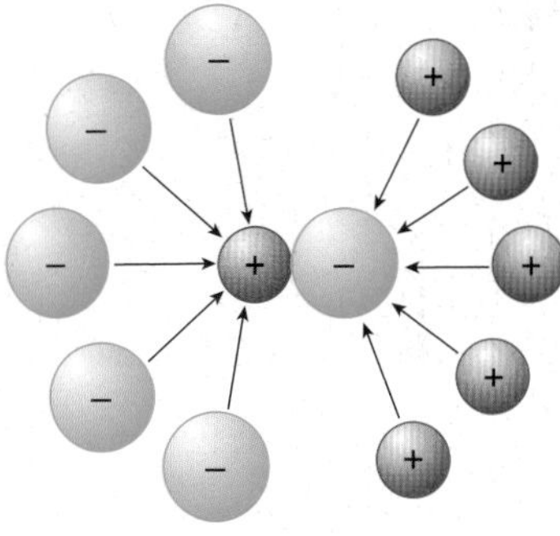

2 Another five Cl^- ions can fit around the Na^+, making six in total, and another five Na^+ ions can fit around the Cl^- ion, making six in total.

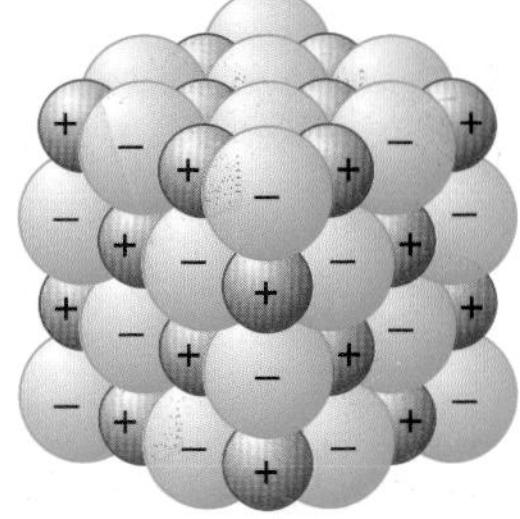

3 Each of these ions then attracts other ions of the opposite charge, and the process continues until millions and millions of oppositely charged ions are all packed closely together.

How a sodium chloride crystal forms.

Carbon minerals – hard and soft

Find out about

- diamond, graphite, and quartz
- giant covalent structures

Diamonds are transparent, and are cut so they reflect light and sparkle.

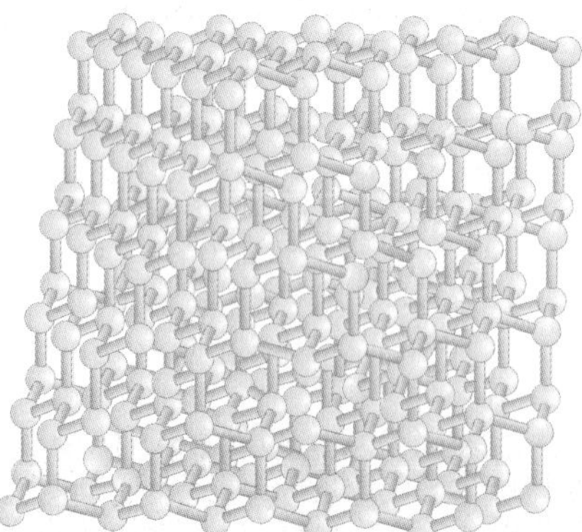

Model of the structure of diamond. Each carbon atom inside the structure is covalently bonded to four others in a huge 3D arrangement.

Diamond and graphite – non-identical twins

If you heat a **diamond** strongly, it will eventually burn and make carbon dioxide. **Graphite** will do the same. This is because both these minerals are made from carbon and nothing else. They both burn, but in many ways they could not be more different. How the carbon atoms are arranged and stuck together in diamond and in graphite is very different, but is easy to picture. Diamond and graphite are good examples of how the behaviour of a solid depends on its structure and bonding.

Geologists think diamonds formed about 3 billion years ago, at very high temperatures and pressures, in the Earth's mantle. Volcanic eruptions forced them up to the surface. Graphite was made in the Earth's crust, where the decayed remains of living things were squashed in metamorphic rocks. The two different sets of conditions have produced two very different minerals.

Diamond and graphite head-to-head

Property	Diamond	Graphite
hardness (1 = softest, 10 = hardest)	10	1–2
melting point (°C)	3560	3650
boiling point (°C)	4830	4830
solubility in water	insoluble	insoluble
electrical conductivity	low	high

Some properties of diamond and graphite.

Carbon atoms in diamond and graphite

In diamond, covalent bonds join each carbon atom to its four nearest neighbours. The four bonds are evenly spaced in *three dimensions* around each carbon atom, making pyramid (tetrahedron) shapes. The arrangement is called a **giant covalent structure**, as it repeats over and over again until you have billions and billions of atoms covalently bonded together.

Graphite has a giant covalent structure, too. Covalent bonds join each carbon atom to its three nearest neighbours. The three bonds are evenly spaced in *two dimensions*, making flat sheets of hexagons that go on and on to include billions and billions of atoms. Each atom has one outer-shell electron that has not been used to make a covalent bond. These electrons drift around freely in the gaps between the layers of atoms. They help stick the layers together, but only weakly.

Different structures – different properties – different uses

Diamond and graphite both have high melting and boiling points. This is because many strong covalent bonds would need to be broken for them to melt. They are also both insoluble in water. This is because of their many strong covalent bonds, and also because, unlike ionic compounds, they are not made up of ions. Ionic compounds dissolve because the ions are attracted to the water molecules.

Strong covalent bonds in three dimensions make diamond the hardest known material. Drills used in mining have bits with diamond tips.

In graphite, weak forces between the layers make it easy for them to slide over one another. This slipperiness makes graphite useful as a lubricant. The electrons drifting between the layers are free to move and take their charge with them. This allows graphite to conduct electricity – this is very unusual for a giant covalent structure. In contrast, diamond has no charged particles that are free to move, so it is an insulator.

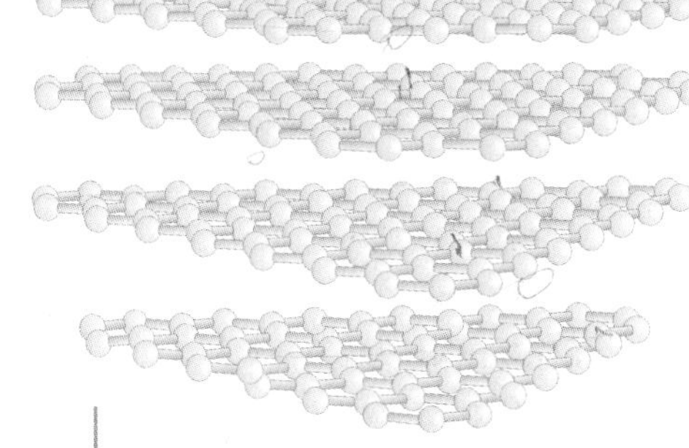

Model of the structure of graphite. Each carbon atom inside the structure is covalently bonded to three others, making flat sheets of hexagons with weak forces between them.

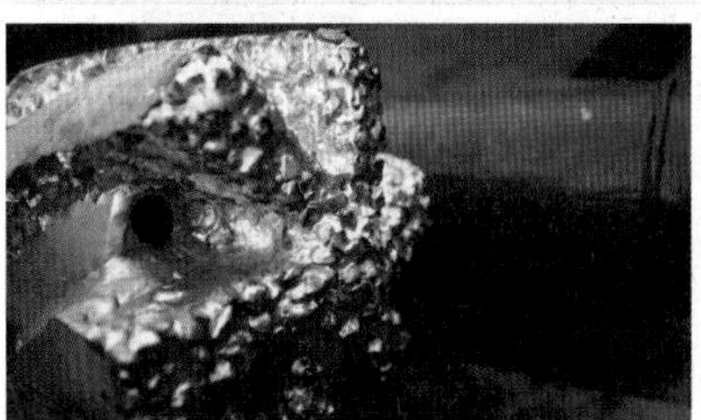

A diamond-tipped drill. The extreme hardness of diamond allows it to cut through anything.

Quartz – a diamond-like mineral

Sandstone is mostly made up of the mineral quartz. Quartz is a common crystalline form of silicon dioxide (SiO_2). It has a giant covalent structure similar to diamond. The strong bonds between the silicon and oxygen atoms make it a strong and rigid mineral. It, too, is hard, so it is suitable for use as an abrasive in sandpaper, for example. Sandstone is a traditional building material – its insoluble quartz helps it to stay weatherproof.

Pure quartz crystals are transparent and very hard.

Pencil lead is made of graphite and clay. As you slide the pencil across the paper, layers of graphite flake off and leave a mark.

Questions

1 Look back at C4 Section J. Draw a table to compare the properties of giant ionic substances with those of giant covalent substances. You should include melting and boiling points, solubility, and electrical conductivity.

2 Explain why graphite is soft but diamond is hard.

Key words

- graphite
- diamond
- giant covalent structure

Metals from the lithosphere

Find out about

- metal ores
- extracting metals
- chemical quantities
- electrolysis

Gold is so unreactive that it occurs uncombined in the lithosphere. Most metals occur as compounds.

An open-pit copper mine in Utah, USA. Mining on this scale has a big impact on the environment.

Metal ores

The wealth of societies has often depended on their ability to extract and use metals. Mining and quarrying for metal ores takes place on a large scale and can have a major impact on the environment.

All metals come from the lithosphere, but most metals are too reactive to exist on their own in the ground. Instead, they exist combined with other elements as compounds. Like other compounds found in the lithosphere, they are called minerals.

Rocks that contain useful minerals are called ores. The valuable minerals are very often the **oxides** or sulfides of metals.

Metal	Name of the ore	Chemical in the mineral
aluminium	bauxite	aluminium oxide, Al_2O_3
copper	copper pyrite	copper iron sulfide, $CuFeS_2$
gold	gold	gold, Au
iron	haematite	iron oxide, Fe_2O_3
sodium	rock salt	sodium chloride, NaCl

Because it occurs in an uncombined state, gold has been used by humans for more than 5000 years. More **reactive metals** like iron were not used by humans until methods for **extracting** them from their ores had been developed. Extraction methods include **reduction** reactions that separate the metal from minerals where it is combined with other elements.

Mineral processing

Over hundreds of millions of years, rich deposits of ores have built up in certain parts of the Earth's crust. But even the richest deposits do not contain pure mineral. The valuable mineral is mixed with lots of useless dirt and rock, which have to be separated off as much as possible. This is called concentrating the ore.

Some ores are already fairly concentrated when they are dug up – iron ore is often over 85% pure Fe_2O_3. But other ores are much less concentrated – copper ore usually contains less than 1% of the pure copper mineral.

Extracting metals: some of the issues

There is a range of factors to weigh up when thinking about the method for extracting a metal.

- **How can the ore be reduced?**
  The more reactive the metal, the harder it is to reduce its ore. The table on the right compares the methods used to reduce different ores.
- **Is there a good supply of ore?**
  Metals ores are mined in different parts of the world. If ore is not very pure, it may not be worth using – the cost of concentrating the ore may be too great. The more valuable the metal, the lower the quality of ore that can be used.
- **What are the energy costs?**
  It takes energy to extract metals, as well as a good supply of ore. This is especially true if the metal is extracted by electrolysis. For example, a quarter of the cost of making aluminium is the cost of electricity.
- **What is the impact on the environment?**
  Metals like iron and aluminium are produced on a huge scale. Millions of tonnes of ore are needed. Mining this ore can have a big environmental impact. This is why it is important to recycle metals. It takes about 250 kg of copper ore to make 1 kg of copper. So recycling 1 kg of copper means that 250 kg of ore need not be dug up.

	Metal	Method
MORE REACTIVE ↑	potassium sodium calcium magnesium aluminium	electrolysis of molten ores
	zinc iron tin lead copper	reduction of ores using carbon
LESS REACTIVE	silver gold	metals occur uncombined

Hot metal can be poured when molten.

Question

1 Suggest explanations for these facts:
  a The Romans used copper, iron, and gold, but not aluminium.
  b Iron is cheap compared with many other metals.
  c Gold is expensive, even though it is found uncombined in nature.
  d About half the iron we use is recycled, but nearly all the gold is recycled.
  e The tin mines in Cornwall have closed, even though there is still some tin ore left in the ground.

Key words

- ✓ **oxides**
- ✓ **reactive metals**
- ✓ **extracting (a metal)**
- ✓ **reduction**

Key words

- ✓ **reducing agent**
- ✓ **oxidised**

Extracting metals from ores

Zinc is a metal that can be extracted from its oxide. Zinc is found in the lithosphere as ZnS, called zinc blende. This can be easily turned to ZnO by heating it in air.

The task is then to remove the oxygen from the zinc oxide, to convert ZnO to Zn. Removing oxygen in this way is called reduction. The process needs a **reducing agent** that will remove oxygen. In this case, the reducing agent is carbon. In removing oxygen, the reducing agent is **oxidised**.

zinc oxide + carbon ⟶ zinc + carbon monoxide

Zn loses O to C and gets reduced...

$$ZnO + C \longrightarrow Zn + CO$$

...C takes O from Zn and gets oxidized

Reducing zinc oxide to zinc using carbon.

Further oxidation of the carbon forms carbon dioxide. Carbon is often used as a reducing agent to extract metals. Carbon, in the form of coke, can be made cheaply from coal. At high temperatures, carbon has a strong tendency to react with oxygen, so it is a good reducing agent. What is more, the carbon monoxide formed is a gas, so it is not left behind to make the zinc impure.

Carbon is also used to extract iron and copper. These reactions can be summarised as:

iron oxide + carbon ⟶ iron + carbon dioxide

copper oxide + carbon ⟶ copper + carbon dioxide

How much metal?

Chemists often ask the 'How much?' question. It's useful to know, say, how much iron it's possible to get from 100 kg of pure iron ore, Fe_2O_3.

Questions

2 a Write an equation for the reaction of zinc sulfide with oxygen to make zinc oxide and sulfur dioxide.

b What problems might arise from the formation of sulfur dioxide on a large scale?

c What might be done to deal with this problem?

3 Why do oxidation and reduction always go together when carbon extracts a metal from a metal oxide?

Relative atomic masses

Chemists need to know the relative masses of the atoms involved to answer questions such as 'How much iron, Fe, could you get from 100 kg of iron oxide, Fe_2O_3?'

Atoms are far too small to weigh directly. For example, it would take around a million million million million hydrogen atoms to make one gram.

Instead of working in grams, chemists find the mass of atoms relative to one another. Chemists can do this using an instrument called a mass spectrometer.

Values for the **relative atomic masses** of elements are shown in the periodic table on page 47. The relative mass of the lightest atom, hydrogen, is 1.

One Mg atom weighs twice as much as one C atom.

Key words

- **relative atomic mass**
- **relative formula mass**

Formula masses

If you know the formula of a compound, you can work out its **relative formula mass** by adding up the relative atomic masses of all the atoms in the formula:

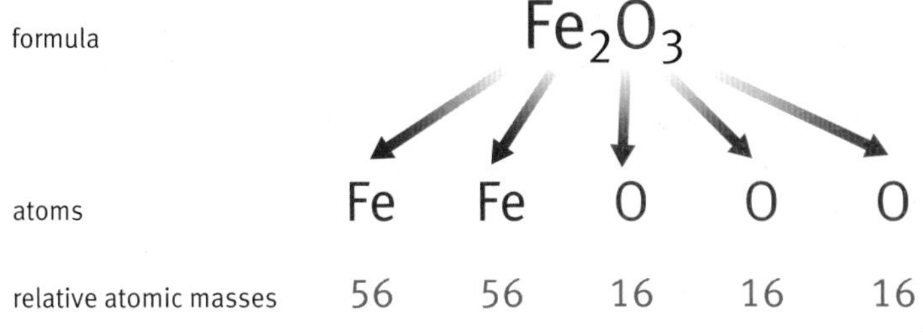

Finding the formula mass of Fe_2O_3.

Worked example

How much Fe could you get from 100 kg of Fe_2O_3?

Fe_2O_3 has a relative formula mass of **160**.

In this formula, there are two atoms of Fe.
2Fe has relative mass 2 × 56 = **112**.

This means that, in 160 kg of Fe_2O_3, there must be 112 kg of Fe.

So 1 kg of Fe_2O_3 would contain 112/160 kg of Fe.

So 100 kg of Fe_2O_3 would contain 100 kg × 112/160 of Fe = 70 kg.

Another way of saying this is that the percentage of Fe in Fe_2O_3 is 70%.

Questions

Look up relative atomic masses in the periodic table.

4 What is the relative formula mass of carbon dioxide?

5 What mass of Al could be made from 1 tonne of Al_2O_3?

6 What mass of Na could be made from 2 tonnes of NaCl?

7 The main ore of chromium is $FeCr_2O_4$. What is the mass of Cr in 100 kg of $FeCr_2O_4$?

8 1000 tonnes of copper ore are dug out of the ground. Only 1% of this is the pure mineral, $CuFeS_2$. What mass of the Cu could be made from 1000 tonnes of the ore?

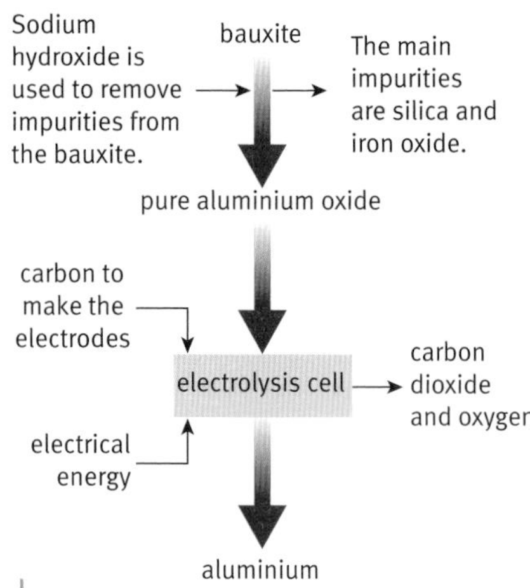

The processing of bauxite to produce aluminium.

Extracting aluminium

Some reactive metals, such as aluminium, hold on to oxygen so strongly that they cannot be extracted using carbon as a reducing agent. To extract these metals, the industry has to use **electrolysis**. Electrolysis is a process that uses an electric current to split a chemical up into its elements.

Aluminium is the most abundant metal in the lithosphere. Much of the metal is in aluminosilicates. It is very hard to separate the aluminium from these minerals.

The main ore of aluminium is bauxite. This consists mainly of aluminium oxide, Al_2O_3. There is some iron oxide in bauxite, which has to be removed before extraction of aluminium.

The diagram below shows the equipment used to extract aluminium by electrolysis. The process takes place in steel tanks lined with carbon. The carbon lining is the negative **electrode** and conducts electricity.

The **electrolyte** is hot, molten Al_2O_3, which contains Al^{3+} and O^{2-} ions.

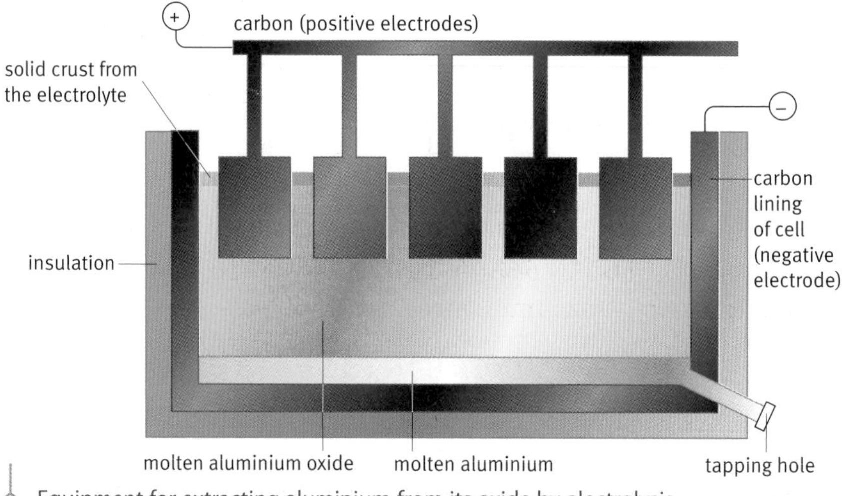

Equipment for extracting aluminium from its oxide by electrolysis.

When an ionic compound, such as Al_2O_3, melts, the ions become free to move around independently, and an electric current can pass through it. The electric current is used to decompose the electrolyte.

Aluminium forms at the negative electrode lining the tank. Because it is very hot in the tank, the aluminium is a liquid and forms a pool of molten metal at the bottom of the tank.

The positive electrodes are blocks of carbon dipping into the molten aluminium oxide. Oxygen forms at the positive electrodes. Some of this combines with the carbon to make carbon dioxide.

Key words

- electrolysis
- electrode
- electrolyte

Ions into atoms and molecules

Electrolysis turns ions back into atoms. Metal ions are positively charged, so they are attracted to the negative electrode. It is a flow of electrons from the power supply into this electrode that makes it negative.

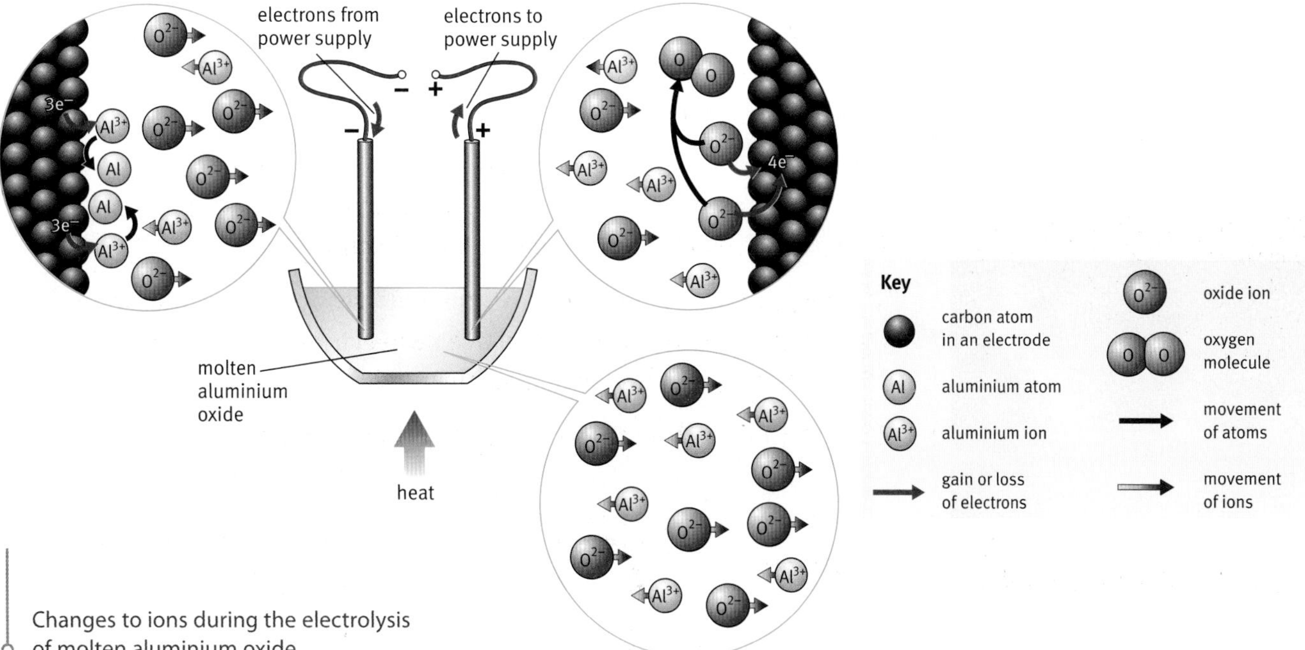

Changes to ions during the electrolysis of molten aluminium oxide.

Positive metal ions gain electrons from the negative electrode and turn back into atoms. During the electrolysis of molten aluminium oxide, the aluminium ions turn into aluminium atoms:

$$Al^{3+} + 3e^- \longrightarrow Al$$

ion; electrons supplied by the negative electrode; atom

Non-metal ions are negatively charged, so they are attracted to the positive electrode. This electrode is positive because the power supply pulls electrons away from it.

Negative ions give up electrons to the positive electrode and turn back into atoms. During the electrolysis of molten aluminium oxide, the oxide ions turn into oxygen atoms, which pair up to make oxygen molecules:

$$O^{2-} \longrightarrow O + 2e^-$$

ion; atom; electrons removed by the positive electrode

$$O + O \longrightarrow O_2$$

atom; atom; molecule

Questions

9 Draw a diagram to show the electron arrangements in:
- a an aluminium atom
- b an aluminium ion.

10 Sodium is extracted by the electrolysis of molten sodium chloride.
- a What are the two products of the process?
- b Use words and symbols to describe the changes at the electrodes during this process.

Structure and bonding in metals

Find out about

- the properties of metals
- the structure of metals
- bonding in metals

Metal properties

Metals have been part of human history for thousands of years. Our lives still depend on metals, despite the development of new materials, including all the different plastics. The varied uses of metals reflect their properties.

Technologists have learnt to use new metals and mixtures of metals called **alloys** so that, as well as steel and aluminium, other metals such as titanium and magnesium can now be used in engineering.

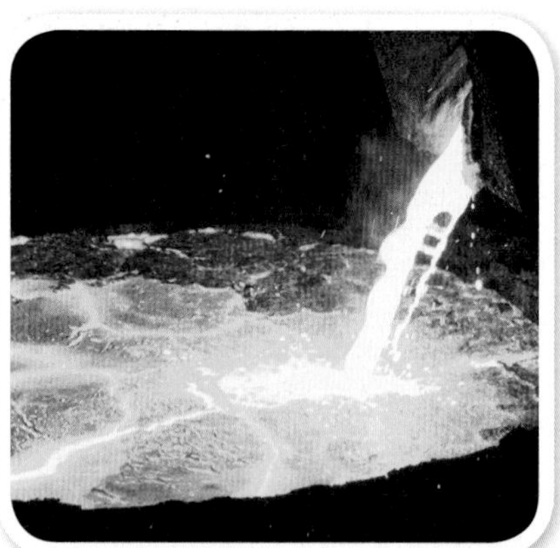

Most metals have high melting points.

Many metals are strong. The titanium hull of this research submarine is strong enough to withstand the pressure at a depth of 6 km. Titanium is also used to make hip joints and racing cars.

Metals can be bent or pressed into shape. They bend without breaking. They are malleable. Aluminium sheet can be moulded under pressure to make cans.

Metals conduct electricity. Copper and aluminium are commonly used as conductors.

Metallic structures

Data can be collected about the properties of a metal, for example, the melting point or tensile strength. Most metals have high melting points and are strong, but to explain *why* they have these properties, materials scientists need to know something about their structure.

Scientists use models to describe what they have discovered about the structures of metals. The models show that metals are made up of atoms that are:

- tiny spheres
- arranged in a regular pattern
- packed closely together in each crystal as a giant structure

This model was thought up creatively to explain the properties of metals. A model is a good one if it explains all the data.

Key words

- metallic bonding
- alloys

The diagram on the right shows the arrangement of atoms in copper, a typical metal. You can see how closely together the atoms of copper are packed – as close together as it is possible to be. Every atom inside the structure has 12 others touching it, the maximum number possible. The atoms are held to each other by strong metallic bonding. Because the bonding is strong, copper is strong and difficult to melt.

The arrangement of atoms in copper. Because the metallic bonding is strong, copper is strong and difficult to melt. Because the bonding is flexible, copper is malleable – the atoms can be moved around without shattering the structure.

Metallic bonding

Metals have a special kind of bonding – not ionic, nor covalent, but metallic. **Metallic bonding** is strong, but flexible, allowing the atoms to slide into new positions.

Metal atoms tend to share the electrons in their outer shell easily. In the solid metal, the atoms lose these shared electrons and become positively charged. The electrons, no longer held by the atomic nuclei, drift freely between the metal atoms. The attraction between the 'sea' of negative electrons and the positively charged metal atoms holds the structure together.

Overall, a metal crystal is not charged. This is because the total negative charge on the electrons balances the total positive charge on the metal atoms. The charged atoms are often described as metal ions but in their chemical reactions metals behave as collections of atoms.

The electrons can move freely through the giant structure, which explains why metals conduct electricity well. When an electric current flows through a metal wire, the free electrons drift from one end of the wire towards the other. Although the electrons are free, the metal atoms themselves are packed closely together in a regular lattice.

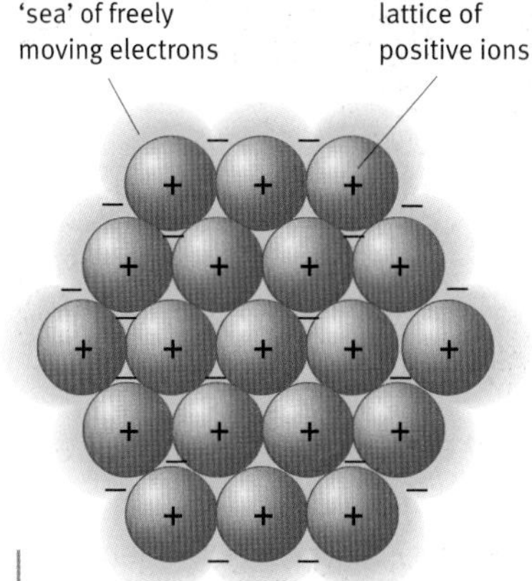

A model of metallic bonding. This model pictures the atoms sharing outer electrons, which hold the charged atoms (metal ions) together.

Questions

1 Give five examples of metals used for their strength. For each metal, give an example of a use that depends on the strength of the metal.

2 Someone looking at a model showing the arrangement of atoms in a copper crystal might think that the following statements are true. Which of these ideas are correct? Which ideas are false? What would you say to put someone right who believed the false ideas?

- a Copper crystals are shaped like cubes because the atoms are packed in a cubic pattern.
- b There is air between the atoms in a crystal of copper.
- c Copper is dense because the atoms are closely packed.
- d The atoms in a copper crystal are not moving at room temperature.
- e Copper has a high melting point because the atoms are strongly bonded in a giant structure.
- f Copper melts when strongly heated because the atoms melt.

3 There are positive metal ions in a metal crystal, but a metal is not an ionic compound. Explain.

The life cycle of metals

Find out about

- ✔ **impacts of extracting, using, and disposing of metals**

Mining, mineral processing, and metal extraction produce many valuable metal products, but these activities can be hazardous and can also have a serious impact on the environment. There can be a conflict between those who want to build up profitable industries and create jobs and those whose aim is to protect the natural world.

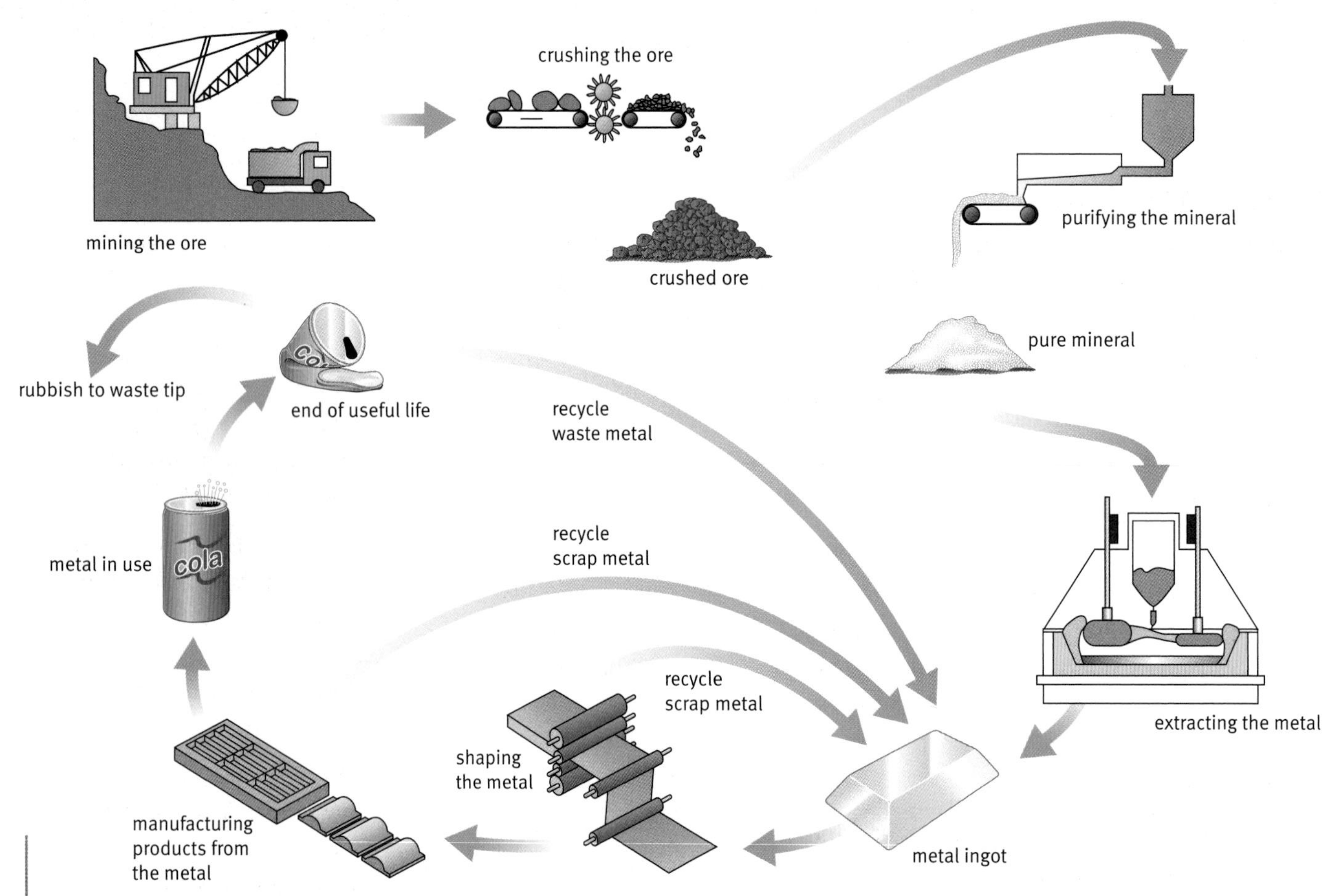

The life cycle of a metal article.

Mining

Mines can be on the surface (open-cast) or underground. Both types of mine produce large volumes of waste rock and can leave very large holes in the ground. Miners use explosives to blast the rock. This is noisy and produces dust.

Miners working in underground mines are at risk from dust, heat, and the possibility of the mine caving in. Modern mines have ventilation shafts and fans to provide fresh air and sensors to monitor the air quality. Walls and ceilings are braced with strong boards to prevent cave-ins.

Mining is much safer than it has been in the past but accidents still happen it is impossible to make mining completely safe, but most countries now have regulations to ensure that mining companies reduce risks.

Processing ores

Many metal ores are high value but low grade. The ore in an open-cast copper mine may contain as little as 0.4% of the metal and still be profitable. This means that 99.6% of the rock dug from the ground becomes waste. Near any mine there are waste tips, which might be large ponds. These can be hazardous if they contain traces of toxic metals such as lead or mercury.

Metal extraction

All the stages of metal extraction and metal fabrication need energy, use large volumes of water, and give off air pollutants. Higher expectations from society and tighter regulation mean that industries have to do more to prevent harmful chemicals escaping into the environment. Economic pressures favour the development of equipment and procedures that minimise the use of energy, water, and other resources.

Metals in use

Careful choice of metals can reduce the environmental impact of our life style. In transport, for example, lighter cars, trucks, and trains mean less fuel consumption and fewer emissions, as well as less wear and tear on roads and tracks. Vehicles can be designed to be lighter by replacing steel with lighter metals, such as aluminium, or with plastics.

Recycling

Recycling is well established in the metal industries. Scrap metal from all stages of production is routinely recycled. Much metal is also recycled at the end of the useful life of metal products.

Recycled steel is as good as new after reprocessing. The scrap is fed to a furnace and melted with fresh metal to make new steel. For every tonne of steel recycled, there is a saving of 1.5 tonnes of iron ore and half a tonne of coal. There is also a big reduction in the total volume of water needed, since large quantities of water are used in mineral processing.

Recycling aluminium is particularly cost-effective because so much energy is needed to extract the metal from its oxide by electrolysis. Recycling also reduces the impact on the environment by cutting the use of raw materials, and the associated mining and processing.

A large pond in Jamaica used to contain the waste from a bauxite mine. Bauxite is impure aluminium oxide. The main impurity is iron oxide, which ends up in the rusty-looking waste.

Questions

1. Why are recycling rates for metal waste from manufacturing higher than recycling rates of metal after use?
2. Draw up a table to list three groups of people affected by metal mining, production, and use. Show the benefits and costs to each group.
3. The price of metals can go up and down. A mining company is considering opening a new copper mine. The price of copper has just fallen – how might this affect the company's decision? Explain your answer.
4. Known bauxite reserves will last for hundreds of years. The aluminium industry claims this makes aluminium a sustainable material. Do you agree?

Science Explanations

Theories of structure and bonding can help to explain the physical properties and chemical reactions of the chemicals we find in the atmosphere, hydrosphere, and lithosphere.

You should know:

- that the elements and compounds in the Earth's atmosphere consist of small molecules
- that the Earth's hydrosphere consists mainly of water with some ionic salts in solution
- that silicon, oxygen, and aluminium are very abundant in the Earth's lithosphere
- why chemicals that are molecular are often liquids or gases at room temperature, and do not conduct electricity
- why ionic compounds have high melting and boiling points and conduct electricity when molten or dissolved in water, but not when solid
- that the formulae of salts show that the total charge on the metal positive ions is balanced by the total charge on the negative ions
- how chemists use precipitation reactions to detect which ions are present in an ionic compound
- how to use ionic equations to describe precipitation reactions
- why chemicals such as silicon dioxide and diamond have very high melting points, do not dissolve in water, and do not conduct electricity
- why metals are strong and conduct electricity
- why the methods used to extract metals from their ores are related to their reactivity
- how to calculate formula masses and reacting masses based on equations, to find the mass of the metal that can be extracted from a metal compound
- why electrolysis turns ions back into atoms and splits ionic compounds into their elements
- how to use equations to show what happens to ions when electrolysis is used to extract metals.

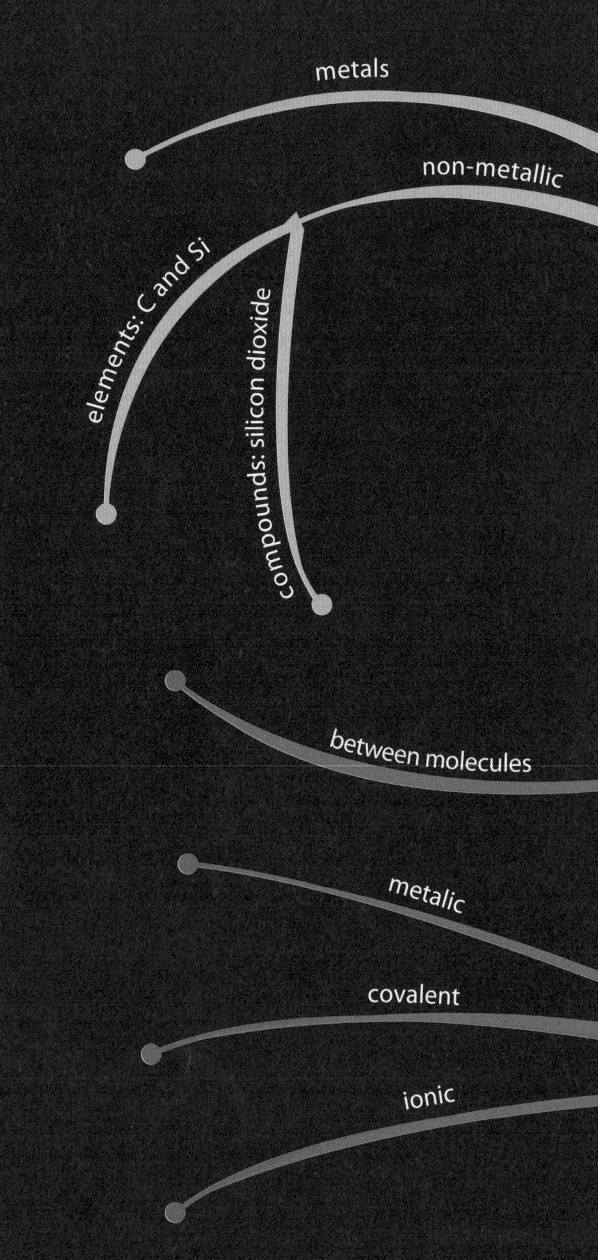

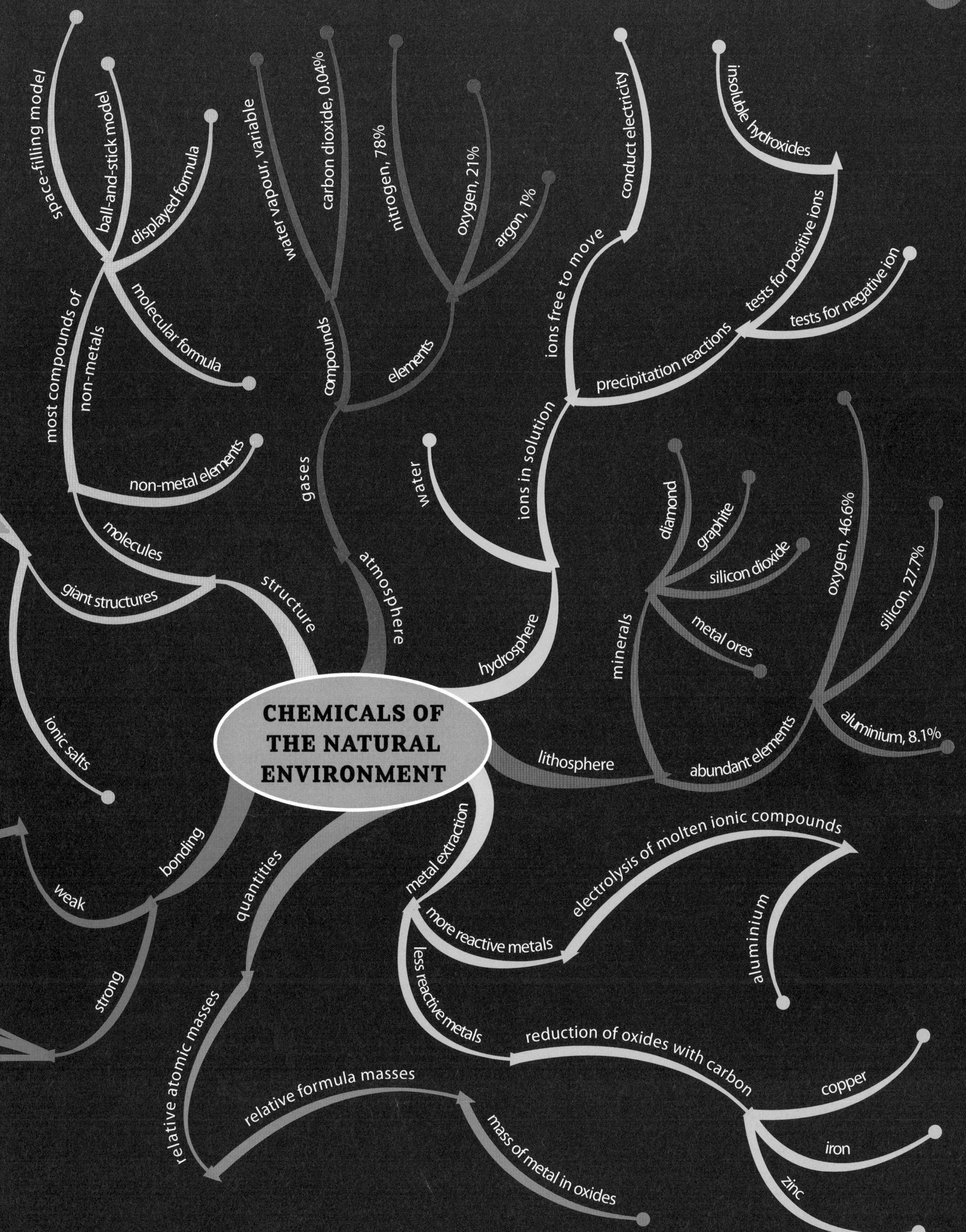

CHEMICALS OF THE NATURAL ENVIRONMENT
structure
molecules
giant structures
most compounds of non-metals
non-metal elements
space-filling model
ball-and-stick model
displayed formula
molecular formula
ionic salts
atmosphere
gases
compounds
elements
water vapour, variable
carbon dioxide, 0.04%
nitrogen, 78%
oxygen, 21%
argon, 1%
hydrosphere
water
ions in solution
ions free to move
conduct electricity
precipitation reactions
tests for positive ions
tests for negative ion
insoluble hydroxides
lithosphere
minerals
diamond
graphite
silicon dioxide
metal ores
abundant elements
oxygen, 46.6%
silicon, 27.7%
aluminium, 8.1%
bonding
weak
strong
quantities
relative atomic masses
relative formula masses
mass of metal in oxides
metal extraction
more reactive metals
electrolysis of molten ionic compounds
aluminium
less reactive metals
reduction of oxides with carbon
copper
iron
zinc

Ideas about Science

Scientific explanations are based on data but they go beyond the data and are distinct from them. An explanation has to be thought up creatively to account for the data. In the context of the theories of structure and bonding you should be able to:

- give an account of scientific work and distinguish statements that report data from statements of explanatory ideas (hypotheses, explanations, and theories)
- recognise data, such as measures of the properties of elements and compounds, that is accounted for by explanations based on theories of structure and bonding.

New technologies and processes based on scientific advances sometimes introduce new risks. Some people are worried about the effects arising from the extraction and use of metals.

You should be able to:

- explain why nothing is completely safe
- identify examples of risks that arise from mining and metal extraction
- suggest ways of reducing a given risk.

Some applications of science, such as the extraction and use of metals, can have unintended and undesirable impacts on the quality of life or the environment. Benefits need to be weighed against costs. You should be able to:

- identify the groups affected and the main benefits and costs of a course of action for each group
- suggest reasons why different decisions on the same issue might be appropriate in view of differences in social and economic context
- identify and suggest examples of unintended impacts of human activity on the environment, such as the production of large volumes of waste by mining, mineral processing, and metal extraction based on low-grade ores
- explain the idea of sustainability, and apply it to the methods used to obtain, use, recycle, and dispose of metals.

Some forms of scientific work have ethical implications that some people will agree with and others will not. When an ethical issue is involved, you need to be able to:

- state clearly what the issue is
- summarise the different views that people might hold.

When discussing ethical issues, common arguments are that:

- the right decision is the one that leads to the best outcome for the majority of the people involved
- certain actions are right or wrong whatever the consequences.

Review Questions

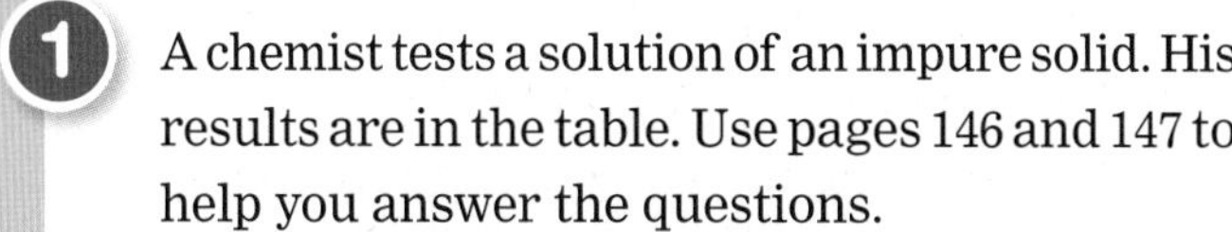

1 A chemist tests a solution of an impure solid. His results are in the table. Use pages 146 and 147 to help you answer the questions.

Test number	Test	Observation
1	acidify then add silver nitrate solution	white precipitate
2	acidify then add barium nitrate solution	white precipitate
3	add dilute sodium hydroxide solution	light-blue precipitate

a Which test shows that the powder contains copper ions?

b Which test shows that the powder contains chloride ions?

c What other ion was found to be present?

2 Zinc metal can be extracted from its oxide by heating with carbon. The equation for the reaction is:

$$ZnO + C \longrightarrow Zn + CO$$

a In the equation above:

- **i** name the element that is oxidised
- **ii** name the reducing agent.

b Calculate:

- **i** the mass of zinc oxide needed to make 1 kg of zinc (relative atomic masses: Zn=65, O=16, C=12)
- **ii** the mass of carbon monoxide produced when 1 kg of zinc is made.

c Explain why recycling helps to make the use of zinc metal more sustainable.

3 Aluminium metal is extracted from its oxide, Al_2O_3, by electrolysis.

a Explain why aluminium cannot be extracted from its oxide by heating with carbon.

b Explain why aluminium oxide conducts electricity when molten.

c Write an equation to summarise what happens at the negative electrode during the electrolysis.

4 The table gives some boiling points.

Chemical	Boiling point (°C)	Solubility in water
nitrogen	–210	insoluble
potassium chloride	770	soluble
silicon dioxide (quartz)	1610	insoluble

a Use ideas about structure and bonding to give reasons for the differences in melting point.

b Why is silicon dioxide found in the lithosphere and not in the hydrosphere or the atmosphere?

P5 Electric circuits

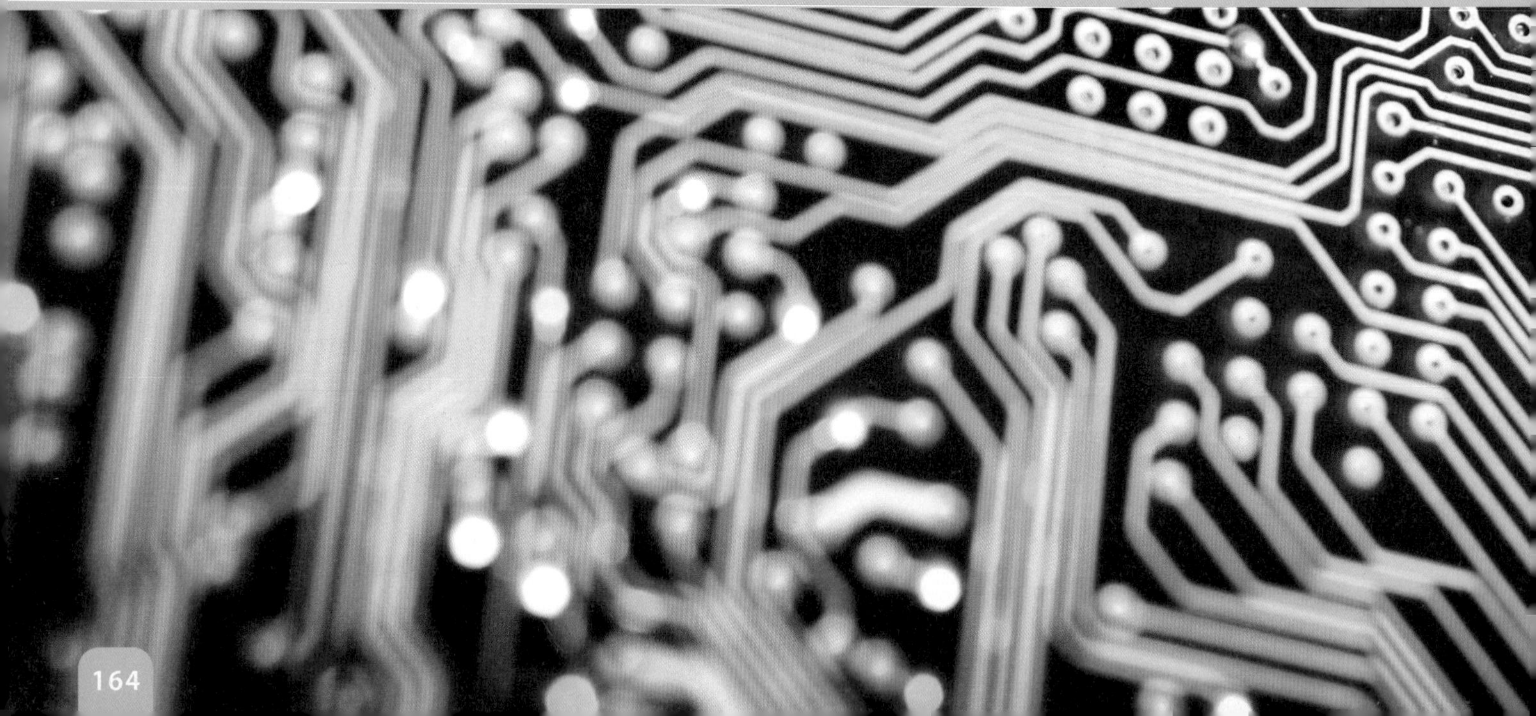

Why study electric circuits?

Imagine life without electricity – rooms lit by candles or oil lamps, no electric cookers or kettles, no radio, television, computers, or mobile phones, no cars or aeroplanes. Electricity has transformed our lives, but you need to know enough to use it safely. More fundamentally, electric charge is one of the basic properties of matter – so anyone who wants to understand the natural world around them needs to have some understanding of electricity.

What you already know

- Electric current is not used up in a circuit.
- Electric current transfers energy from the battery to components in the circuit.
- The power rating of a device is measured in watts and power = voltage × current.
- There is a magnetic field near a wire carrying a current. This can be used to make an electromagnet.
- In a power station a turbine drives a generator to produce electricity.

Find out about

- the idea of electric charge, and how moving charges result in an electric current
- how models that help us 'picture' what is going on in an electric circuit can be used to explain and predict circuit behaviour
- electric current, voltage, and resistance
- energy transfers in electric circuits, and how mains electricity is generated and distributed
- electric motors.

The Science

The particles that atoms are made of carry an electric charge. An electric current is a flow of charges. A useful model of an electric circuit is to imagine the wires full of charges, being made to move around by the battery. The size of current depends on the voltage and the resistance of the circuit. A voltage can be produced by moving a magnet near a coil.

Ideas about Science

Electrical properties of materials are used in many measuring instruments. To be sure that the reading gives a true value it is essential that the external conditions do not affect the reading.

A Electric charge

Find out about

- electric charge and how it can be moved when two objects rub together
- the effects of like and unlike charges on each other
- the connection between charge and electric current

Many of the properties of matter are obvious; matter has mass and it takes up space. But there is another property of matter that is less obvious. The most dramatic evidence of it is lightning. Other evidence comes from the shock you sometimes feel when you touch a car door handle after you get out. It is the property we call **electric charge**. Two hundred years ago, scientists were just beginning to understand electric charge, and to learn how to control it. Their work led to technological developments that have transformed our everyday lives.

Why is it called 'charge'?

In the late 1700s, experimenters working on electricity thought that the effect you produced when you rubbed something was a bit like preparing a gun. In those days, you had to prime a gun by pushing down an explosive mixture into the barrel. This was called the 'charge'. When you had done this, the gun was 'charged'. It could be 'discharged' by firing.

Charging by rubbing

Electrical effects can be produced by rubbing two materials together. If you rub a balloon against your jumper, the balloon will stick to a wall. If you rub a plastic comb on your sleeve, the comb will pick up small pieces of tissue paper – they are attracted to the comb. In both cases, the effect wears off after a short time.

When you rub a piece of plastic, it is changed in some way; it can now affect objects nearby. The more it is rubbed, the stronger the effect. It seems that something is being stored on the plastic. If a lot is stored, it may escape by jumping to a nearby object, in the form of a spark. We say that the plastic has been charged. Charging by rubbing also explains the example discussed at the top of the page. When you get out of a car, you slide across the seat, rubbing your clothes against it.

Two types of charge

If you rub two identical plastic rods and then hold them close together, the rods push each other apart – they **repel**. The forces they exert on each other are very small, so you can only see the effect if one of the rods can move freely.

Lightning is the most striking and dramatic evidence of the electrical properties of matter. A lightning strike occurs when electric charges move at high speed from a thundercloud to the ground or vice versa.

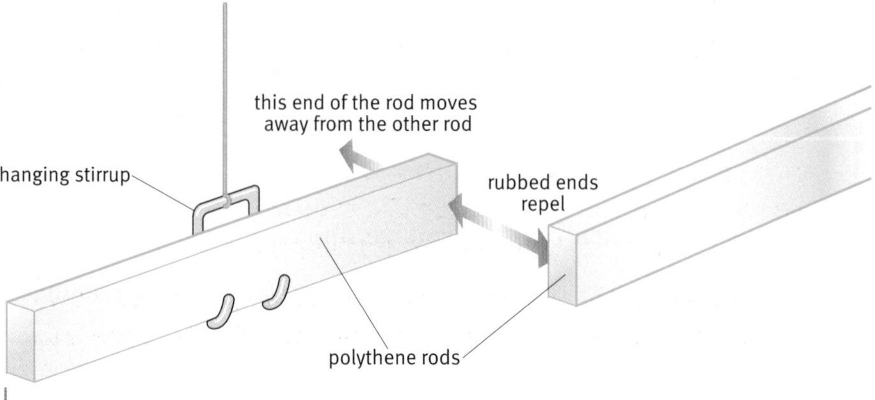

Two rubbed polythene rods repel each other. The hanging rod moves away from the other one.

If you try this with two rods of different plastics, however, you can find some pairs that **attract** each other. Scientists' explanation of this is that there are two types of electric charge. If two rods have the same type of charge, they repel each other. But if they have charges of different types, they attract. The early electrical experimenters called the two types of charge **positive** and **negative**. These names are just labels. They could have called them red and blue, or A and B.

Where does charge come from?

Scientists believe that charge is not *made* when two things are rubbed together it is just *moved around*. If you rub a plastic rod with a cloth, both the rod and the cloth become charged. (To see this, you need to wear a polythene glove on the hand holding the cloth, otherwise the charge will escape through your body.) Each object gets a different charge; if the rod has a positive charge, the cloth has a negative one. Rubbing does not make charge. It separates charges that were there all along.

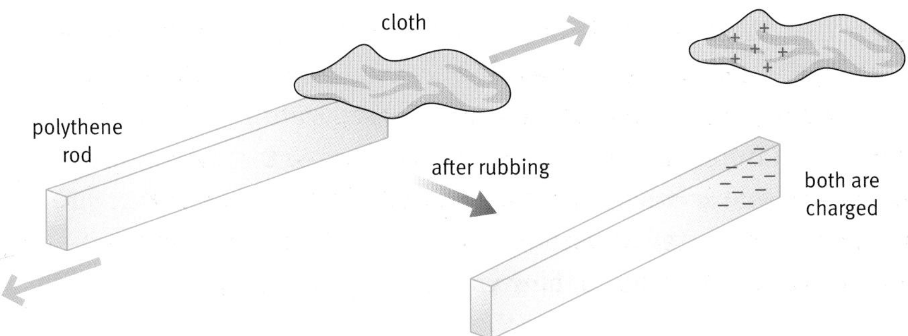

After it has been rubbed, the rod has a negative charge and the cloth has a positive charge. A possible explanation is that some electrons have been transferred from the cloth to the rod.

Questions

1. Imagine that you have two plastic rods that you know get a positive and a negative charge when you rub them with a cloth. Now you are given a third plastic rod. Explain how you could test whether it gets a positive or a negative charge when rubbed.
2. Some picture frames are made with plastic rather than glass. If you clean the plastic with a duster, it may get dusty again very quickly. Use the ideas on these pages to explain why this happens.
3. Use the ideas in the box on the right to explain why after you have rubbed a balloon on your jumper, the balloon will stick to the wall.

Attracting light objects

An object with a positive charge attracts another object with a negative charge. But why does a charged rod also attract light objects, such as little pieces of paper? The reason is that there are charges in the paper itself. Normally these are mixed up together, with equal amounts of each. So a piece of paper is uncharged. If a negatively charged rod comes near, it repels negative charges in the paper to the far end. This leaves a surplus of positive charges at the near end. The attraction between these positive charges and the rod is stronger than the repulsion between the negative charges (at the far end) and the rod. So the little piece of paper is attracted to the rod.

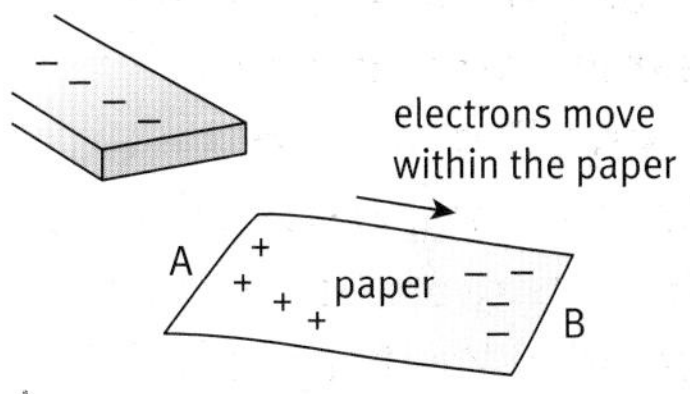

A charged rod separates the charges in a piece of paper nearby – and the paper is attracted to the rod.

Static electricity

The effects discussed here are often called electrostatic effects, and are said to be due to **static electricity**. The word 'static' indicates that the charges are fixed in position and do not move (except when there is a spark).

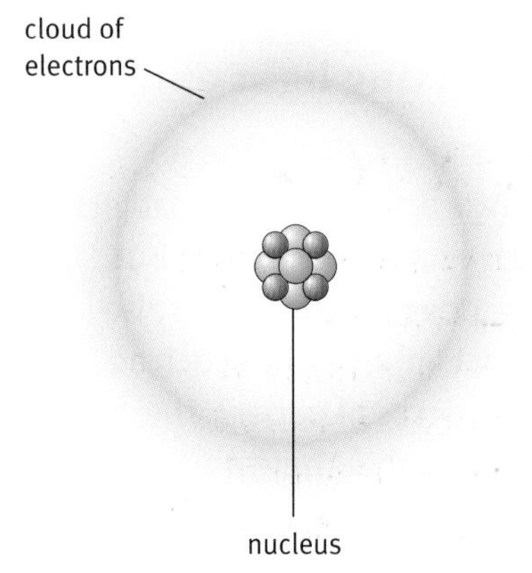

What is electric charge?

Charge is a basic property of matter, which cannot be explained in terms of anything simpler. All matter is made of atoms, which in turn are made out of protons (positive charge), neutrons (no charge), and **electrons** (negative charge). In most materials there are equal numbers of positive and negative charges, so the whole thing is neutral. When you charge something, you move some electrons to it or from it.

The atom has a tiny nucleus made up of protons and neutrons. This has a positive charge. It is surrounded by a cloud of electrons, which have negative charge. As the electrons are on the outside of the atom they can be 'rubbed off', on to another object.

Although scientists cannot explain charge, they have developed useful ideas for predicting its effects. One is the idea of an **electric field**. Around every charge there is an electric field. In this region of space, the effects of the charge can be felt. Another charge entering the field will experience a force.

Moving charge = current

A van de Graaff generator is a machine for separating electric charge. When it is running, charge collects on its dome. As the charge builds up, the electric field around the dome gets stronger. The charge may 'jump' to another nearby object, in the form of a spark. If you hold a mains-testing screwdriver close to the dome and touch its metal end-cap with your finger the neon bulb inside the screwdriver lights up. There is an **electric current** through the lamp, making it light up. Charge on the dome is escaping across the air gap, through the neon bulb, and through you to the Earth. This (and other similar observations) suggests that an electric current is a flow of charge.

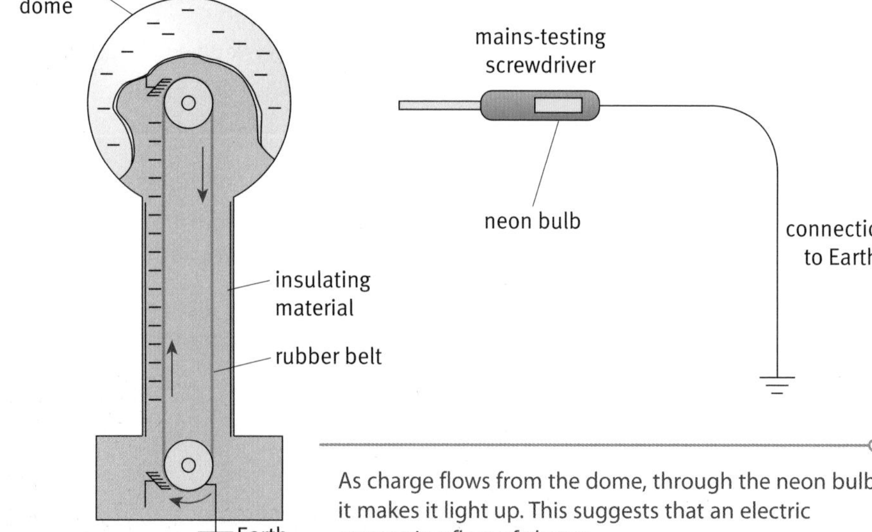

As charge flows from the dome, through the neon bulb, it makes it light up. This suggests that an electric current is a flow of charge.

Key words

- electric charge
- repel, attract
- positive, negative
- electron
- electric field
- electric current

Electric currents in circuits

A closed loop

The diagram on the right shows a simple **electric circuit**. If you make a circuit like this, you can quickly show that:

- if you make a break *anywhere* in the circuit, *everything* stops.

This suggests that something has to go all the way round an electric circuit to make it work. This 'something' is electric charge.

You will also notice that:

- both bulbs come on *immediately* when the circuit is completed. And they go off *immediately* if you make a break in the circuit.

This might seem to indicate that electric charge moves round the circuit very fast. Even with a large electric circuit and long wires, it is impossible to detect any delay. But there is another possible explanation. Perhaps there are **charges** (tiny particles with electric charge) in all the components of the circuit (wires, lamp filaments, batteries) *all the time.* Closing the switch allows these charges to move. They all move together, so the effect is immediate, even if the charges themselves do not move very fast.

The diagram below shows a model that is useful for thinking about how a simple electric circuit works. The power source is the hamster in the treadmill. As the treadmill turns, it pushes the peas along the pipe. If the pipe is full of peas all the time, then they will start moving everywhere around the circuit – immediately. The paddle wheel at the bottom will turn as soon as the hamster runs. The turning paddle wheel could be used to lift a mass. The hamster loses energy – and the mass gains energy. The hamster does work on the treadmill to make it turn and set the peas moving – and the moving peas do work on the paddle wheel.

Find out about

- ✔ how simple electric circuits work
- ✔ models that help explain and predict the behaviour of electric circuits
- ✔ how to measure electric current

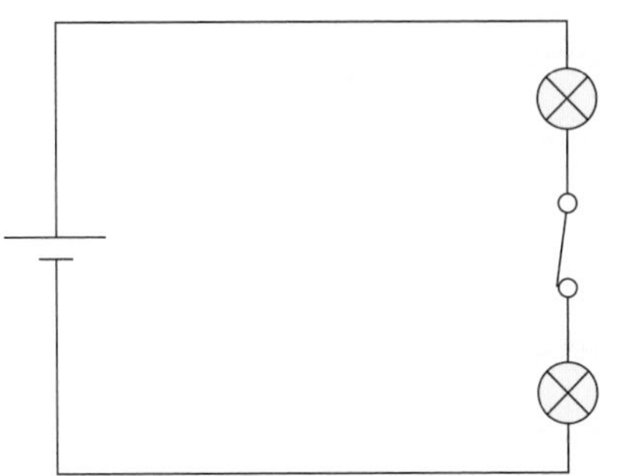

When you open the switch in this electric circuit, both bulbs go off.

Electric circuit model.

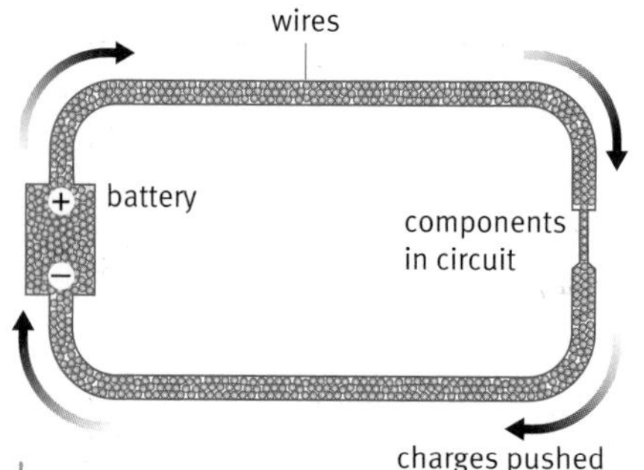

You can think of an electric current as a flow of charges, which are present in all materials (and free to move in conductors), moving round a closed conducting loop, pushed by the battery.

conventional current

electron flow

An electric current is a flow of electrons through the wires of the circuit. You can think of it equally well as a 'conventional current' of positive charges going the other way.

Key words

- ✓ electric circuit
- ✓ series circuit
- ✓ ammeter
- ✓ charges

An electric circuit model

A model is a way of thinking about how something works. The diagram on the right summarises a useful scientific model of an electric circuit.

The key ideas are:

- charges are present throughout the circuit all the time
- when the circuit is a closed loop, the battery makes the charges move
- all of the charges move round together

The battery makes the charges move in the following way. Chemical reactions inside the battery have the effect of separating electric charges, so that positive charge collects on one terminal of the battery and negative charge on the other. If the battery is connected to a circuit, the charges on the battery terminals set up an electric field in the wires of the circuit. This makes free charges in the wire drift slowly along. However, even though the charges move slowly, they all begin to move at once, as soon as the battery is connected. So the effect of their motion is immediate. Notice too that the flow of charge is continuous, all round the circuit. Charge also flows through the battery itself.

Conventional current, electron flow

In the model above, the charges in the circuit are shown moving away from the positive terminal of the battery, through the wires and other components, and back to the negative terminal of the battery. This assumes that the moving charges are positive. In fact there is no simple way of telling whether the moving charges are positive or negative, or which way they are moving. Long after this model was first proposed and had become generally accepted, scientists came to think that the moving charges in metals were electrons, which have negative charge. In all metals, the atoms have some electrons that are only loosely attached to their 'parent' atom and are relatively free to wander through the metal. It is these that move, in the weak electric field that the battery sets up in the wire. Insulators have few charges free to flow.

To explain and predict how electric circuits behave, it makes no difference whether you think in terms of a flow of electrons in one direction or positive charges in the other. Although scientists now believe it is electrons that flow in metals, in this course we use the model of conventional current going the other way, as most physicists and engineers do.

Electric current

An electric current is a flow of charge. You cannot see a current, but you can observe its effects. The current through a torch bulb makes the fine wire of the filament heat up and glow. The bigger the current through a bulb, the brighter it glows (unless the current gets too big and the bulb 'blows').

To measure the size of an electric current we use an **ammeter**. The reading (in amperes, or amps (A) for short) indicates the amount of charge going through the ammeter every second.

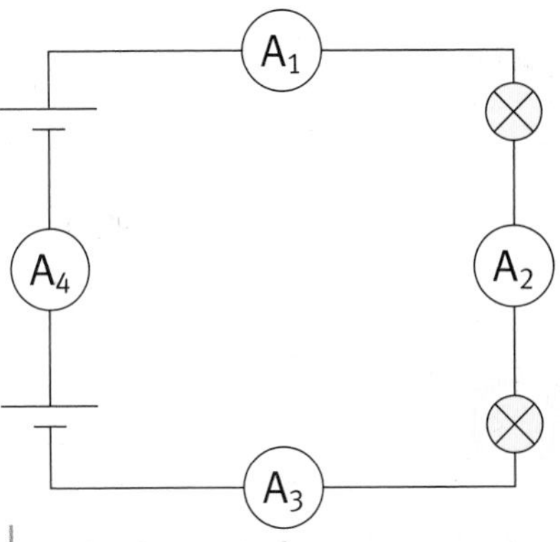

The current is the same size at all these points – even between the batteries. Current is not used up to make the bulbs light. This is a **series** circuit.

Current around a circuit

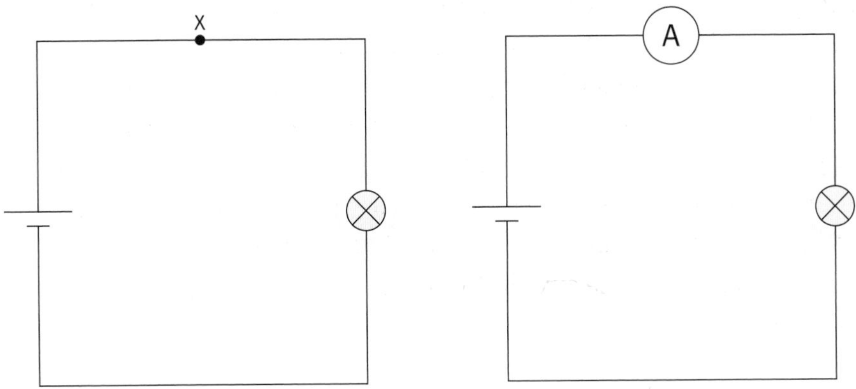

To measure the current at point X, you have to make a gap in the circuit at X and insert the ammeter in the gap, so that the current flows through it.

If you use an ammeter to measure the size of the electric current at different points around a circuit, you get a very important result.

- The current is the same everywhere in a simple (single-loop) electric circuit.

This may seem surprising. Surely the bulbs must use up current to light. But this is not the case. Current is the movement of charges in the wire, all moving round together like dried peas in a tube. No charges are used up. The current at every point round the circuit must be the same.

Of course, *something* is being used up. It is the energy stored in the battery. This is getting less all the time. The battery is doing work to push the current through the filaments of the light bulbs, and this heats them up. The light then carries energy away from the glowing filament. So the circuit constantly transfers energy from the battery, to the bulb filaments, and then on to the surroundings (as light). The current enables this energy transfer to happen. But the current itself is not used up.

Questions

1 Look at the electric circuit model on page 169. What corresponds to:
   i the battery, ii the electric current, iii the size of the electric current?
   a Suggest one thing in a real electric circuit that might correspond to the paddle wheel.
   b How might you model a switch?

2 How would you change the electric circuit model to explore what happens in a series circuit with two identical bulbs? Use the model to explain:
   a why both bulbs go on and off together when the circuit is switched on and off
   b why both bulbs light immediately when the circuit is switched on
   c why both bulbs are equally bright.

Branching circuits

In this portable MP3 player, the battery has to run the motor that turns the hard drive, the head that reads the disk, and the circuits that decode and amplify the signals.

Often, we want to run two or more things from the same battery. One way to do this is to put them all in a single loop, one after the other. Components connected like this are said to be **in series**. All of the moving charges then have to pass through each of them.

Another way is to connect components **in parallel**. In the circuit on the left, the two bulbs are connected in parallel. This has the advantage that each bulb now works independently of the other. If one burns out, the other will stay lit. This makes it easy to spot a broken one and replace it.

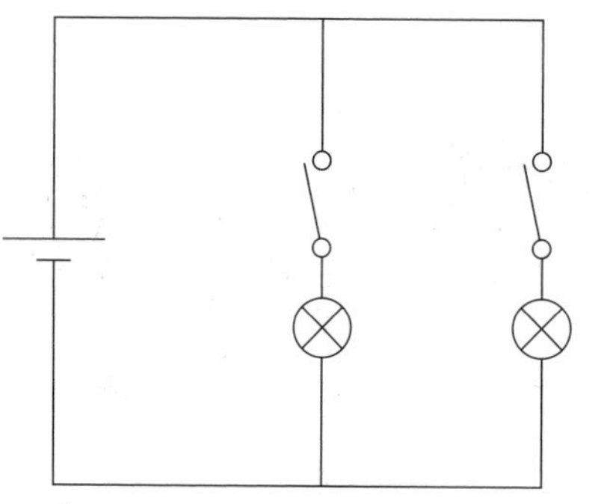

An advantage of connecting components in parallel is that each can be switched on and off independently.

Currents in the branches

Look at the circuit below, which has a motor and a buzzer connected in parallel. A student, Nicola, was asked to measure the current at points a, b, c, and d. Her results are shown in the table below on the right.

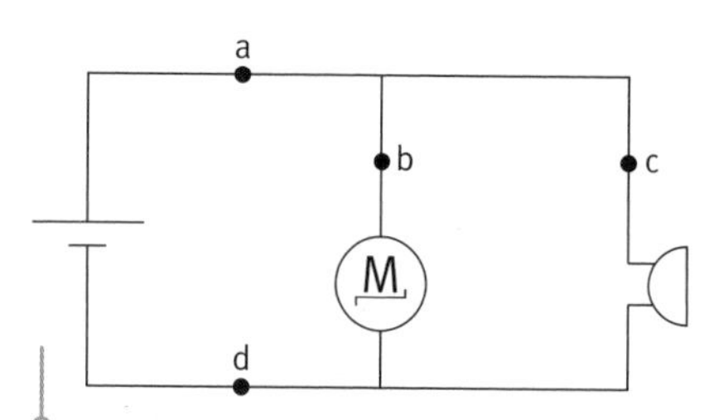

Point in circuit	Current (mA)
a	230
b	150
c	80
d	230

Measuring currents in a circuit with two parallel branches.

Nicola noticed that the current is the same size at points a and d: 230mA. When she added the currents at b and c, the result was also 230mA. This makes good sense if you think about the model of charges moving round. At the junctions, the current splits, with some charges flowing through one branch and the rest flowing through the other branch. Current is the amount of charge passing a point every second. So the amounts in the two branches must add up to equal the total amount in the single wire before or after the branching point.

Key words

- in series
- in parallel

Questions

1 When we want to run several things from the same battery, it is much more common to connect them in parallel than in series. Write down three advantages of parallel connections.

2 Look at the circuit above on the left. If you wanted to switch both bulbs on and off together, where would you put the switch? Draw two diagrams showing two possible positions of the switch that would do this.

3 In the circuit above with the motor and buzzer, what size is the electric current:
- a in the wire just below the motor?
- b in the wire just below the buzzer?
- c through the battery itself?

Controlling the current

D

The scientific model of an electric circuit imagines a flow of charge round a closed conducting loop, pushed by a battery. This movement of charge all round the circuit is an electric current.
The size of the current is determined by two factors:

- the **voltage** of the battery
- the **resistance** of the circuit components.

Find out about

- how the battery voltage and the circuit resistance together control the size of the current
- what causes resistance
- the links between battery voltage, resistance, and current

Battery voltage

Batteries come in different shapes and sizes. They usually have a voltage measured in volts (V), marked clearly on them, for example, 1.5 V, 4.5 V, 9 V. To understand what voltage means and what this number tells you, look at the following diagrams, which show the same bulb connected first to a 4.5 V battery and then to a 1.5 V battery. Voltage is measured using a voltmeter.

All the batteries on the front row are marked 1.5 V – but are very different sizes. The three at the back are marked 4.5 V, 6 V, and 9 V.

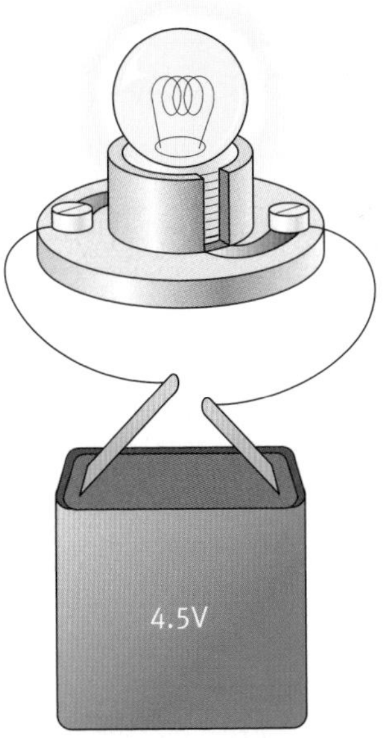

With a 4.5 V battery, this bulb is brightly lit.

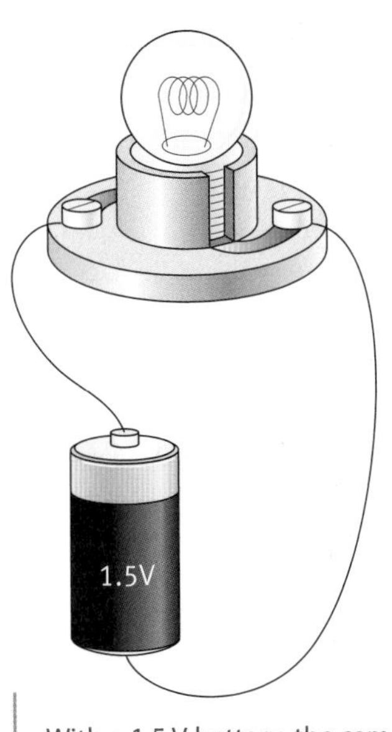

With a 1.5 V battery, the same bulb is lit, but very dimly.

The bigger the current through a light bulb, the brighter it will be (up to the point where it 'blows'). So the current through the bulb above is bigger with the 4.5 V battery. You can think of the voltage of a battery as a measure of the 'push' it exerts on the charges in the circuit. The bigger the voltage, the bigger the 'push' – and the bigger the current as a result.

The battery voltage depends on the choice of chemicals inside it. To make a simple battery, all you need are two pieces of different metal and a beaker of salt solution or acid. The voltage quickly drops, however. The chemicals used in real batteries are chosen to provide a steady voltage for several hours of use.

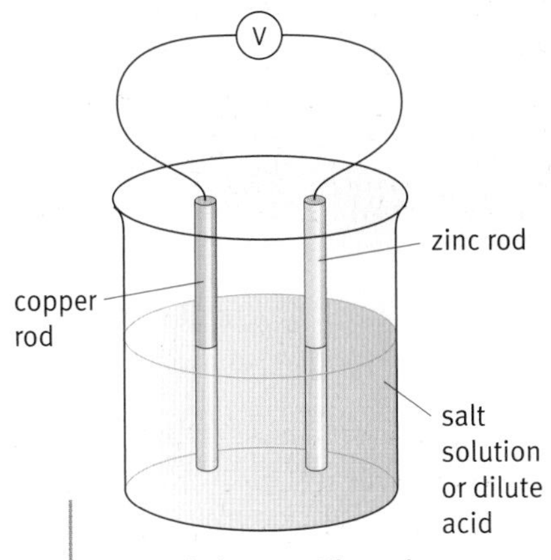

A simple battery. The voltage depends on the metals and the solution you choose.

Resistance

The size of the current in a circuit also depends on the resistance that the components in the circuit provide to the flow of charge. The battery 'pushes' against this resistance. You can see the effect of this if you compare two circuits with different resistors. Resistors are components designed to control the flow of charge.

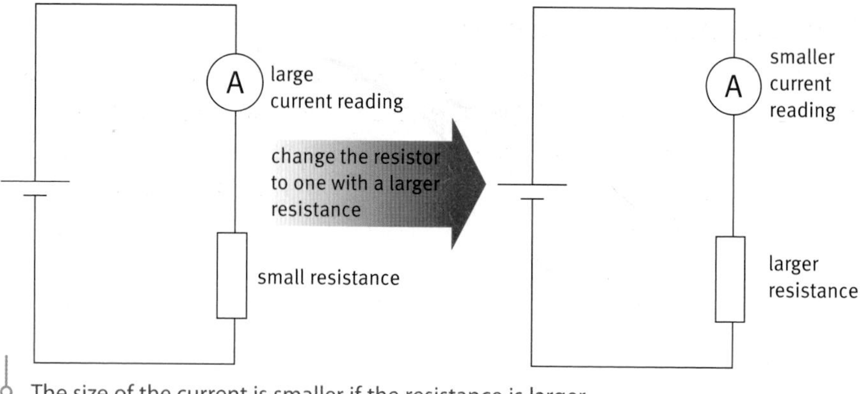

The size of the current is smaller if the resistance is larger.

Changing resistance changes the size of the current. The bigger the resistance, the smaller the current.

What causes resistance?

Everything has resistance, not just special components called resistors. The resistance of connecting wires is very small, but not zero. Other kinds of metal wire have larger resistance. The filament of a light bulb has a lot of resistance. This is why it gets so hot and glows when there is a current through it. A heating element, like that in an electric kettle, is a resistor.

Why the temperature of a wire rises when a current flows through it.

All metals get hot when charge flows through them. In metals, the moving charges are free electrons. As they move round, they collide with the fixed array, or lattice, of ions in the wire. These collisions make the ions vibrate a little more, so the temperature of the wire rises. In some metals, the ions provide only small targets for the electrons, which can get past them relatively easily. In other metals, the ions present a much bigger obstacle – and so the resistance is bigger.

Questions

1 Look back at the electric circuit model in Section B. How would you change the model to show the effect of:
   a a bigger battery voltage?
   b an increase in resistance?

   (You might be able to think of two different ways of doing this.)

   Does the model correctly predict the effect these would have on the current?

2 Suggest two different ways in which you could change a simple electric circuit to make the electric current bigger.

Key relationships in an electric circuit: A summary

The size of the electric current (I) in a circuit depends on the battery voltage (V) and the resistance (R) of the circuit.

- If you make V bigger, the current (I) increases.
- If you make R bigger, the current (I) decreases.

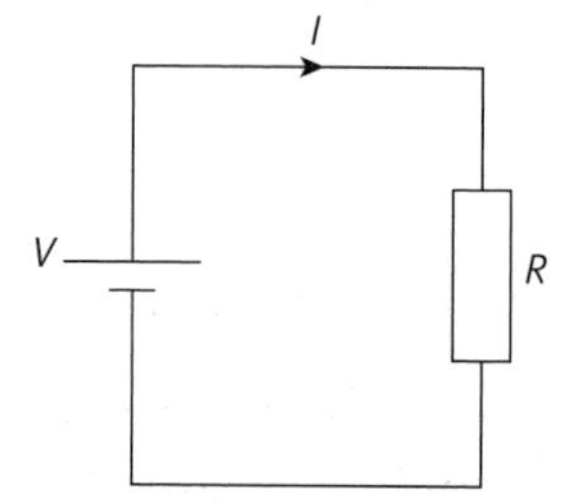

Measuring resistance

The relationship between battery voltage, resistance, and current leads to a way of measuring and defining resistance. The first step is to explore the relationship between voltage and current in more detail. Keiko, a student, did this by measuring the current through a coil of wire with different batteries.

1. Keiko connected a coil of resistance wire to a 1.5 V battery and an ammeter. She noted the current.

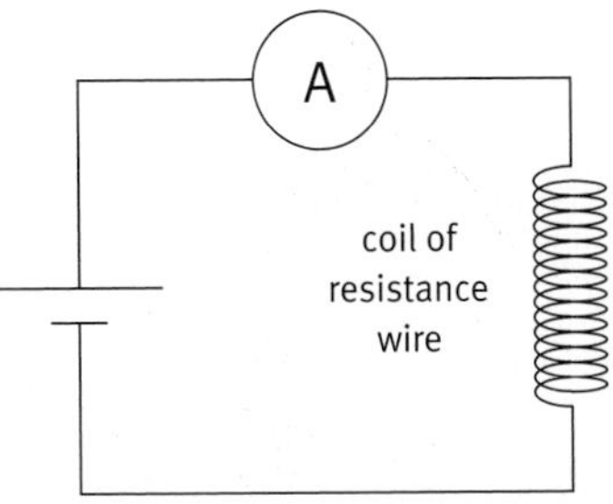

2. She then added a second battery in series. She noted the current again.

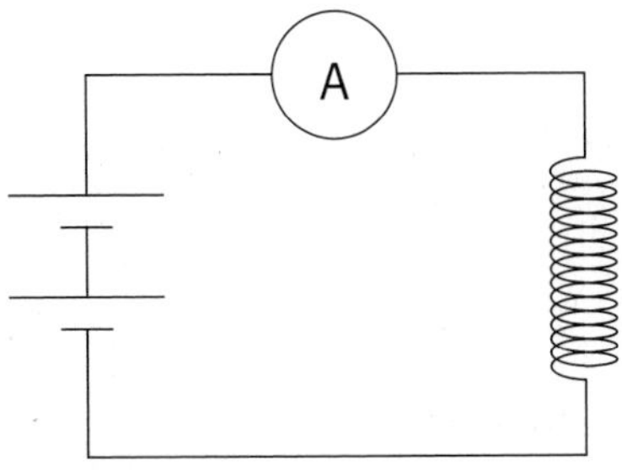

3. Keiko repeated this with 3, 4, 5, and 6 batteries, to get a set of results:

4. Finally, she drew a graph of current against battery voltage:

Number of 1.5 V batteries	Battery voltage (V)	Current (mA)
1	1.5	75
2	3.0	150
3	4.5	225
4	6.0	300
5	7.5	375
6	9.0	450

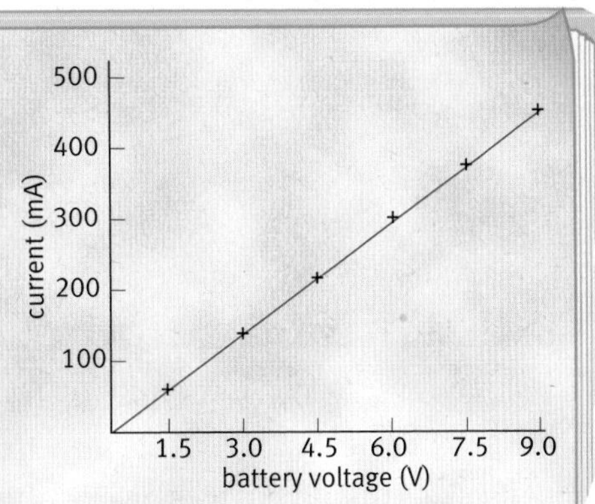

The straight-line graph means that the current in the circuit is proportional to the battery voltage. This result is known as **Ohm's law**. The number you get if you divide voltage by current is the same every time. The bigger the number, the larger the resistance. This is how to measure resistance:

$$\text{resistance of a conductor} = \frac{\text{voltage across the conductor}}{\text{current through the conductor}}$$

$$R = \frac{V}{I}$$

The units of resistance are called ohms (Ω).

Rearranging this equation gives $I = V/R$. You can use this to calculate the current in a circuit, if you know the battery voltage and the resistance of the circuit.

Questions

3 In Keiko's investigation:
  a how many 1.5 V batteries would she need to use to make a current of 600 mA flow through her coil?
  b what is the resistance of her coil, in ohms?

4 In a simple series circuit, a 9 V battery is connected to a 45 Ω resistor.

  What size is the electric current in the circuit?

Ohm's law

Ohm's law says that the current through a conductor is **proportional** to the voltage across it – provided its temperature is constant.
It only applies to some types of conductor (such as metals). An electric current itself causes heating, which complicates matters. For example, the current through a light bulb is not proportional to the battery voltage. The *I–V* graph is curved. The reason is that bulb filament heats up – and its resistance increases with temperature.

Key words

- voltage
- resistance
- Ohm's law

Variable resistors

Resistors are used in electric circuits to control the size of the current. Sometimes we want to vary the current easily, for example, to change the volume on a radio or CD player. A variable resistor is used. Its resistance can be steadily changed by turning a dial or moving a slider.

The circuit diagram on the left shows the symbol for a variable resistor. As you alter its resistance, the brightness of the light bulb changes, and the readings on *both* ammeters increase and decrease together. The variable resistor controls the size of the current everywhere round the circuit loop. (Is this what the circuit model in Section B would predict?)

A_1 A_2 R

Using a variable resistor to control the current in a series circuit.

Each of these sliders adjusts the value of a variable resistor.

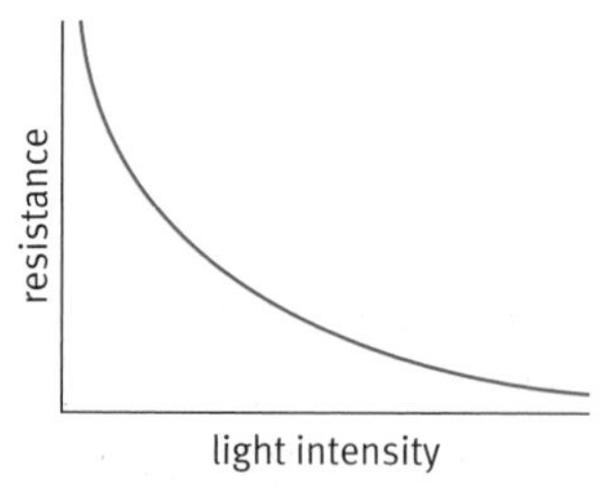

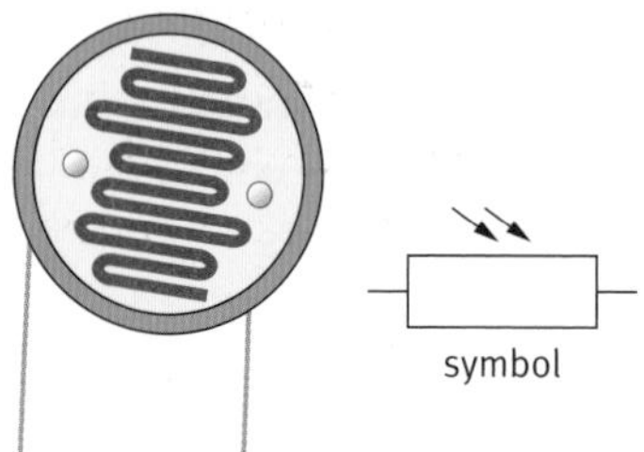

The resistance of a light-dependent resistor decreases as the light intensity increases.

Some useful sensing devices are really variable resistors. For example, a light-dependent resistor (LDR) is a semiconductor device whose resistance is large in the dark but gets smaller as the light falling on it gets brighter. An LDR can be used to measure the intensity (brightness) of light or to switch another device on and off when the intensity of the light changes. For example, it could be used to switch an outdoor light on in the evening and off again in the morning.

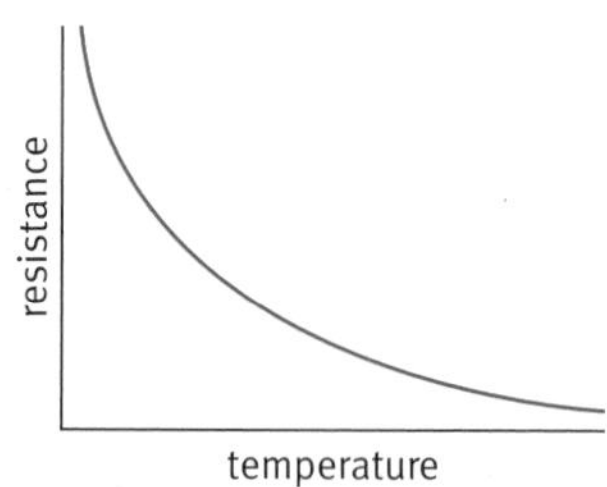

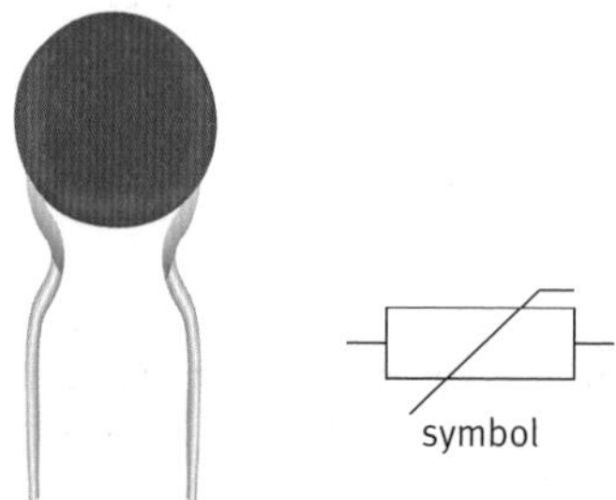

The resistance of a thermistor decreases as the temperature increases.

A thermistor is another device made from semiconducting material. Its resistance changes rapidly with temperature. The commonest type has a lower resistance when it is hotter. Thermistors can be used to make thermometers (to measure temperature) or to switch a device on or off as temperature changes. For example, a thermistor could be used to switch an immersion heater on when the temperature of water in a tank falls below a certain value and off again when the water is back at the required temperature.

Combinations of resistors

Most electric circuits are more complicated than the ones discussed so far. Circuits usually contain many components, connected in different ways. There are just two ways of connecting circuit components: in series or in parallel.

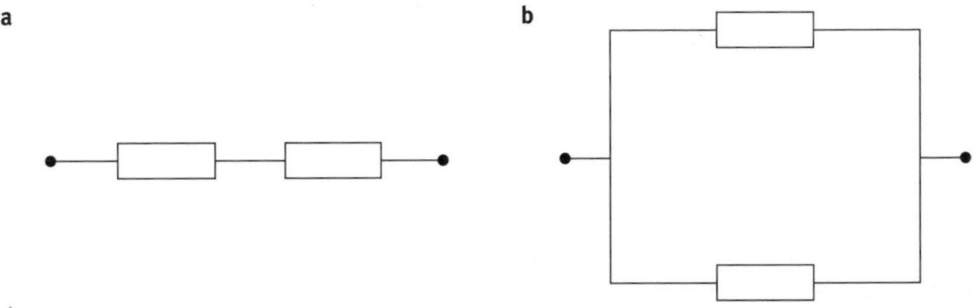

Two resistors connected **a** in series, **b** in parallel.

This circuit board from a computer contains a complex circuit, with many components.

Two resistors in series have a larger resistance than one on its own. The battery has to push the charges through both of them. But connecting two resistors in parallel makes a smaller total resistance. There are now two paths that the moving charges can follow. Adding a second resistor in parallel does not affect the original path but adds a second equivalent one. It is now easier for the battery to push charges round, so the resistance is less.

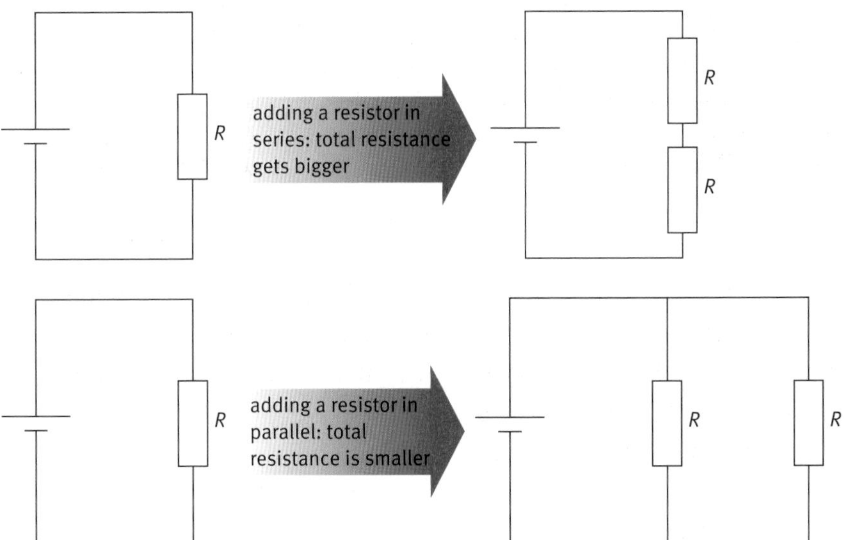

Different ways of adding a second resistor: how you do it makes a difference.

Question

5 All the resistors in the three diagrams below are identical. Put the groups of resistors in order, from the one with the largest total resistance to the one with the smallest total resistance.

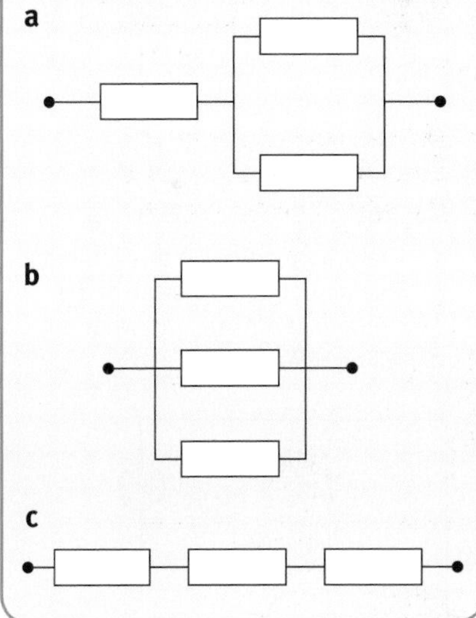

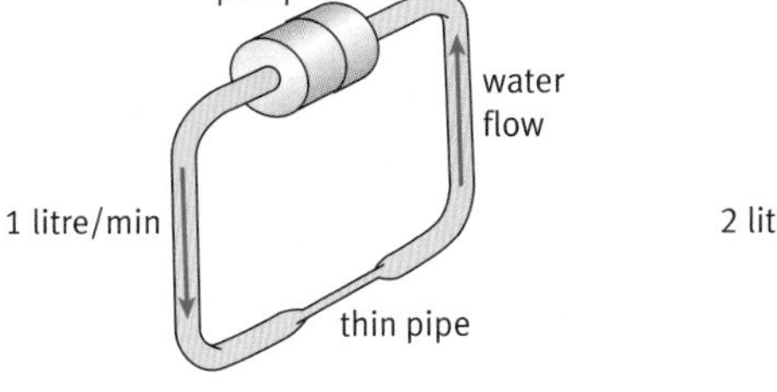

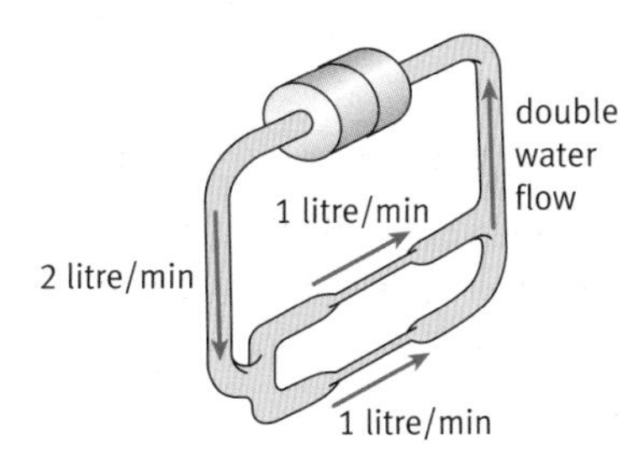

A water-flow model shows how the total resistance gets less when a second parallel path is added.

Potential difference

Find out about

- how voltmeters measure the potential difference between two points in a circuit
- how height provides a useful model for thinking about electrical potential
- how current splits between parallel branches
- how potential dividers are used

You can think of the voltage of a battery as a measure of the 'push' it exerts on the charges in a circuit. But a voltmeter also shows a reading if you connect it across a resistor or bulb in a working circuit as in the circuit on the left. Resistors and bulbs do not 'push'. So the **voltmeter** reading must be indicating something else.

A useful picture is to think of the battery as a pump, lifting water up to a higher level. The water then drops back to its original level as it flows back to the inlet of the pump. The diagram below shows how this would work for a series circuit with three resistors (or three lamps). The pump does **work** on the water to raise it to a higher level. This increases the gravitational potential energy stored. The water then does work on the three water wheels as it falls back to its original level. If we ignore energy losses, this is the same as the amount of work done by the pump.

In the electric circuit, the battery does work on the electric charges, to lift them up to a higher 'energy level'. They then do work (and transfer energy) in three stages as they drop back to their starting level. A voltmeter measures the difference in 'level' between the two points it is connected to. This is called the potential difference between these points. **Potential difference (p.d.)** is measured in volts (V).

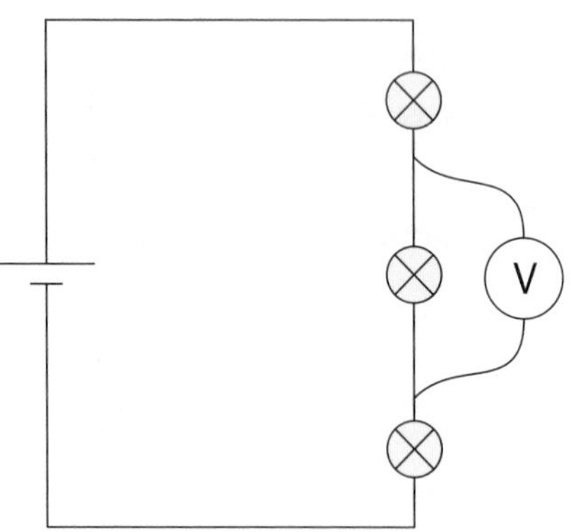

There is a reading on this voltmeter. This cannot be a measure of the strength of a 'push' – so what is it telling us?

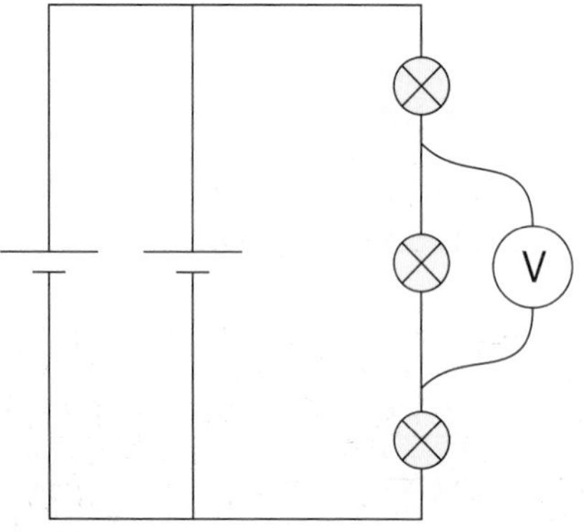

Adding a second battery in parallel does not change the potential difference across the lamps. So the current through the lamps does not change.

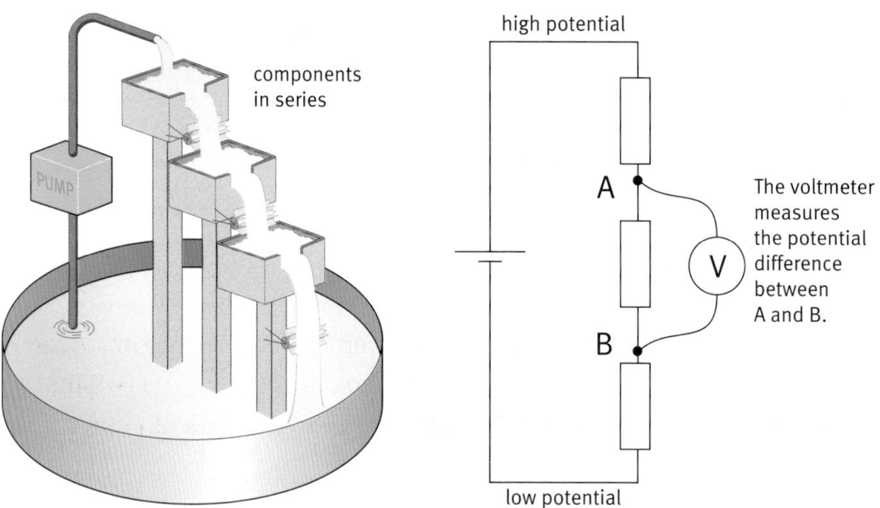

The voltage of a battery is the potential difference between its terminals. If you put a battery with a larger voltage into the circuit above, this would mean a bigger potential difference across its terminals. The potential difference across each lamp (or resistor) would also now be bigger. Going back to the water-pump model, this is like changing to a stronger pump that lifts the water up to a higher level. The three downhill steps then also have to be bigger, so that the water ends up back at its starting level.

The same idea works for a parallel circuit. In this case, the water divides into three streams. Each loses all its energy in a single step.

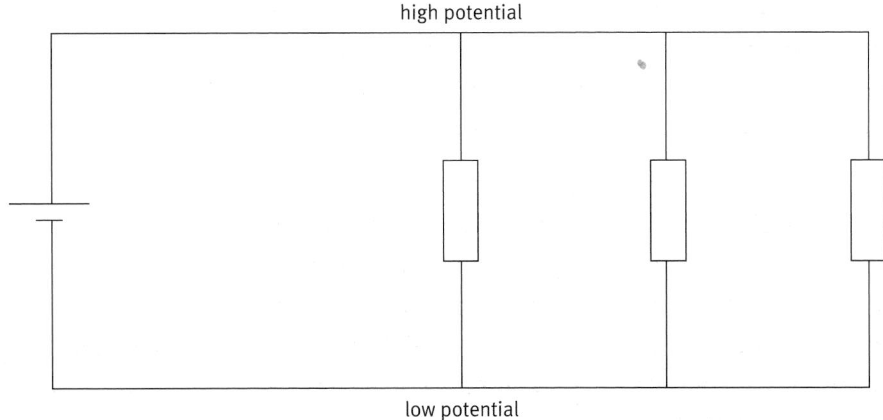

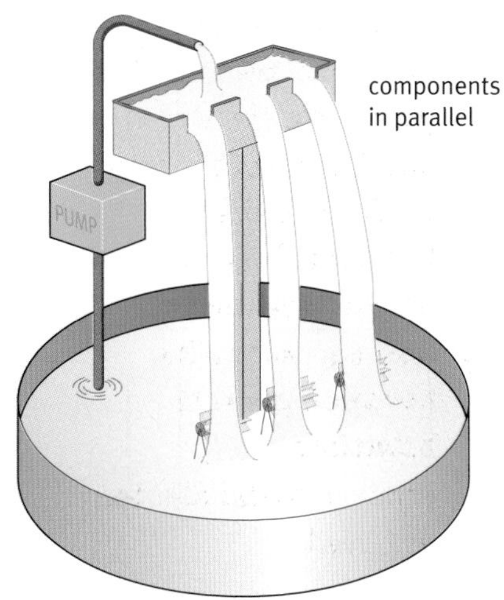

Voltmeter readings across circuit components

This water-pump model helps to explain and to predict voltmeter readings across resistors in different circuits. If several resistors are connected in parallel to a battery, the potential difference across each is the same. It is equal to the battery voltage.

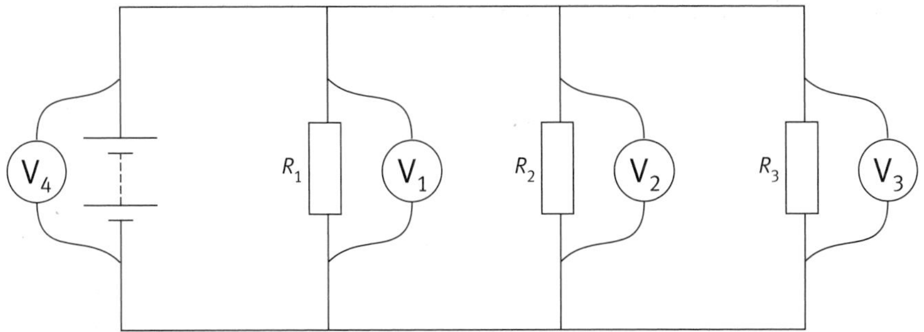

All of these voltmeters will have the same reading, even if R_1, R_2, and R_3 are different.

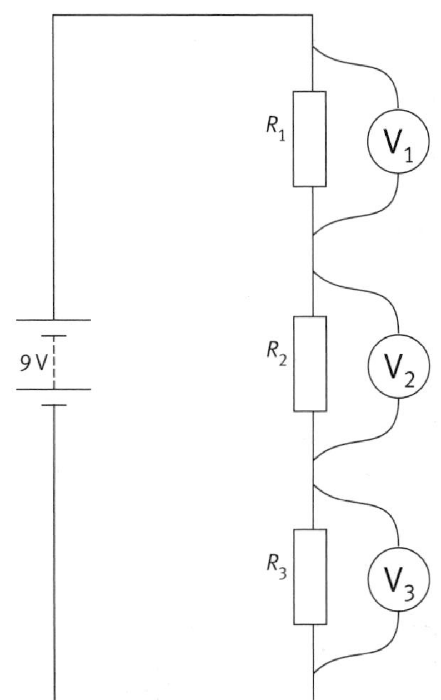

The voltages are in proportion to the resistances, and their sum is equal to the battery voltage.

If the resistors are connected in series (as on the right), the sum of the potential differences across them is equal to the battery voltage. This is exactly what you would expect from the 'waterfall' picture on the previous page.

In the series circuit, the potential difference across each resistor depends on its resistance. The biggest voltmeter reading is across the resistor with biggest resistance. Again this makes sense. More work has to be done to push charge through a big resistance than a smaller one.

If several identical batteries are connected in parallel with a single resistor, the potential difference across the resistor remains the same as for one battery. As the potential difference across the resistor does not change, the current through the resistor does not change.

Key words

- ✓ voltmeter
- ✓ potential difference (p.d.)
- ✓ work

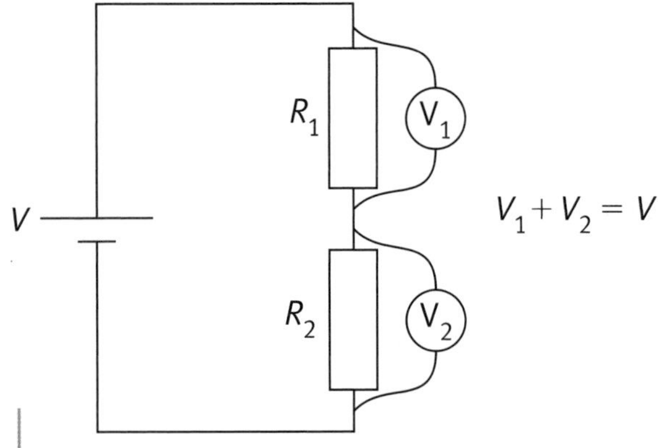

Two resistors in series make a potential divider.

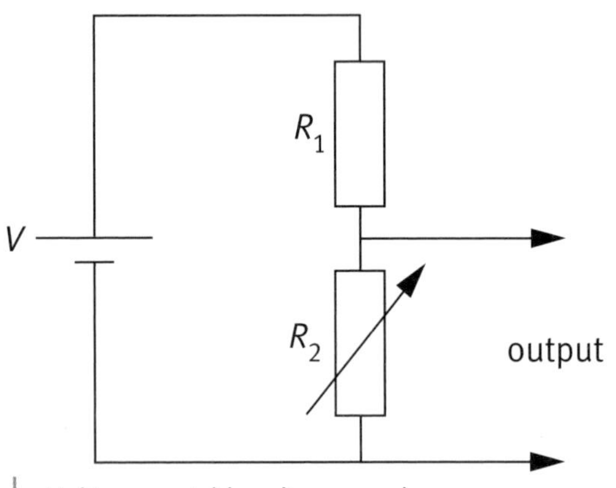

Making a variable voltage supply.

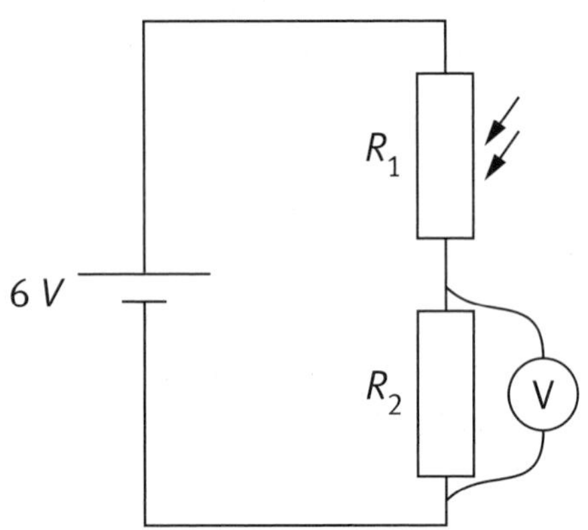

A simple light intensity meter.

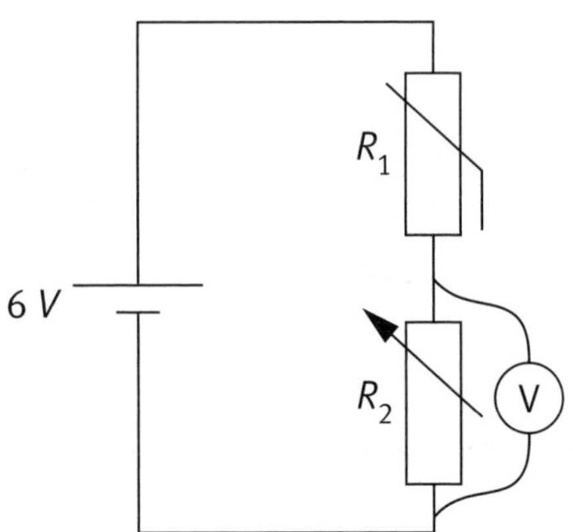

Potential divider

A circuit with two resistors in series has many important applications, particularly in electronics. The sum of the potential differences across the two resistors is equal to the p.d. across the battery (the battery voltage). The resistors divide up the battery voltage into two parts. For this reason, this kind of circuit is often called a **potential divider**. The two parts are often unequal. The bigger p.d. is always across the bigger resistance.

One application is to make a variable voltage supply. If the resistance of the variable resistor R_1 is turned right down to zero, then the potential difference across the output leads will be zero. The p.d. across the fixed resistor R_2 will be equal to the battery voltage. But if the resistance of R_1 is now increased, a larger share of the total will appear across it. The potential difference across the output leads will increase. Unlike a battery with a fixed voltage, we have a power supply with a voltage that we can change by turning a control knob.

Sensors also use the potential divider principle. A light sensor uses a light dependent resistor (LDR) in series with a fixed resistor. In the dark, the resistance of the LDR is large – much bigger than the resistance of R_2. So the reading on the voltmeter is very small, close to zero.

But in bright light, the resistance of the LDR falls dramatically until it is very small. Now the resistance of R_2 is much bigger than the resistance of the LDR. So the p.d. across R_2 is almost all of the battery voltage – and the reading on the voltmeter is close to 6 V.

So the reading on the voltmeter indicates the brightness of the light falling on the LDR.

Question

1 Describe and explain the reading on the voltmeter in this circuit when the thermistor is:
   a cold
   b hot.

   Explain how the circuit could be used as an electrical thermometer.

Currents in parallel branches

The potential difference across resistors R_1, R_2, and R_3 in the parallel circuit earlier is exactly the same for each. It is equal to the p.d. across the battery itself. But the *currents* through the resistors are not necessarily the same. This will depend on their resistances. The current through the biggest resistor will be the smallest. There are two ways to think of this.

1 Imagine water flowing through a large pipe. The pipe then splits in two, before joining up again later. If the two parallel pipes have different diameters, more water will flow every second through the pipe with the larger diameter. The wider pipe has less resistance than the narrower pipe to the flow of water. So the current through it is larger.

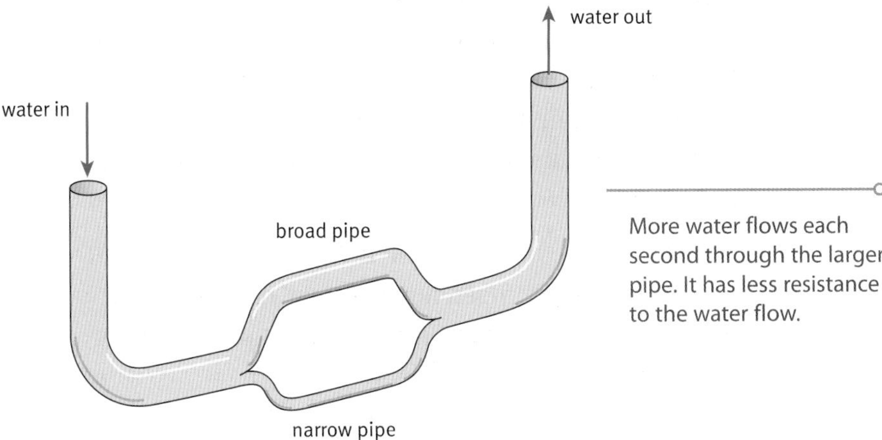

More water flows each second through the larger pipe. It has less resistance to the water flow.

2 Think of two resistors connected in parallel to a battery as making two separate simple-loop circuits that share the same battery. The current in each loop is independent of the other. The smaller the resistance in a loop, the bigger the current. Some wires in the circuit are part of both loops, so here the current will be biggest. The current here will be the sum of the currents in the loops.

Question

2 Imagine removing the red resistor from the circuit below leaving a gap. What would happen to:

a the current through the purple resistor?

b the current from (and back to) the battery?

Explain your reasoning each time.

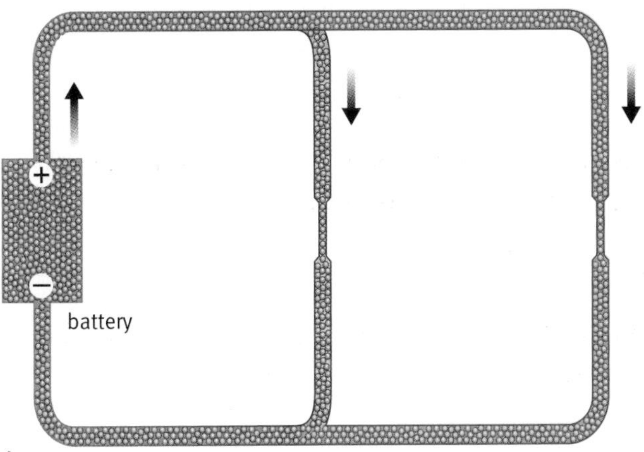

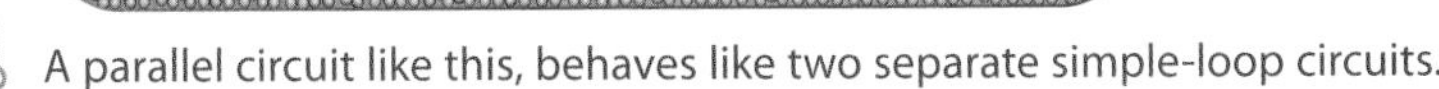
A parallel circuit like this, behaves like two separate simple-loop circuits.

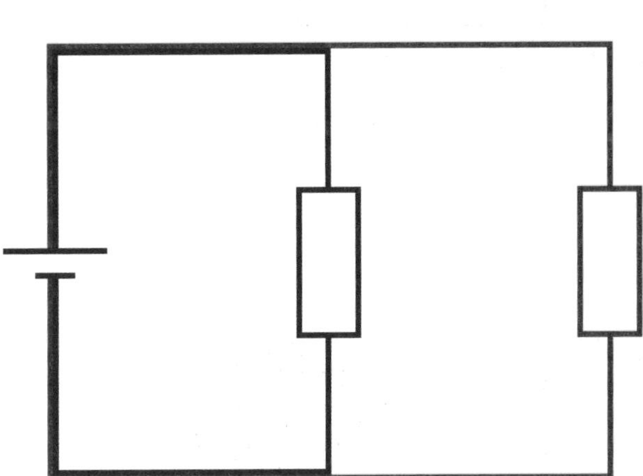

Electrical power

Find out about

- how the power produced in a circuit component depends on both current and voltage

An electric circuit is primarily a device for doing work of some kind. It transfers energy initially stored in the battery to somewhere else. A key feature of any electric circuit is the rate at which work is done on the components in the circuit – that is, the rate at which energy is transferred from the battery to the other components and on into the environment. This is called the **power** of the circuit.

Measuring the power of an electric circuit

Imagine starting with a simple battery and bulb circuit and trying to double, and treble, the power. You could do this in two ways.

- One is to add a second bulb, and then a third, in parallel with the first. In the circuits down the left-hand side of the diagram below, the p.d. is the same, but the current supplied by the battery doubles and trebles. The power is proportional to the current.
- Another is to add a second bulb, and then a third, in series with the first. Now you need to add a second battery, and then a third, to keep the brightness of the bulbs the same each time. In these circuits (across the diagram below), the current is the same, but the p.d. of the battery doubles and trebles. The power is proportional to the voltage.

This is summarised in the box in the bottom right-hand corner.

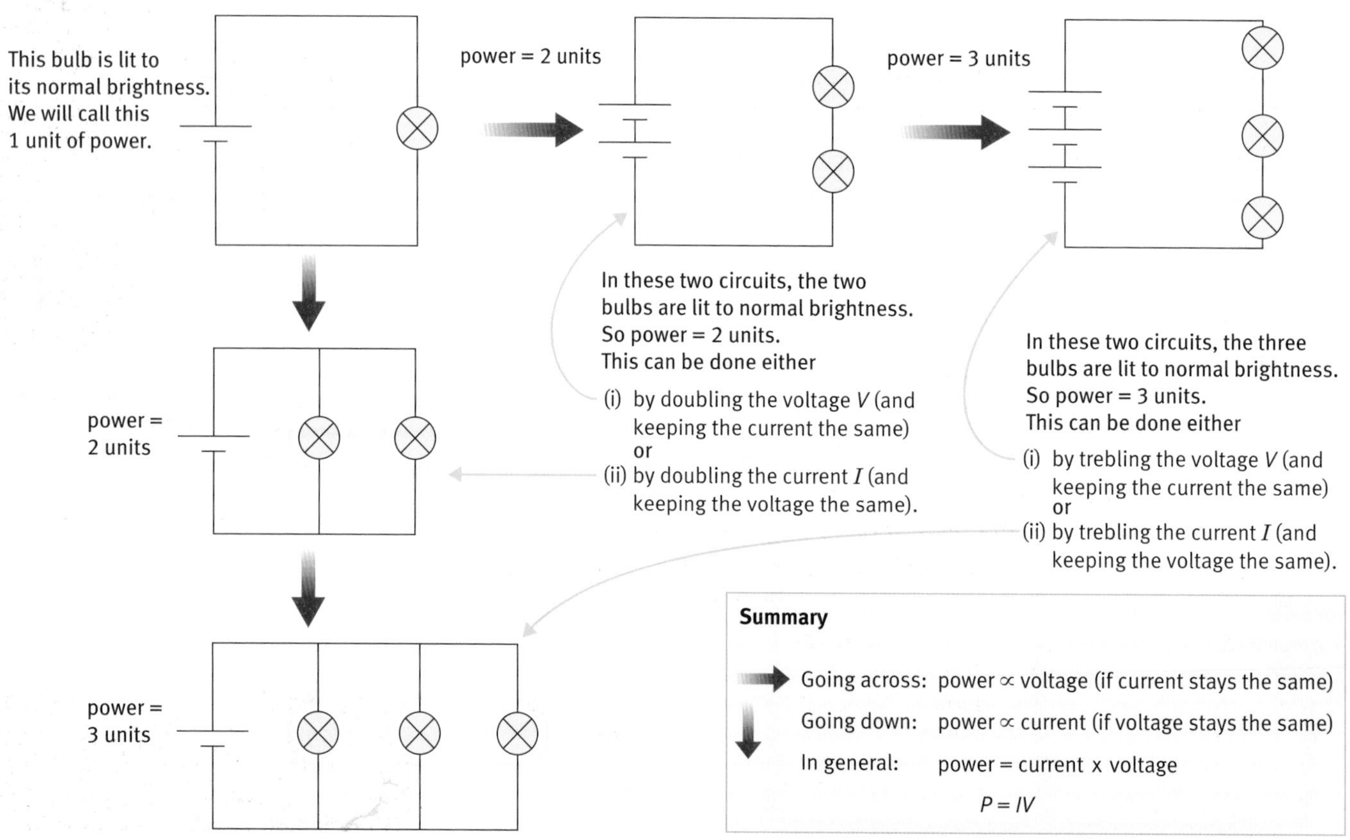

In general, the power dissipated in an electric circuit depends on both the current and the voltage:

power	=	current	×	voltage
P	=	I	×	V
(watt, W)		(ampere, A)		(volt, V)

The unit of power is the watt (W). One watt is equal to one joule per second.

To see how this equation for power makes sense, look back at the explanation of resistance and heating on page 174. If the battery voltage is increased, the electric field in the wires gets bigger. So the free charges (the electrons) move faster. When a charge collides with an atom in the wire, more energy is transferred in the collision. These collisions also happen more often, simply because the charges are moving along faster. So when the voltage is increased, collisions between electrons and the lattice of atoms are *both* harder *and* more frequent.

If you know the power, it is easy to calculate how much work is done (or how much energy is transferred) in a given period of time:

work done (or energy transferred)	=	power	×	time
(joule, J)		(watt, W)		(second, s)

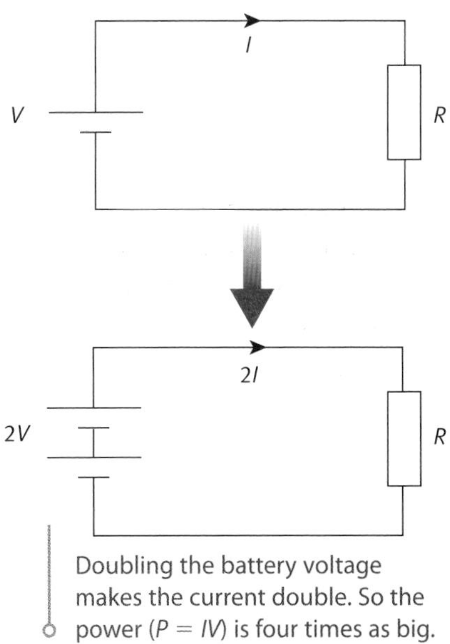

Doubling the battery voltage makes the current double. So the power ($P = IV$) is four times as big.

The power of the electric motor in this tube train is much greater than the power of the strip light above the platform. Both the voltage and the current are bigger.

Questions

1 In these two circuits, resistor R_1 has a large resistance and resistor R_2 has a small resistance. If each circuit is switched on for a while, which resistor will get hotter? Explain your answers.

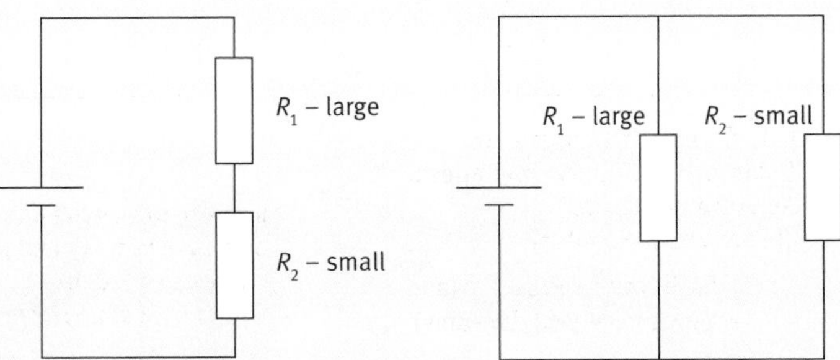

2 In circuit A, a battery is connected to a resistor with a small resistance. In circuit B, the resistor has a large resistance. The two batteries are identical. Which will go 'flat' first? Explain your answer.

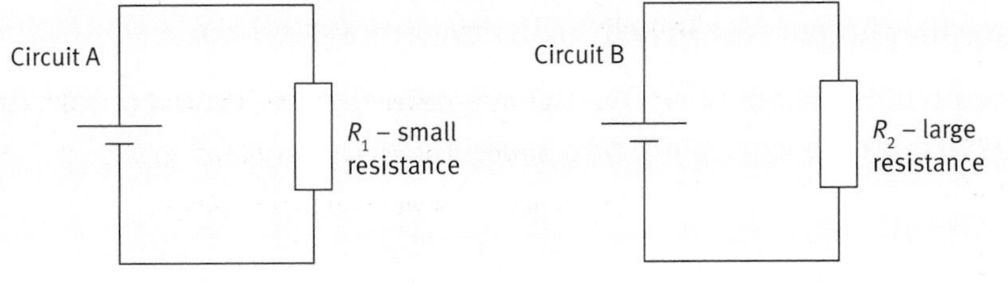

Key word

✔ power

Magnets and motors

Find out about

- the force on a current-carrying wire in a magnetic field
- why an electric motor spins

Magnetic effects

In 1819, the Danish physicist Hans-Christian Oersted noticed that the needle of a magnetic compass moved every time he switched on an electric current in a nearby wire. He investigated this further, and showed that there was a link between electricity and magnetism. When there is an electric current in a wire, there is a **magnetic field** in the region around the wire. The compass needle is a magnet and experiences a force because it is in the magnetic field caused by the electric current.

Winding a wire into a coil makes the magnetic field stronger. This is because the fields of each turn of the coil add together. It can be strengthened further by putting an iron core inside the coil to make an **electromagnet**.

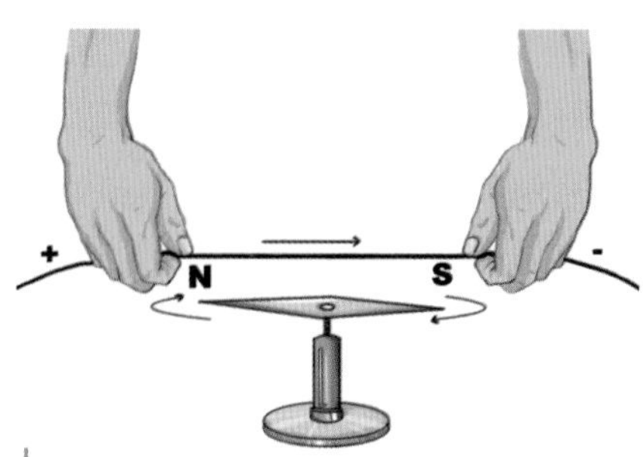

The wire is above the compass needle. The needle moves when the electric current is switched on.

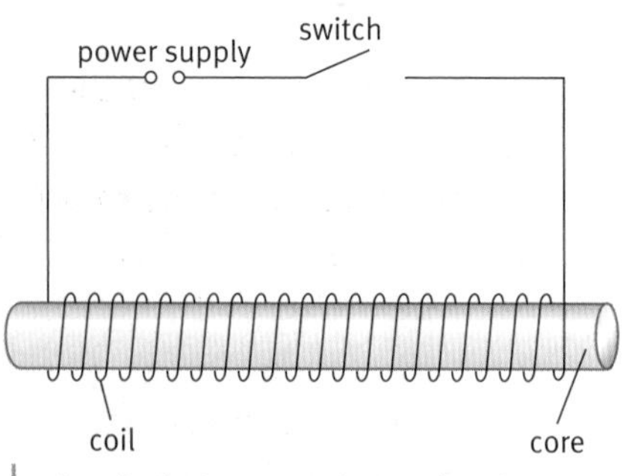

A coil of wire wound round an iron core makes an electromagnet – a magnet that can be switched on and off.

Magnetic forces

A permanent magnet, such as a compass needle, experiences a force if it is placed in the magnetic field near a wire that is carrying an electric current. What if we keep the magnet fixed and allow the wire to move? The diagram shows one way of doing this. The two long parallel wires, and the wire 'rider' that is laid across them, are all made of copper wire with no insulation. The 'rider' is sitting in the magnetic field between the two flat magnets on the metal holder. If you switch on the power supply, the 'rider' slides sideways. The force acting on it is at right angles to both the magnetic field lines and to the electric current.

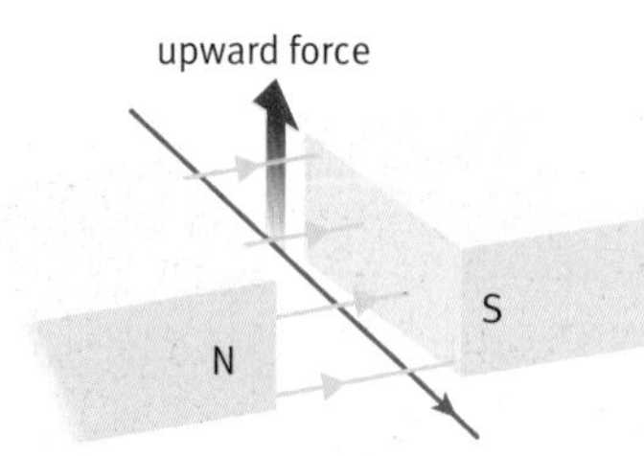

The magnetic force on the wire is at right angles to both the magnetic field lines and the electric current.

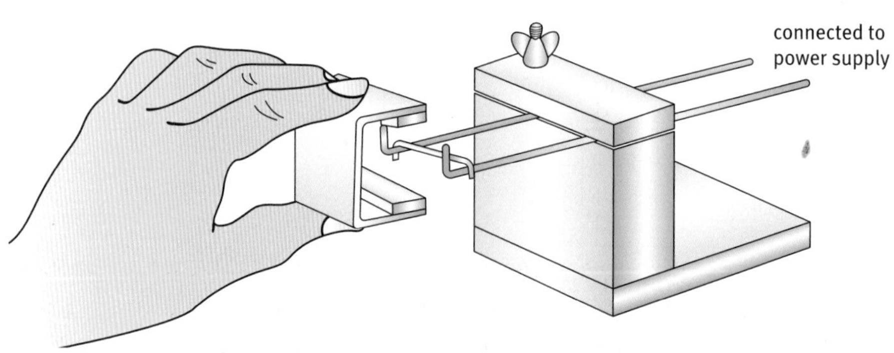

When the current is switched on, the rider moves sideways.

If we turn the magnets, so that the magnetic field is parallel to the wire rider, it does not experience any magnetic force.

Turning effect on a coil

Magnetic forces can make a wire coil turn when a current flows. The diagram below shows a square coil of wire in a magnetic field. There will be no forces on the two ends of the coil, because the currents in these wires are parallel to the magnetic field lines. There will, however, be forces on the two sides of the coil, because the currents in these are at right angles to the magnetic field lines. The forces will be at right angles to both the field and the current. One force is up and the other down, because the currents in the two sides of the coil are in opposite directions. The effect of these forces is to make the coil rotate around the dotted line. If the coil is made with several turns of wire, this will make the turning forces stronger.

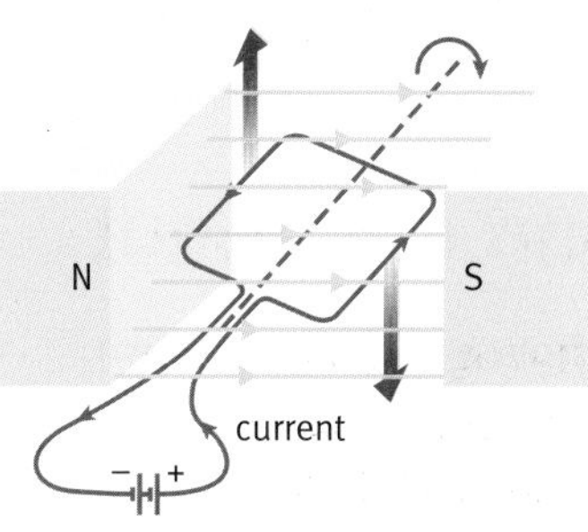

Turning effect of the magnetic forces on a flat coil

An electric motor

The coil in the diagram above would only turn through 90° before stopping. But if we could reverse the direction of the current in the coil at this point, this would reverse the direction of the forces on each side – and keep it turning for a further half-turn. If we could then reverse the current direction again, we could make the coil turn continuously. This is how a simple electric **motor** works.

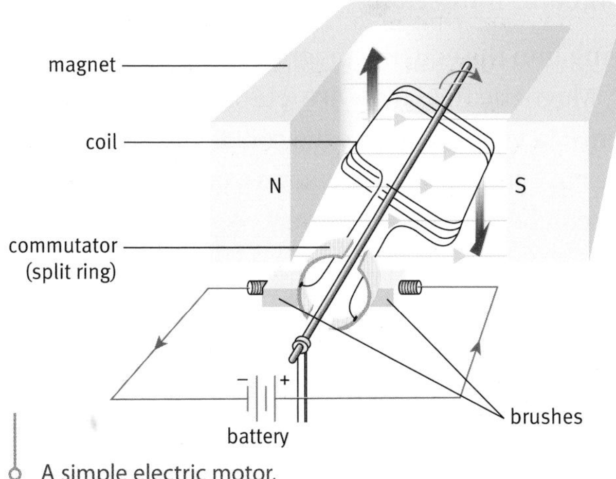

A simple electric motor.

Rather than having fixed wires, the electric circuit now includes a pair of brushes that rub against a split-ring **commutator**. This is fixed to the coil and rotates with it. As the coil rotates, each half of the split ring touches one brush for half a turn, and the other brush for the next half-turn. So the current direction in the coil is reversed twice every turn – changing the direction of the magnetic forces, and keeping the coil rotating.

Questions

1 Look at the diagram of the flat coil between the magnets. Explain why there is no rotating force on the coil when it is vertical.

2 Explain how the commutator ensures that the current in the coil changes direction even though it is connected to a battery with a direct current.

Key words

- **magnetic field**
- **electromagnet**
- **motor**
- **commutator**

H Generating electricity

Find out about

- how a magnet moving near a coil can generate an electric current
- the factors that determine the size of this current
- how this is used to generate electricity on the large scale

There is a connection between electricity and magnetism. An electric current generates a magnetic field, which can then be used to cause motion, as in an electric motor. But can we do this in reverse? Is it possible to generate an electric current by moving a wire near a magnet?

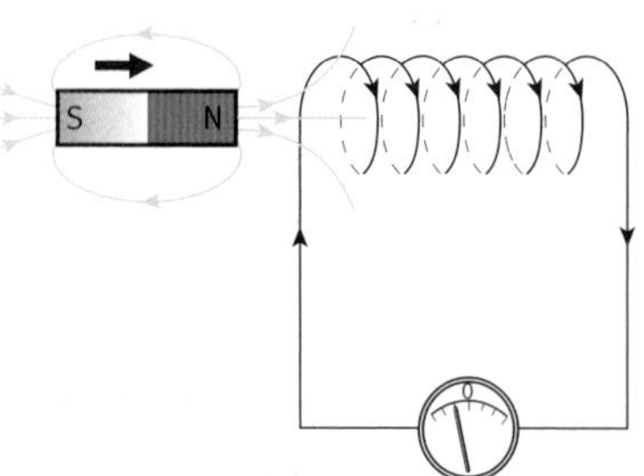

1 While the bar magnet is moving into the coil, there is a small reading on the sensitive ammeter.

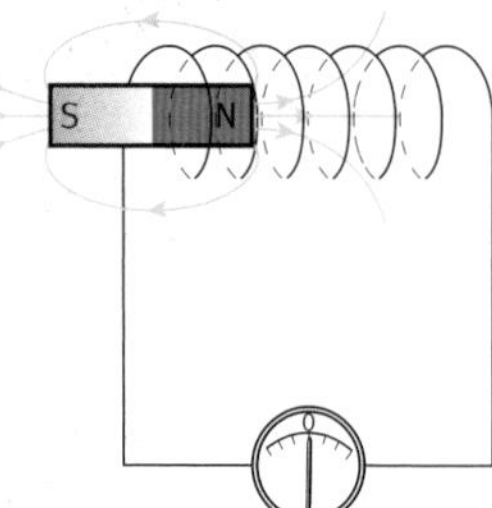

2 There is no current while the magnet is stationary inside the coil.

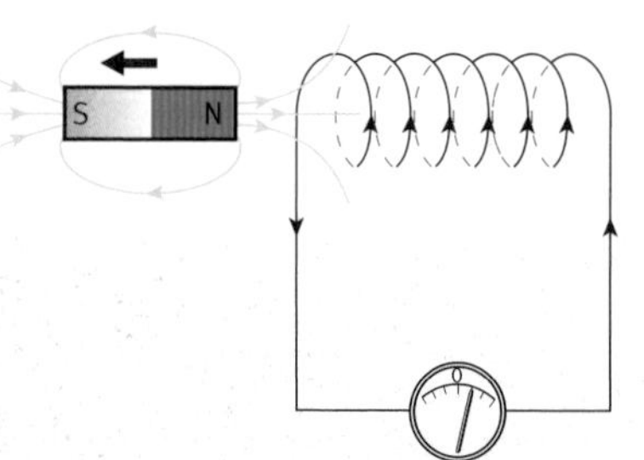

3 While the magnet is being removed from the coil, there is again a small current, but now in the opposite direction.

Moving a magnet into, or out of, a coil generates a current.

Electromagnetic induction

Michael Faraday discovered electromagnetic induction.

In the 1830s, the English physicist Michael Faraday did a series of investigations with magnets, wires, and coils. He found that he could generate an electric current by moving a magnet into a coil of wire. There was a current only while the magnet was moving, not while it was stationary inside the coil. When he pulled the magnet out of the coil again, there was another current but now in the opposite direction. This effect is called **electromagnetic induction**.

While the magnet is moving into the coil, the magnetic field around the wires of the coil changes. Magnetic field lines are 'cutting' the coil. This causes a potential difference (a voltage) across the coil. If the coil is connected into a complete circuit, this voltage causes a current. While the magnet is moving, the coil behaves like a battery.

There is no voltage (and hence no current) while the magnet is stationary inside the coil. It is *changes* in the magnetic field, not the field itself, that cause the induced voltage. When the magnet is pulled out again, the magnetic field changes again. There is an induced voltage (and current) in the other direction.

The size of the induced voltage can be increased by:

- moving the magnet in and out more quickly
- using a stronger magnet
- using a coil with more turns (there is an induced voltage in each turn of the coil, and these add together).

Making a generator

We can make a simple **generator** by rotating a magnet near one end of a coil. The effect is more noticeable if we put an iron core into the coil. This greatly increases the strength of the magnetic field inside the coil. As the magnet rotates, the magnetic field around the coil is constantly changing. This induces a voltage across the ends of the coil, which

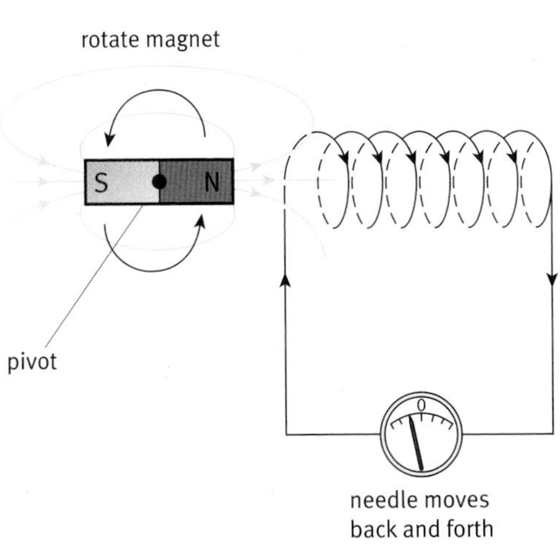

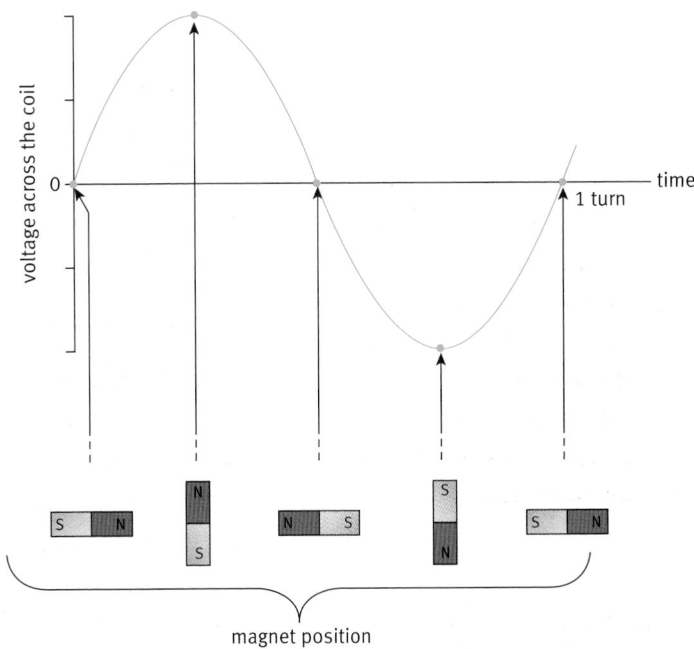

When the magnet rotates, it induces an alternating voltage across the coil.

causes an electric current in the circuit. The voltage and current change direction every half-turn of the magnet. This arrangement generates an **alternating current (a.c.)** in the circuit. For many applications, a.c. works just as well as the **direct current (d.c.)** produced by a battery. A direct current is one that flows in just one direction.

The size of the alternating voltage produced by a generator of this sort can be increased by:

- using a stronger rotating magnet or electromagnet
- rotating the magnet or electromagnet faster (though this also affects the frequency of the a.c. produced)
- using a fixed coil with more turns
- putting an iron core inside the fixed coil (this makes the magnetic field a lot bigger – as much as 1000 times).

In a typical power station generator, an electromagnet is rotated inside a fixed coil. As it spins, a.c. is generated in the coil. In power stations in the UK, the rate of turning is set at 50 cycles per second. The generator is turned by a turbine, which is driven by steam. The steam is produced by burning gas, oil, coal, or by the heating effect of a nuclear reaction.

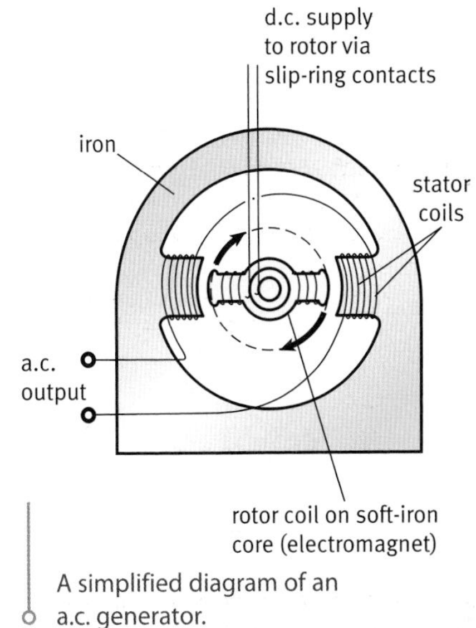

A simplified diagram of an a.c. generator.

Questions

1 Describe the difference between alternating current produced by a power station and the direct current from a battery.

2 Look at the graph. Explain why the voltage across the coil changes with time as the magnet rotates.

Key words

- ✓ **generator**
- ✓ **electromagnetic induction**
- ✓ **alternating current (a.c)**
- ✓ **direct current (d.c)**

Transformers

An electric current can be generated by moving a magnet into (or out of) a coil of wire. The moving magnet could be replaced by an electromagnet. If a coil is wound around an iron core, it becomes quite a strong magnet when a current flows through it.

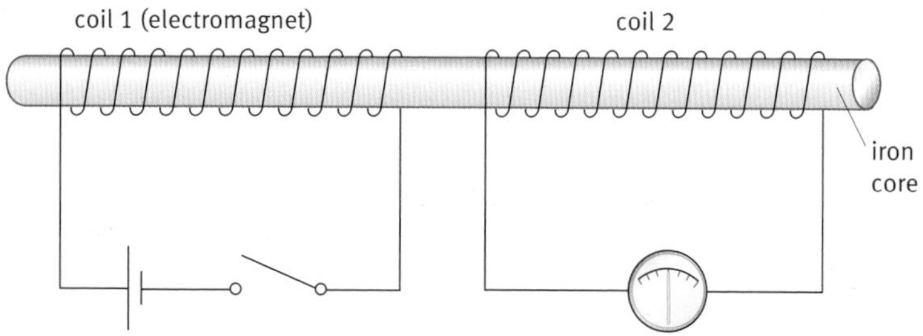

When the current in the electromagnet (coil 1) is switched on, this has the same effect as plunging a bar magnet into coil 2. So a current is generated in coil 2, whilst the current in coil 1 is changing. This arrangement of two coils on the same iron core is called a **transformer**. Changing the current in the primary coil induces a voltage across the secondary coil. If the current in the primary is a.c., it is changing all the time. So an alternating voltage is induced across the secondary coil.

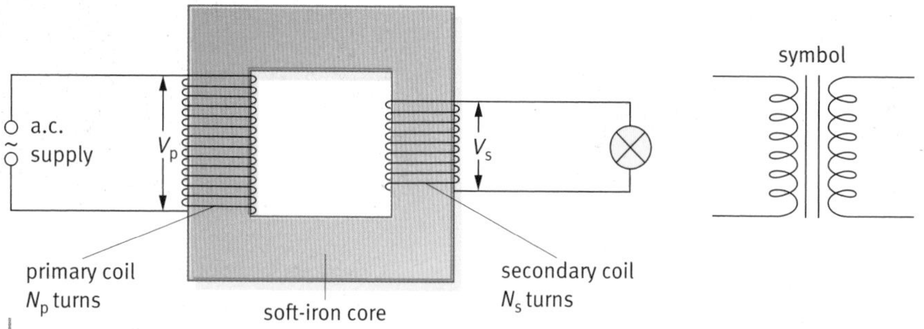

The transformer. When the current in the primary coil is changing, a voltage is induced across the secondary coil. This makes a current flow round the right-hand circuit. Notice that there is no direct electrical connection between the coils of a transformer. The only connection is through the magnetic field.

The output voltage of a transformer depends on the number of turns of wire on the two coils.

$$\frac{\text{voltage across secondary coil } (V_S)}{\text{voltage across primary coil } (V_P)} = \frac{\text{number of turns on secondary coil } (N_S)}{\text{number of turns on primary coil } (N_P)}$$

If there are more turns on the secondary coil, then the induced voltage across this coil is bigger than the applied voltage across the primary coil. However, this is not something for nothing! The current in the secondary will be less, so that the power from the secondary is no greater than the power supplied to the primary (remember: power = IV).

Find out about

- how transformers are used to alter the voltage of a supply
- the main components of the National Grid

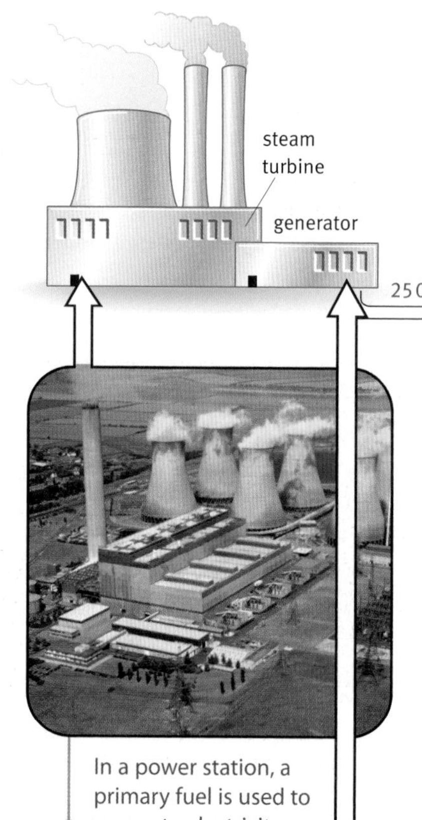

In a power station, a primary fuel is used to generate electricity.

The heart of a power station is a turbine, and as it turns it makes the coil of a generator rotate.

The National Grid

Transformers play an important role in the National Grid system. The National Grid distributes electricity from the power stations to the rest of the country. It does this by means of a long chain of links. These are cables and magnetic fields in transformers.

Transformers are used to step up the output from the power stations and then step down again at the end of the power line. This can only happen because mains electricity supplies an alternating current.

Key word

✓ **transformer**

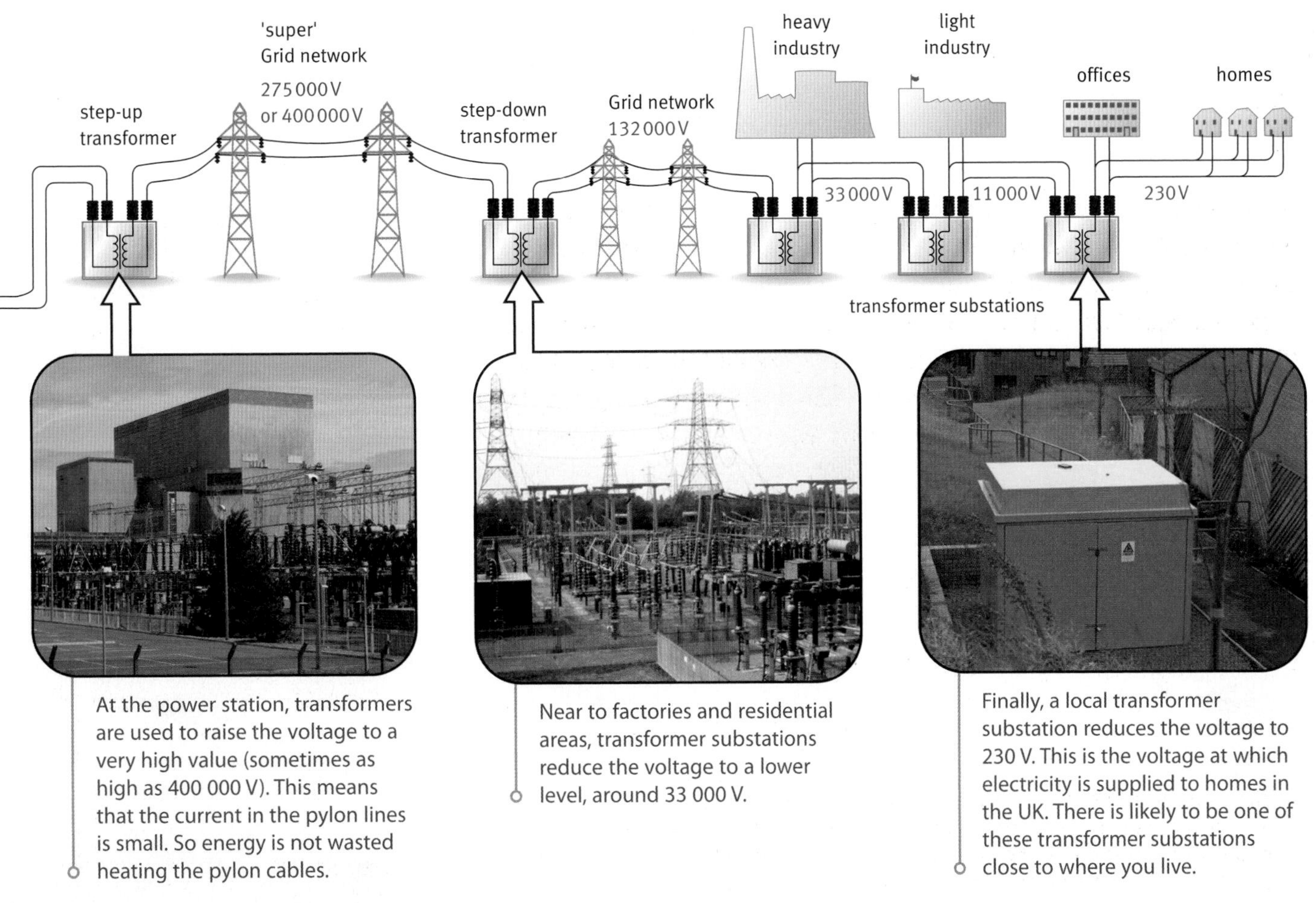

At the power station, transformers are used to raise the voltage to a very high value (sometimes as high as 400 000 V). This means that the current in the pylon lines is small. So energy is not wasted heating the pylon cables.

Near to factories and residential areas, transformer substations reduce the voltage to a lower level, around 33 000 V.

Finally, a local transformer substation reduces the voltage to 230 V. This is the voltage at which electricity is supplied to homes in the UK. There is likely to be one of these transformer substations close to where you live.

Questions

1. A transformer has 100 turns on its primary coil, and 25 turns on its secondary coil. A 12 V a.c. supply is connected to the primary coil. What will be the voltage across the secondary coil?
2. You have a 6 V a.c. supply and want to use it to operate a 12 V bulb. Explain how you could make a simple transformer to do this.
3. In the National Grid, transformers are used to 'step up' the voltage from 25 000 V to 400 000 V. What gets smaller as a result (and stops us getting something for nothing)?

Science Explanations

Electricity is essential to modern-day life. An understanding of electric charge, current, voltage, and resistance in a circuit allows us to use electricity safely and enables power to be generated and distributed.

You should know:

- about electric charge and how positive and negative charges can be separated
- that electric current is a flow of charges already present in the materials of the circuit
- how electric circuits work, and about models that help us understand electric circuits
- that current is not used up as it goes around but it does work on the components it passes through, transferring energy from the battery to other components
- that the voltage of a battery is a measure of the push on the charges
- that the bigger the voltage the bigger the current
- that the components in a circuit resist the flow of charge and how the current depends on the battery voltage and the circuit resistance
- why resistors get hotter when current flows through them and why filament lamps glow
- about components with a variable resistance, including thermistors and light-dependent resistors
- how to measure the voltage between two points using a voltmeter
- that the battery can be thought of as raising charges to a higher level of potential energy and that the charges then lose this energy as they go around the circuit
- that the voltage is also called the potential difference (p.d.)
- about the p.d. across and the current through resistors connected in series and in parallel
- that a circuit with two resistors is sometimes called a potential divider and how this circuit can be useful when used with a variable resistor, thermistor, or LDR
- about the power (energy per second) transferred by an electric circuit
- about the force on a current-carrying wire in a magnetic field and why a motor spins
- about electromagnetic induction, including:
  - a p.d. is induced across the ends of a wire, or coil, in a changing magnetic field
  - if this wire, or coil, is part of a circuit there is an induced electric current in the circuit
  - the magnetic field must be changing, otherwise there is no effect
  - how the effect is increased and how it is used in making an electrical generator
- that electrical generators are used to produce mains electricity
- the difference between a.c. and d.c. electricity
- about the electricity supply system (the National Grid) and why it uses high voltages although the mains voltage to our homes is 230 V.
- about transformers and the effect of changing the number of turns in the coils.

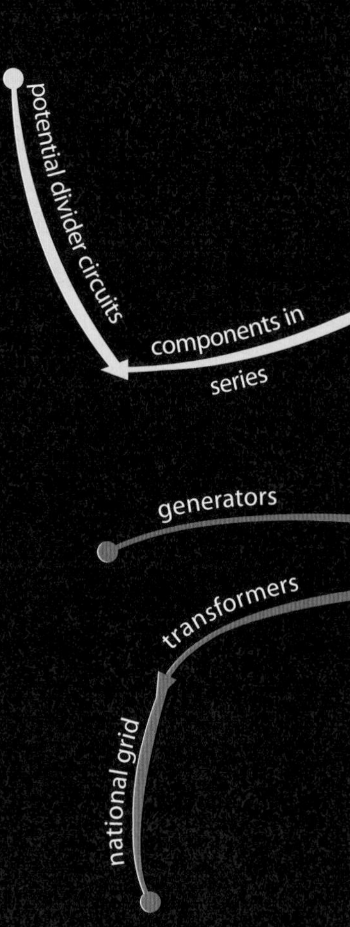

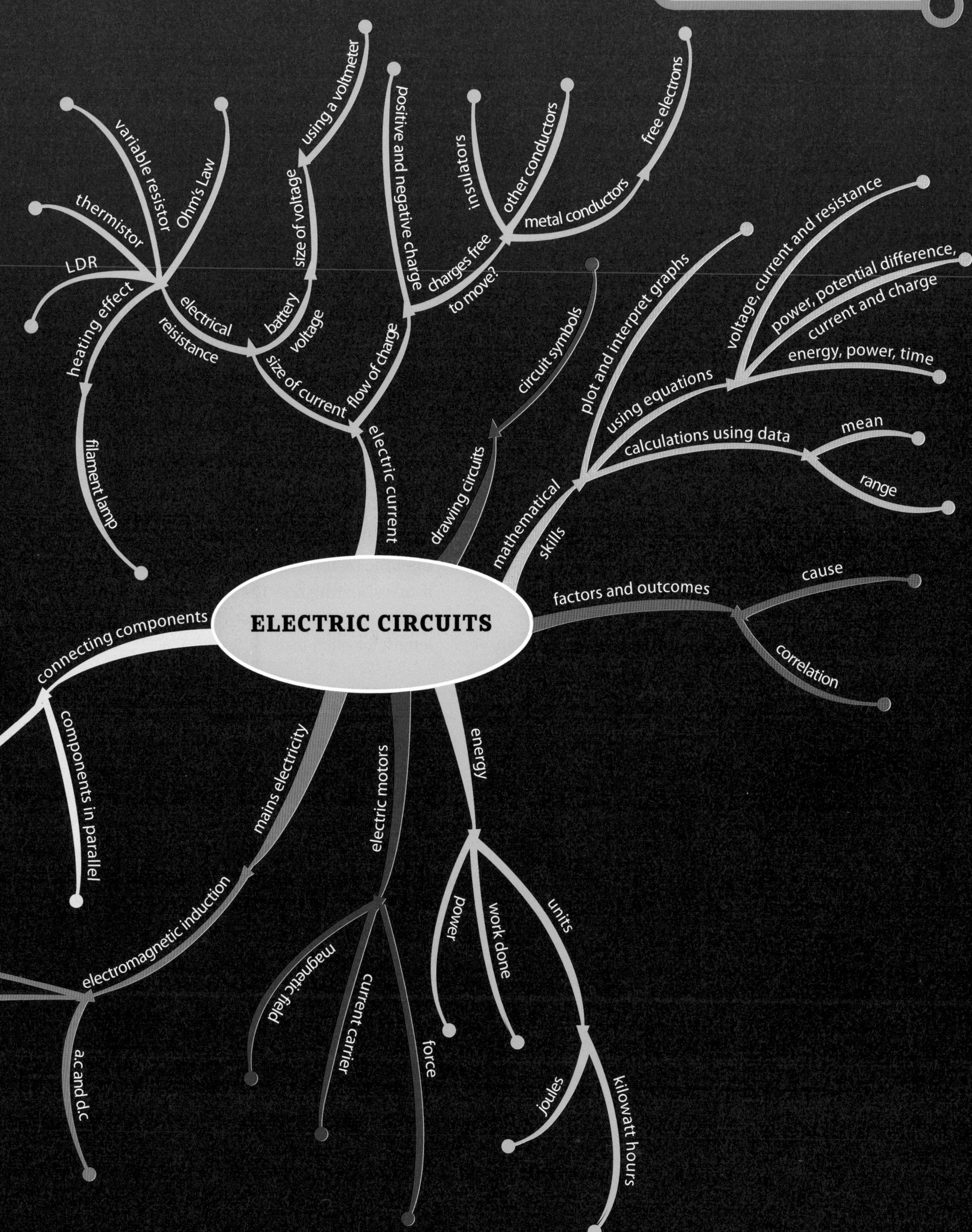
ELECTRIC CIRCUITS
electric current
size of current
electrical reisistance
heating effect
filament lamp
LDR
thermistor
variable resistor
Ohm's Law
voltage
battery
size of voltage
using a voltmeter
flow of charge
positive and negative charge
charges free to move?
insulators
other conductors
metal conductors
free electrons
drawing circuits
circuit symbols
mathematical skills
plot and interpret graphs
using equations
voltage, current and resistance
power, potential difference, current and charge
energy, power, time
calculations using data
mean
range
factors and outcomes
cause
correlation
connecting components
components in parallel
mains electricity
electromagnetic induction
a.c and d.c
electric motors
magnetic field
current carrier
force
energy
power
work done
units
joules
kilowatt hours

Ideas about Science

In addition to developing an understanding of electric circuits, it is important to understand how scientists use data to develop their ideas. Collecting data is often the starting point for a scientific enquiry, but data can never be trusted completely. Data is more reliable if it can be repeated; when making several measurements of the same quantity, the results are likely to vary. This may be because:

- you have measured several individual samples, for example, several samples of a resistance wire
- the quantity you are measuring is varying, for example, the light level in the room is varying as you measure the resistance of an LDR
- there are limitations in the measuring equipment, for example, a poor electrical connection in the circuit.

Usually the best estimate of the true value of a quantity is the mean of several repeated measurements. When there is a spread of values in a set of measurements, the true value is probably in the range between the highest and the lowest values. You should:

- be able to calculate the mean from a set of repeat measurements
- know that a measurement may be an outlier if it is well outside the range of other measurements
- be able to explain whether or not an outlier should be included as part of the data or rejected when calculating the mean.

When comparing sets of data to decide if there is a difference between the two means, it is useful to look at the ranges of the data. You should know:

- if the ranges of two sets of data do not overlap there may be a real difference between the means.

To investigate the relationship between a factor and an outcome, it is important to control all the other factors that might affect the outcome. In a plan for an investigation you should be able to:

- recognise that the control of other factors is a positive feature of an investigation and it is a design flaw if factors are not controlled
- explain why it is necessary to control all the factors that might affect the outcome, other than the factor being investigated, for example, if investigating how the thickness of a wire affects its resistance, use the same material and length for each test.

Factors and outcomes may be linked in different ways, and it is important to distinguish between them. A correlation between a factor and an outcome does not necessarily mean that the factor causes the outcome; both might be caused by some other factor. For example, the more electricity substations there are in an area, the more babies are born in that area. But this is because there are more houses needing an electricity supply where more people live. You should be able to:

- identify a correlation from data, a graph, or a description
- explain why an observed correlation does not necessarily mean that the factor causes the outcome
- explain why individual cases do not provide convincing evidence for or against a correlation.

Review Questions

1 Look at the electric circuit models in this module. Copy and complete the following table.

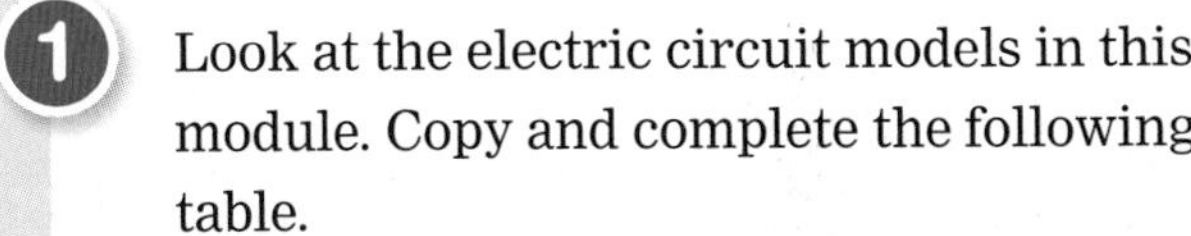

	What corresponds to:		
Model	**the battery?**	**electric current?**	**the resistors or lamps?**
'peas in a pipe'			
'water in a pipe'			

2 In a simple single-loop electric circuit, the current is the same everywhere. It is not used up. How does each of the models above help to account for this?

3 Imagine a simple electric circuit consisting of a battery and a bulb. For each of the following statements, say if it is true or false (and explain why):

a Before the battery is connected, there are no electric charges in the wire. When the circuit is switched on, electric charges flow out of the battery into the wire.

b Collisions between the moving charges and fixed atoms in the bulb filament make it heat up and light.

c Electric charges are used up in the bulb to make it light.

4 In shops, you can buy batteries labelled 1.5 V, 4.5 V, 6 V, or 9 V. But you cannot buy batteries labelled 1.5 A, 4.5 A, 6 A, or 9 A. Explain why not.

5 You are given four 4 Ω resistors. Draw diagrams to show how you could connect all four together to make a resistance of:

a 16 Ω **b** 1 Ω

c 10 Ω **d** 4 Ω

Note that there is more than one possible way to do parts c and d.

6 Peter has a sensor labelled LDR.

a What do the letters LDR stand for?

b What does an LDR detect?

c What does Peter need to measure to work out the resistance?

d Draw a circuit diagram to show how he could measure the quantities in your answer to part b.

7 Copy and complete these sentences:
When a magnet is moved into a coil of wire, a voltage is ______ in the coil. The voltage is produced only when the magnet is ________.
This is used in an a.c. generator, which has an ________ rotating near a fixed coil. To increase the size of the induced voltage, you could use a ______electromagnet, have more ________ on the fixed coil, turn the rotor coil _______, or put a core of _________ inside it.
The current in the external circuit constantly changes direction, so it is called ________ current (__). This is different from the current from a battery, which always goes in one direction and is called current (__).

8 What are the similarities and differences between a motor and a generator?

9 A school laboratory has a set of transformers to demonstrate how power lines work. The transformer has 240 turns on the primary coil and 1200 turns on the secondary coil.

a How will the output voltage be different to the input voltage?

b The input voltage is 2 V. Calculate the output voltage.

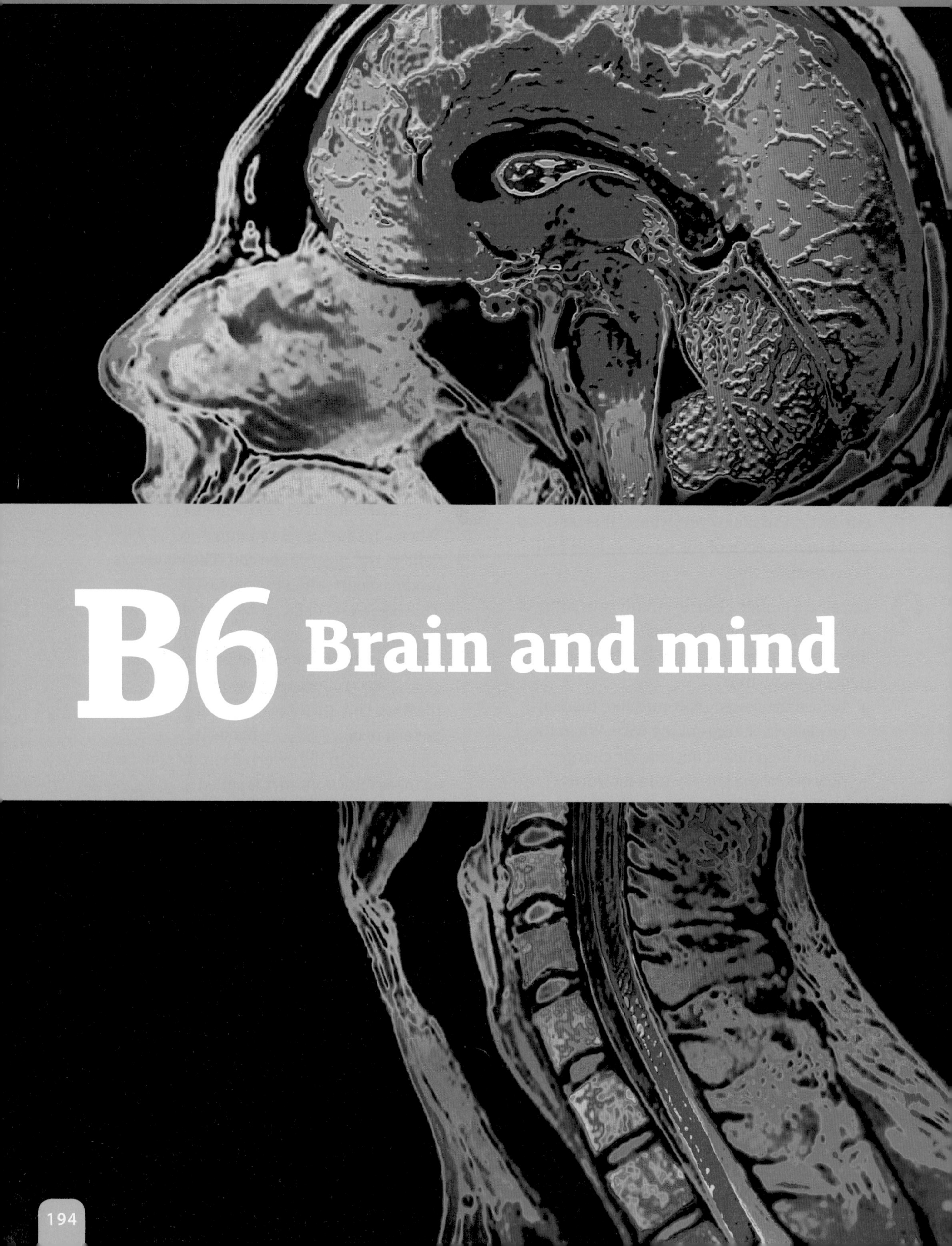

B6 Brain and mind

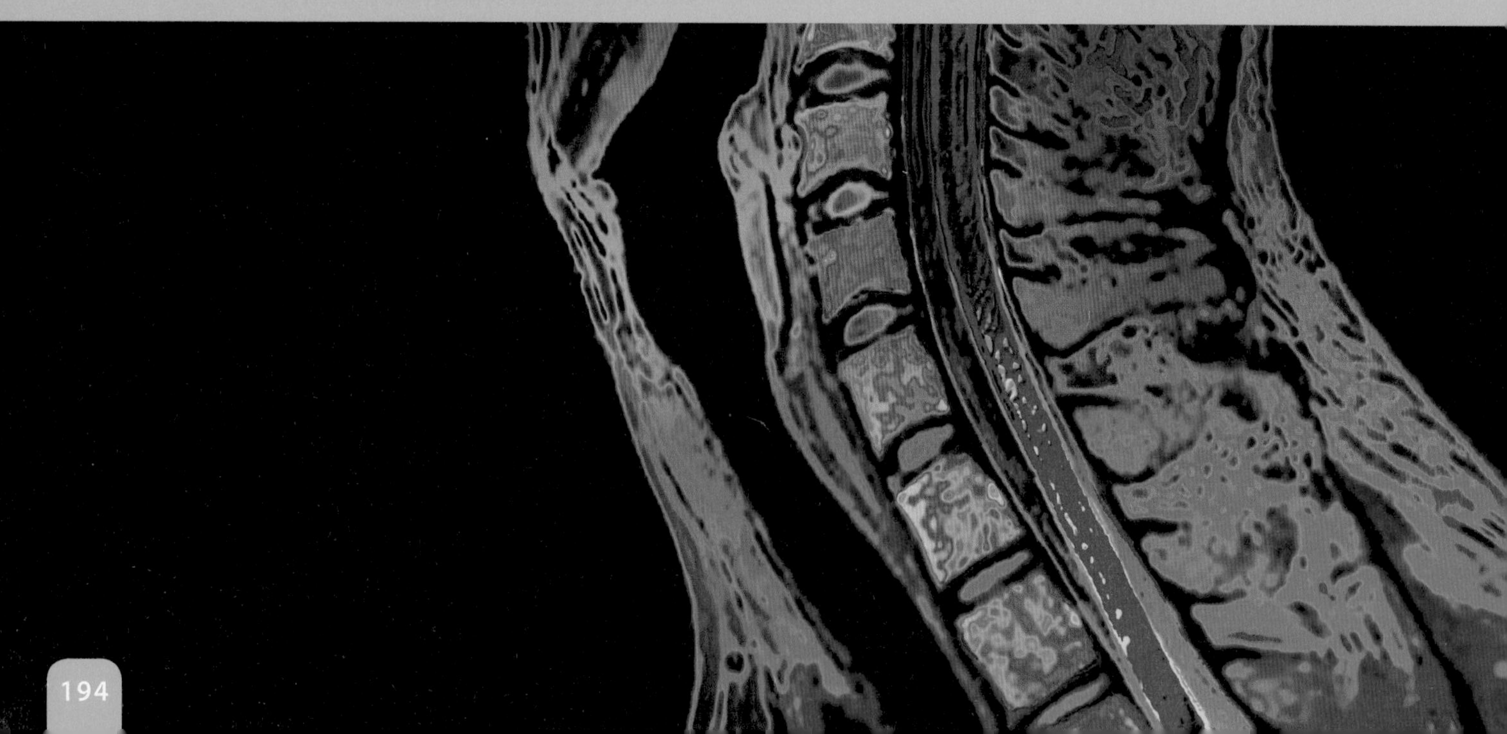

Why study the brain and mind?

The human brain allows our species to survive on Earth. It gives us advantages of intelligence and sophisticated behaviour.

What you already know

- The success of humans is mainly due to the evolution of a large and complex brain.
- Drugs can affect human behaviour.
- Nerves and hormones help you respond to your environment.

Find out about

- how organisms respond to stimuli
- how nerve impulses are passed around your body
- how your brain coordinates your senses
- how you learn new skills
- how scientists are finding out about memory.

The Science

Animals respond to stimuli in order to survive. The central nervous system – the brain and spinal cord – coordinates millions of electrical impulses every second. These impulses determine how we think, feel, and react – our behaviour. Some drugs can affect this.

Ideas about Science

Applications of scientific research can have ethical implications. Some people say that the right decision is the one leading to the best outcome for the most people. Others say that some actions are always unnatural or wrong.

What is behaviour?

Find out about

- what behaviour is
- how simple behaviour helps animals survive

Imagine you are sitting outside. The temperature drops, and you get cold. You start to shiver.

Shivering is a **response** to the change in temperature. A change in your environment, like a drop in temperature, is called a **stimulus**. Eating is a response to the stimulus of hunger. Scratching is a response to an itch. Shivering, eating, and scratching are all examples of **behaviour**.

You can think of behaviour as anything an animal does. The way an animal responds to changes in its surroundings is important for its survival.

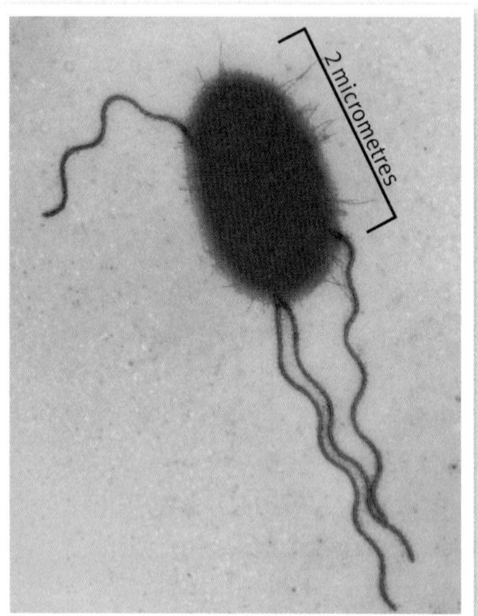

Escherichia coli bacteria are found in the lower gut of warm-blooded animals. They detect the highest concentration of food and move towards it.

Simple behaviour

Simple animals always respond to a stimulus in the same way. For example, woodlice always move away from light. This is an example of a **simple reflex** response. Reflexes are always **involuntary** – they are automatic. Reflexes are important because they increase the animal's chance of survival. The photographs in this section all show reflexes.

Why are simple reflexes important?

Simple reflex behaviour helps an animal to:

- find food, shelter, or a mate
- escape from predators
- avoid harmful environments, for example, extreme temperatures.

Woodlice move away from light, so you are most likely to find them in dark places.

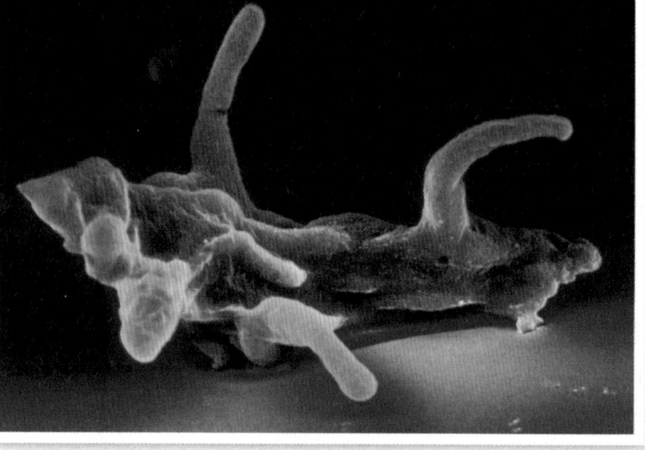

Single-celled *Amoeba* move away from high concentrations of salt, strong acids, and alkalis.

Simple reflexes usually help animals to survive. But animals that only behave with simple reflexes cannot easily change their behaviour, or learn from experience. This is a problem if conditions around them change. Their simple reflexes may no longer be helpful for survival.

When a giant octopus sees a predator, it rapidly contracts its body muscles. This squirts out a jet of water to push the octopus away from danger. The octopus may also release a dark chemical (often called 'ink'), which hides its escape.

When the tail of the sea hare *Aplysia* is pinched, the muscle contracts quickly and strongly. This reflex helps the animal escape from the spiny lobster that preys on it.

Earthworms have some of the fastest reflexes in the animal kingdom. A sharp tap from a beak on its head end is detected in the body wall. Rapid contraction of the worm's muscles pulls it back into its burrow. But this time the bird was too quick.

Have you ever tried to swat a fly? It has a very fast response to any movement that its sensitive eyes detect.

Complex behaviour – a better chance of survival

Complex animals, like mammals, birds, and fish, have simple reflex responses. But a lot of their behaviour is far more complicated. It includes reflex responses that have been altered by experience. Also, much of their behaviour is not involuntary – they make conscious decisions. For example, if it gets very cold, you do not just rely on your reflexes to keep you warm; you decide to put on extra clothes.

Because complex animals can change their behaviour when environmental conditions change, they are more likely to survive.

Key words

- **response**
- **stimulus**
- **behaviour**
- **simple reflex**
- **involuntary**

Questions

1. Write a sentence to explain each key word on this page.
2. Describe an action you did today that:
   - you did not have to think about and you have never had to learn to do
   - you have learnt to do but you can now do without thinking
   - you had to think about while you were doing it.
3. Which of the actions you described is an example of conscious behaviour, and which is most likely to be a simple reflex response?

Simple reflexes in humans

Find out about

- reflexes in newborn babies
- simple reflexes that help you to survive

Behaviour in humans and other mammals is usually very complex. But simple reflexes are still important for survival. For example:

- When an object touches the back of your throat, you gag to avoid swallowing it. This is the gag reflex.
- When a bright light shines in your eye, your pupil becomes smaller. This **pupil reflex** stops bright light from damaging the sensitive cells at the back of your eye.

These types of behaviour are inherited through our genes. This is called **innate** behaviour.

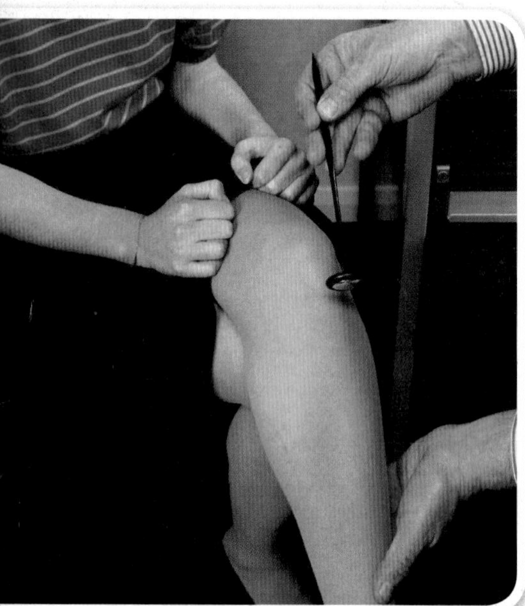

The knee-jerk reflex causes your thigh muscle to contract, so your lower leg moves upwards. Doctors may test this and other reflexes when you have a health check. Try standing still with your eyes closed. You will notice this reflex helping you to balance.

Newborn reflexes

When a baby is born, the nurse checks for a set of **newborn reflexes**. Many of these reflexes are only present for a short time after birth. They are gradually replaced by behaviours learnt from experience. In a few cases these reflexes are missing at birth, or they are still present when they should have disappeared. This may mean that the baby's nervous system is not developing properly.

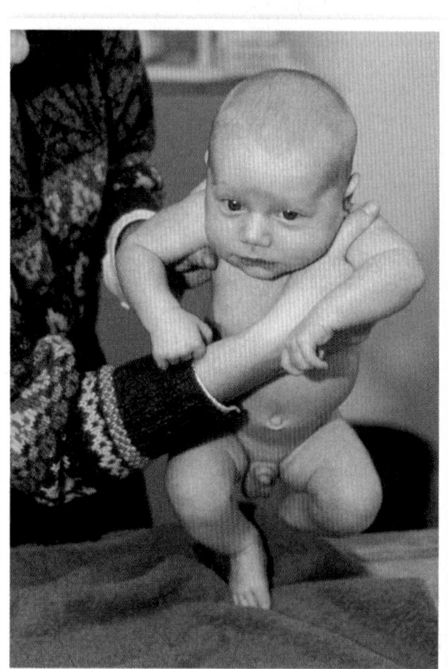

Stepping. If you hold a baby under his arms, support his head, and allow his feet to touch a flat surface, he will appear to take steps and walk. This reflex usually disappears by two–three months after birth. It then reappears as he learns to walk at around 10–15 months.

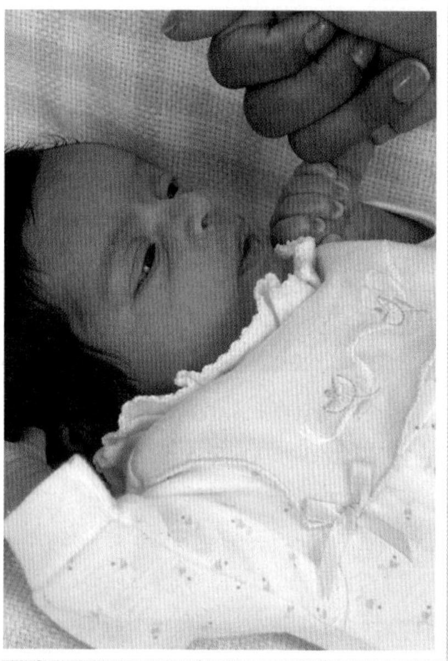

Grasping. When you put your finger in a baby's open palm, the baby grips the finger. When you pull away, the grip gets stronger. This reflex usually disappears by five–six months. If you stroke the underneath of a baby's foot, its toes and foot will curl. This reflex usually disappears by 9–12 months.

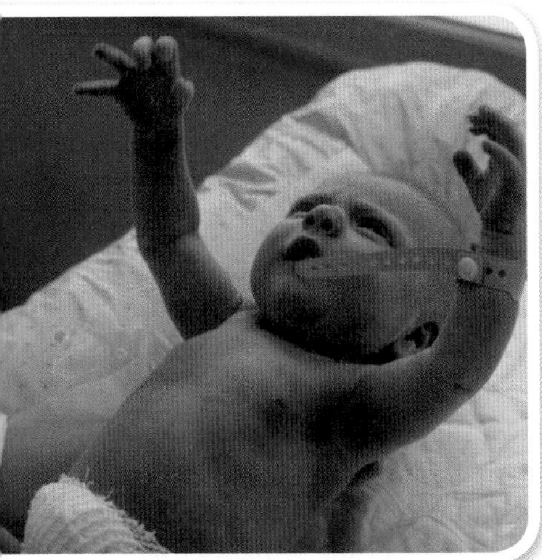

Startle. This is also called the Moro reflex, named in 1687 after the Italian scientist Artur Moro. It usually happens when a baby hears a loud noise or is moved quickly. The response includes spreading the arms and legs out and extending the neck. The baby then quickly brings her arms back together and cries. This reflex usually goes by three–six months.

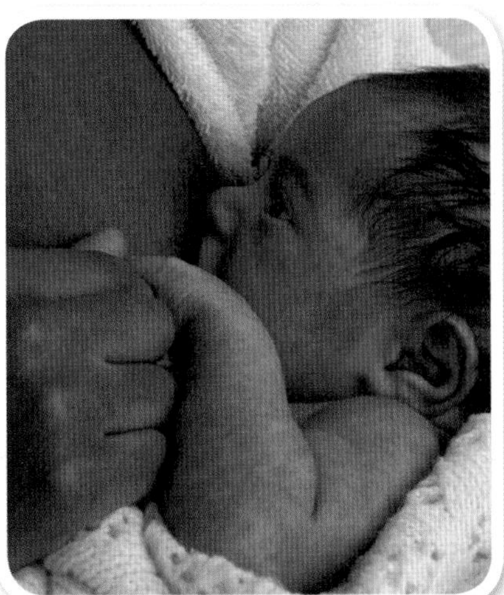

Sucking. Placing a nipple (or a finger) in a baby's mouth causes the sucking reflex. It is slowly replaced by voluntary sucking at around two months.

Rooting. Stroking a baby's cheek makes her turn towards you, looking for food. This reflex helps the baby find the nipple when she is breastfeeding. The rooting reflex is gone by about four months.

Swimming. If you put a baby under six months of age in water, he moves his arms and legs while holding his breath.

Sudden infant death syndrome (SIDS)

Sudden Infant Death Syndrome, or cot death, is tragic and unsolved. In the UK about seven babies a week die from SIDS. This is 0.7 deaths for every 1000 live births.

It is likely that there are many different causes of cot death. Some people think that it could be because a baby's simple reflexes have not matured properly. This is how doctors think this may happen:

- When a fetus detects that oxygen in its blood is low, its reflex response makes it move around less. This makes sense because the less it moves around, the less oxygen it will use up in cell respiration.
- This response changes as the baby matures. When an older baby or child's airways are covered, for example, by a duvet, the baby moves more. He turns his head from side to side. He also pushes the obstruction away. So now the response to low oxygen is more activity, not less.
- If the newborn baby has not grown out of the fetal reflex, he may lie still if his bedcovers cover his airways. He is more likely to suffocate.

Doctors now advise mothers to put babies onto their backs to sleep, and not to use soft bedding like duvets. This way their faces are less likely to become covered.

Key words

- ✓ **pupil reflex**
- ✓ **newborn reflexes**

Questions

1. Describe two reflexes in:
   a. adult humans
   b. newborn babies.
2. How do you think the startle reflex helps a baby to survive?
3. Why are premature babies more at risk from SIDS than babies born at the correct time?

Receptors

You can only respond to a change if you can detect it. **Receptors** inside and outside your body detect stimuli, or changes in the environment.

You can detect many different stimuli, for example, sound, texture, smell, temperature, and light. Different types of receptors each detect a different type of stimulus. Receptors on the outside of your body monitor the external environment. Others monitor changes inside your body, for example, core temperature and blood sugar levels.

Sense organs

Some receptors are made up of single cells, for example, pain receptors in your skin. Other receptor cells are grouped together as part of a complex sense organ, for example, your eye. Vision is very important in humans and most other mammals. Light entering our eyes helps us humans produce a three-dimensional picture of our surroundings. This gives us information about objects such as their shape, movement, and colour.

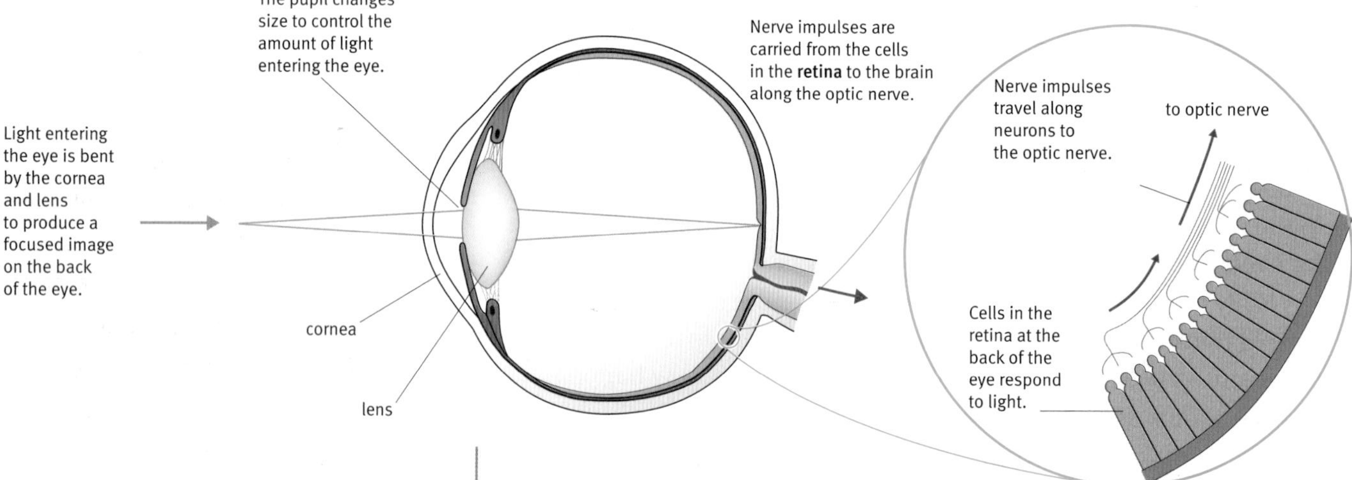

Light is focused by the cornea and lens onto light-sensitive cells at the back of the eye. These cells are receptors. They trigger nerve impulses to the brain.

Effectors

The body's responses to stimuli are carried out by **effector** organs. In multicellular organisms the effectors are either **glands** or **muscles**. Nervous and **hormonal** communication systems have developed as living things have evolved. Multicellular organisms have evolved a complex communication system that co-ordinates short-term and long-term responses.

Stimuli from the environment bring about responses in muscles or glands. Nerve impulses bring about fast, short-lived responses, for example, contractions of muscles. Hormones bring about longer-lasting effects such as an increase in growth rate.

Effectors are either glands or muscles.

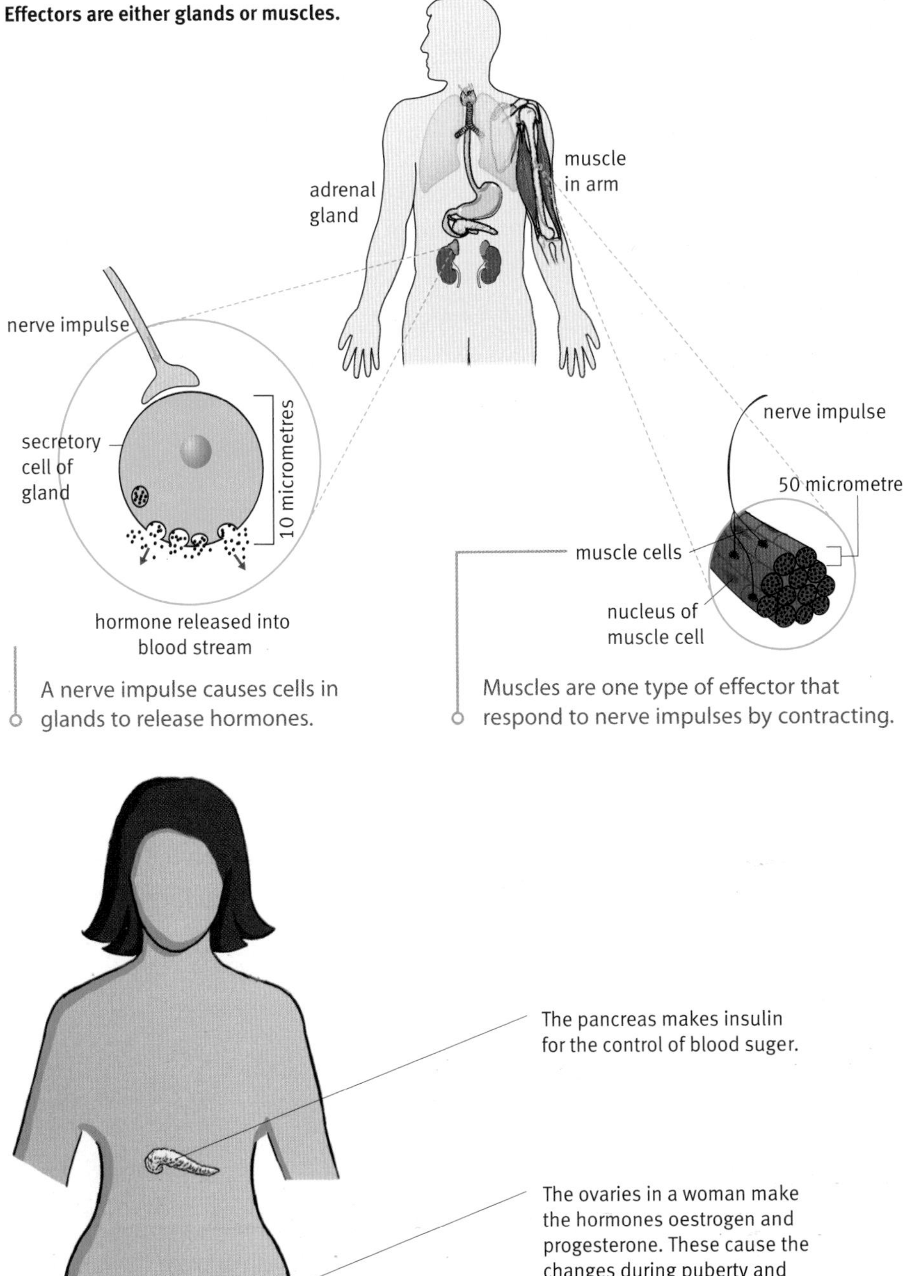

A nerve impulse causes cells in glands to release hormones.

Muscles are one type of effector that respond to nerve impulses by contracting.

Hormones are released when a nerve impulse reaches glands such as the ovaries and pancreas. Hormones have a slower and longer-lasting effect than a nerve impulse.

Key words

- receptor
- retina
- effector
- glands
- muscles
- hormone
- nervous system

Questions

4 Name the two different types of effector and say what they do.

5 Which receptors would you use in order to thread your trainers with new laces?

6 Which effectors are you using when you:
- text a friend?
- cry?
- run a race?

7 Some people suffer from a disease where tiny clusters of light receptor cells in different parts of the eye become damaged. How would this affect what the person sees?

C Your nervous system

Find out about

- ✔ different parts of your nervous system
- ✔ how reflexes are controlled

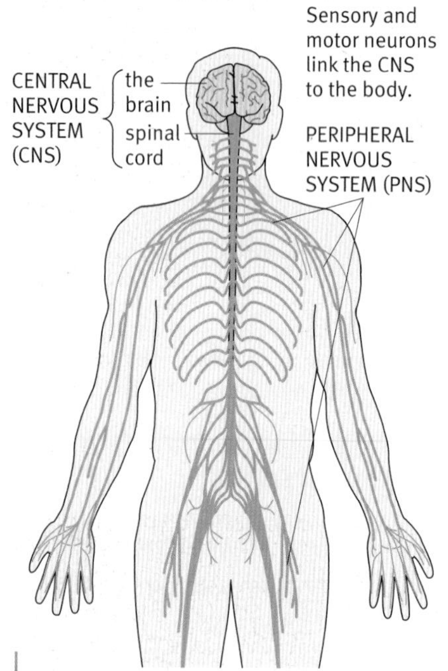

In the mammalian nervous system the brain and spinal cord is connected to the body via the peripheral system.

The process of evolution produced larger and more complex multicellular animals. These animals developed nervous and hormonal systems to allow them to respond to the environment. This gave them a survival advantage over simpler animals.

Walk out from a dark cinema on a bright afternoon and your pupils will become smaller. This pupil reflex prevents bright light damaging your eye. Like all reflexes, this behaviour is coordinated by your nervous system.

Cells in your nervous system carry **nerve impulses**. These nerve impulses allow the different parts of the nervous system to communicate with each other.

Peripheral nervous system

Many nerves link your brain and spinal cord to every other part of your body. These nerves make up the **peripheral nervous system**.

Nerves and neurons

Nerves are bundles of specialised cells called **neurons**. Like most body cells, neurons have a nucleus, a cell membrane, and cytoplasm. They are different from other cells because the cytoplasm is shaped into a very long thin extension. This is called the **axon**, and it is how neurons connect different parts of the body.

Axons carry electrical nerve impulses. Like wiring in an electrical circuit, the axons must be insulated from each other. The insulation for an axon is a **fatty sheath** wrapped around the outside of the cell. The fatty sheath increases the speed that impulses move along the axon.

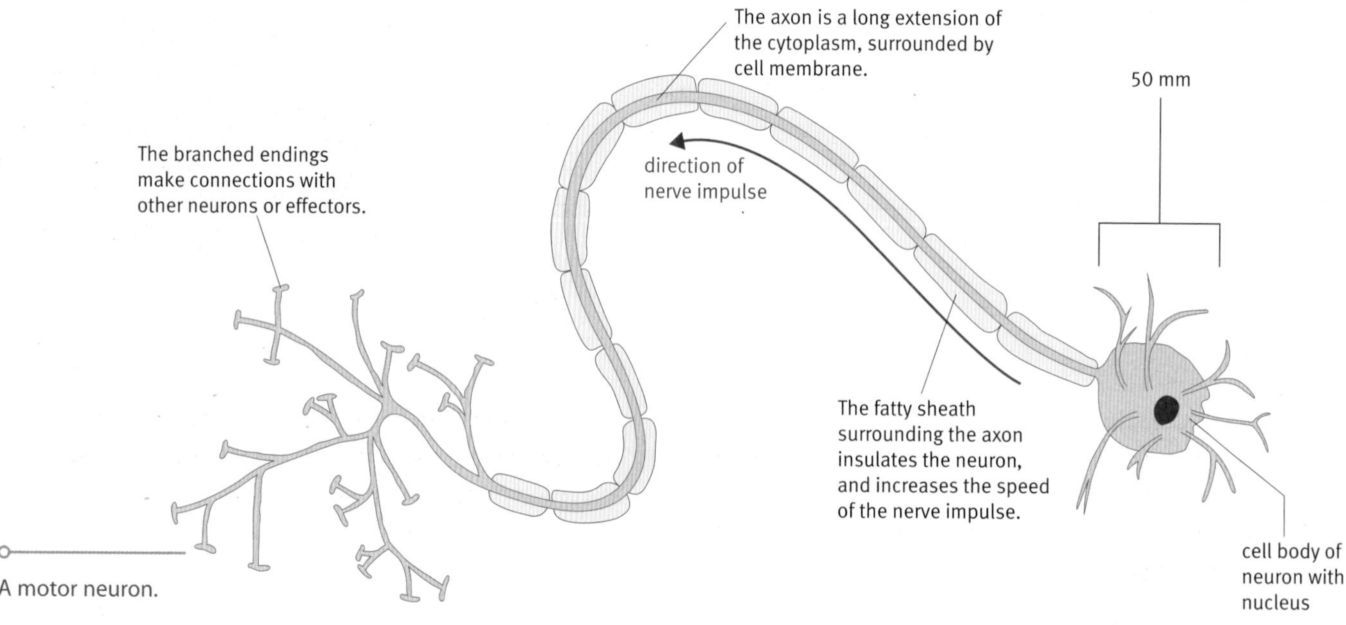

A motor neuron.

The reflex arc

In a simple reflex, impulses are passed from one part of the nervous system to the next in a pathway called a **reflex arc**. The diagram below shows this pathway for a pain reflex. The **relay neurons** in the spinal cord connect the **sensory neuron** to the appropriate motor neuron.

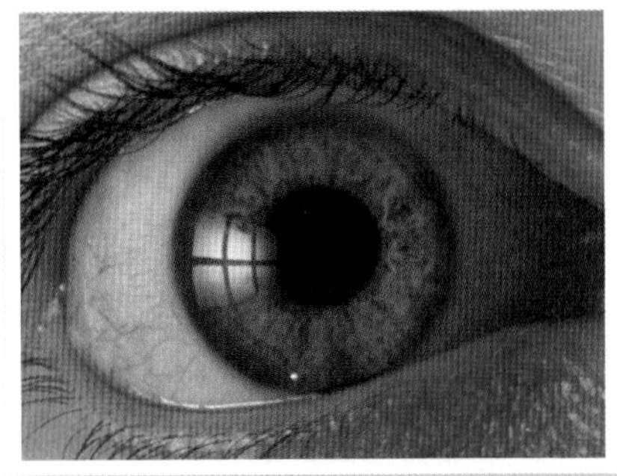

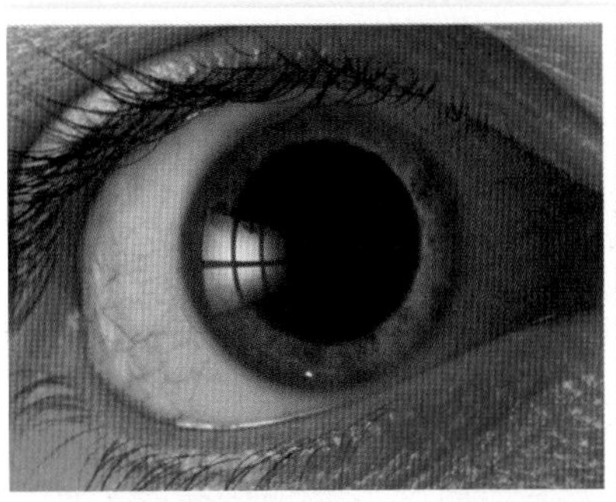

Muscles in the iris cause the pupil to change size depending on the brightness of light entering the eye. (The pupil size controls the amount of light intensity that is the stimulus.)

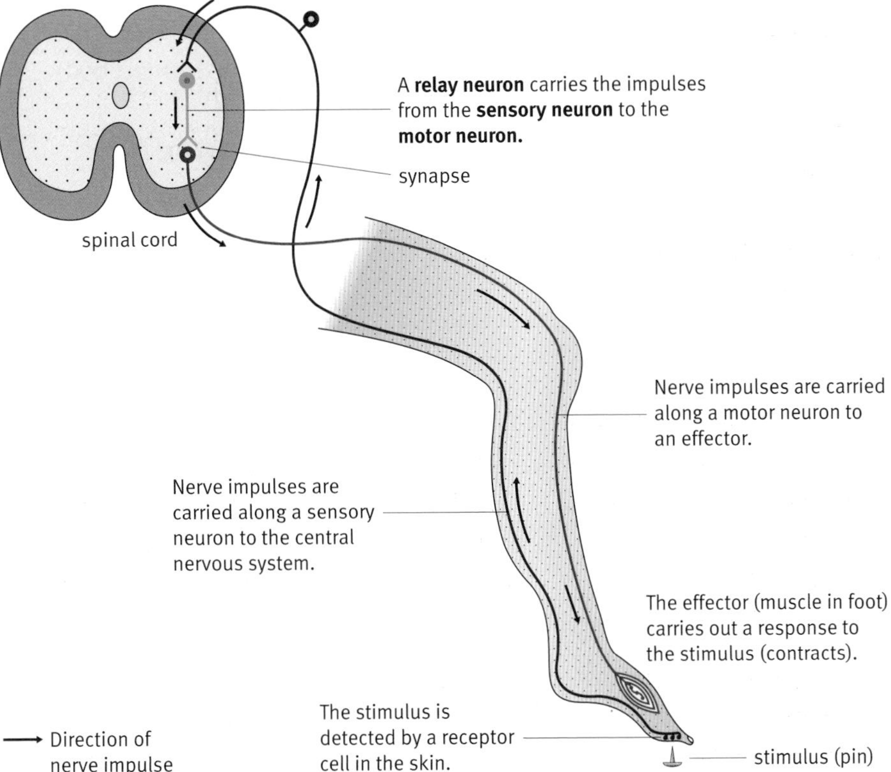

Central nervous system

Your **central nervous system** (CNS) coordinates all the information it receives from your receptors via sensory neurons. Information about a stimulus goes to either your brain or your spinal cord. In a reflex arc the CNS links incoming information from receptors with **motor neurons**. Motor neurons complete the reflex arc by stimulating effectors that carry out the necessary response.

Key words

- **nerve impulses**
- **reflex arc**
- **central nervous system**
- **relay neuron**
- **sensory neuron**
- **motor neuron**
- **peripheral nervous system**
- **neurons**
- **axon**
- **fatty sheath**

Questions

1 Describe the difference between:
- the job of a sensory neuron and a motor neuron
- an axon and a neuron
- the central nervous system and the peripheral nervous system.

2 Draw a labelled diagram to show a reflex arc for the newborn grasp reflex shown on page 198. This reflex is coordinated by the spinal cord.

D Synapses

Find out about

- how nerve impulses pass from one neuron to the next

Quick responses to stimuli are also essential in real life. They help you survive by avoiding danger.

Curare is a very powerful toxin. It is used on the tips of blowpipe darts.

Key words

- synapses
- transmitter substances
- receptor molecules
- serotonin
- Ecstasy

Think about playing a fast sport or computer game. You need very quick reactions to win. Nerve impulses give you fast reactions because they travel along axons at 400 metres per second.

Mind the gap

Neurons do not touch each other. So when nerve impulses pass from one neuron to the next, they have to cross tiny gaps. These gaps are called **synapses**. Some drugs and poisons (toxins) interfere with nerve impulses crossing a synapse. This is how they affect the human body.

How do nerve impulses cross a synapse?

Nerve impulses cannot jump across a synapse. Instead, chemicals called **transmitter substances** are used to pass an impulse from one neuron to the next. The diagram below explains how this works.

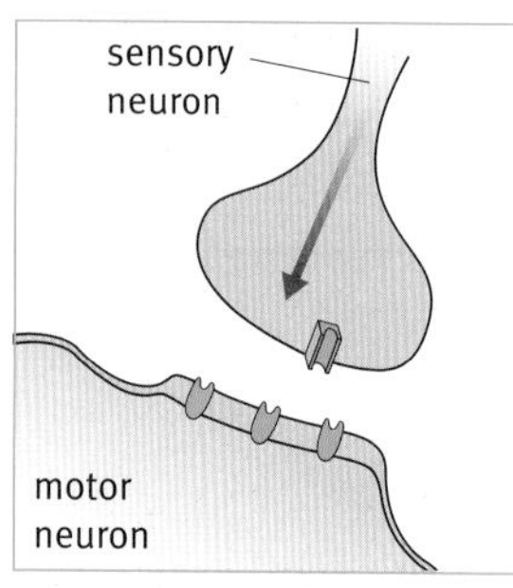

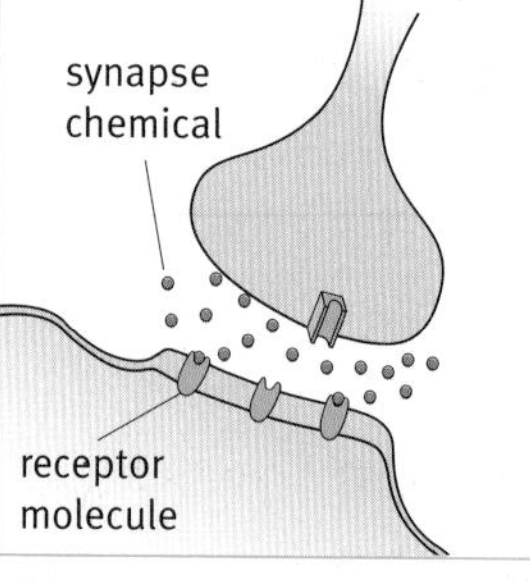

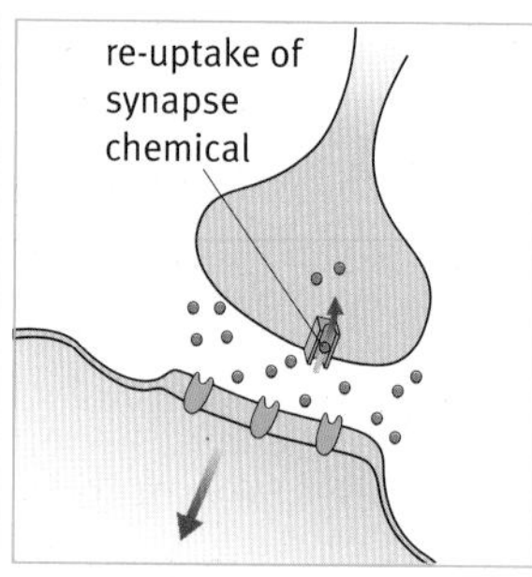

1. A nerve impulse arrives at a synapse. The direction of the impulse is shown by the arrow.

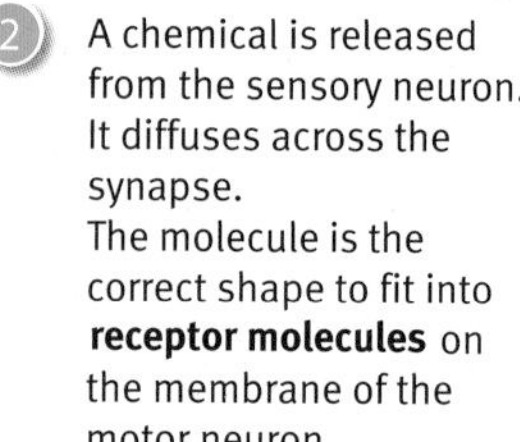

2. A chemical is released from the sensory neuron. It diffuses across the synapse. The molecule is the correct shape to fit into **receptor molecules** on the membrane of the motor neuron.

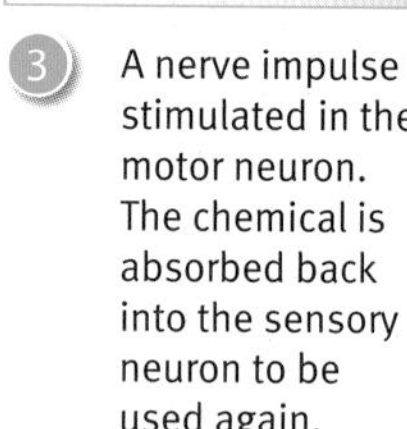

3. A nerve impulse is stimulated in the motor neuron. The chemical is absorbed back into the sensory neuron to be used again.

How a synapse works.

Do synapses slow down nerve impulses?

The gap at a synapse is only about 20 nanometres (nm) wide. The synapse chemical travels across this gap in a very short time. Synapses do slow down nerve impulses to about 15 metres per second. A nerve impulse still travels from one part of your body to another at an incredible speed.

Being human – just chemicals in your brain?

The way we think, feel, and behave does involve a series of chemicals moving across synapses between neurons, but there is more to behaving like a human than chemicals in your brain.

These processes are very complicated. Scientists researching how your nervous system works are only just beginning to understand the brain.

Serotonin

Serotonin is a chemical released at one type of synapse in the brain. When serotonin is released, you get feelings of pleasure. Pleasure is an important response for survival. For example, eating nice-tasting food gives you a feeling of pleasure. So you are more likely to repeat eating, which is essential for survival.

Lack of serotonin in the brain is linked to depression. Depression is a very serious illness. At least one person in five will suffer from a depressive illness at some point in their life. They feel very unhappy for many days on end and often find it difficult to manage normal everyday things like working, studying, or looking after their family.

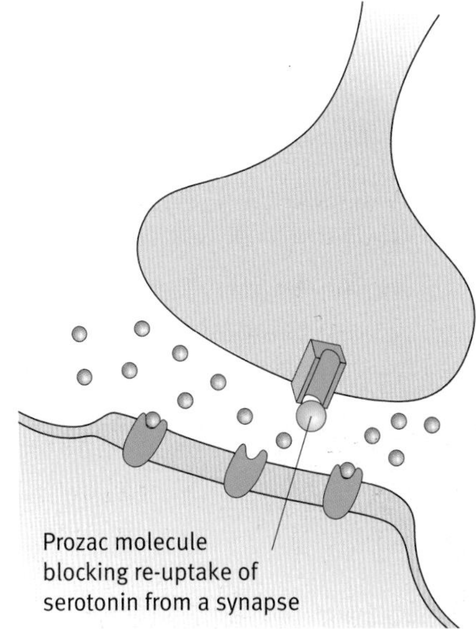

Feelings of depression can be caused by too little serotonin in the brain. Prozac works by blocking re-uptake of serotonin.

How do some drugs affect the brain?

Prozac is the name of an antidepressant drug. Prozac causes serotonin concentration to build up in synapses in the brain. So a person may feel less unhappy. The diagram explains how Prozac works. Like all drugs, Prozac can have unwanted effects.

Ecstasy

Ecstasy is the common name for the drug MDMA. Ecstasy works in a similar way to Prozac. People who have taken Ecstasy say that it can give them feelings of happiness and being very close to other people. Studies on monkeys suggest that long-term use of Ecstasy may destroy the synapses in the pleasure pathways of the brain. Permanent anxiety and depression might result, along with poor attention span and memory. For some people the harmful effects of Ecstasy are more immediate. Ecstasy interferes with the body's temperature control systems. It also slows down production of the hormone ADH in the brain. These effects can be fatal. You can read more about ADH in module B2.

Beta blockers

Some people suffer from severe chest pains called angina. This pain can be triggered when people are stressed or excited. Nerve impulses stimulate the heart to speed up. This can leave the heart muscle painfully starved of oxygen. A doctor might prescribe a drug called a beta blocker to help. Beta blockers reduce the transmission of impulses across nerve synapses, stopping the heart beat from speeding up. Beta blockers also help to control nerve impulses inside the heart, making sure that the heart beats in a regular, controlled way.

Questions

1. Write down a sentence to describe a synapse.
2. Draw a flow diagram to describe what happens when a nerve impulse arrives at a synapse.
3. Explain how the release of serotonin in the brain helps us survive.
4. Some drugs (like curare) block the receptors on the motor neuron at a synapse. Explain how this would affect muscles that the motor neuron is linked to.
5. Prozac is described as a selective serotonin re-uptake inhibitor or SSRI. Explain how an SSRI might help a person with depression.

The brain

Find out about

- the structure of your brain
- how scientists learn about the brain

Think of some things you did today. Getting up, deciding what to have for breakfast, travelling to school, talking, and listening to friends. All of this complex behaviour has been controlled by your brain. But exactly how it happens is still being researched. Scientists who study the brain are called **neuroscientists**. Neuroscience is a fairly 'new' science. This means that scientists have only recently started to investigate how the brain works.

Complex nervous and hormonal communication systems only developed once multicelllular organisms evolved. These organisms have specialised tissues and organs to carry out communication processes.

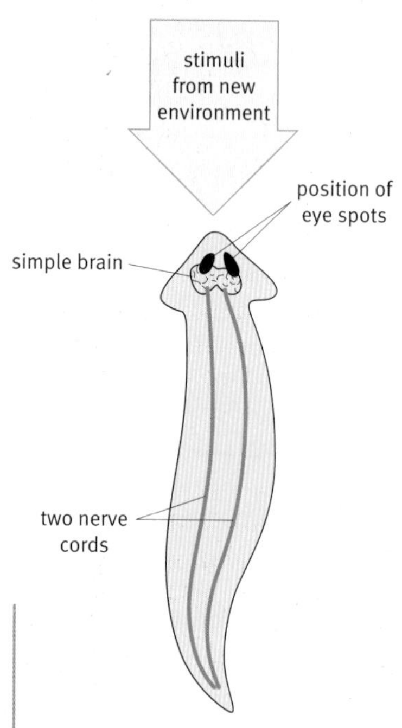

Sense organs on the flatworm head detect light and chemical stimuli. A simple brain processes the response.

Simple animals

Neurons carry electrical impulses around your body. Quite simple animals have a larger mass of neurons at one end of their body – the head end. This end reaches new places first, as the animal moves. These neurons act as a simple brain. They process information coming from the receptors on their head end.

Looking at how the brains of simple animals work can help scientists begin to understand more complicated brains.

Complex animals

More complex behaviour like yours needs a much larger brain. So your brain is made of billions of neurons. It also has many areas, each carrying out one or more specific functions all in the same organ. Your complex brain allows you to learn from experience, for example, how to behave with other people.

This is a diagram called the 'sensory homunculus'. Each body part is drawn so that its size represents the surface area of the sensory cortex that receives nerve impulses from it.

Average human brain statistics
width: 140 mm
length: 167 mm
height: 93 mm
weight: 1.4 kg

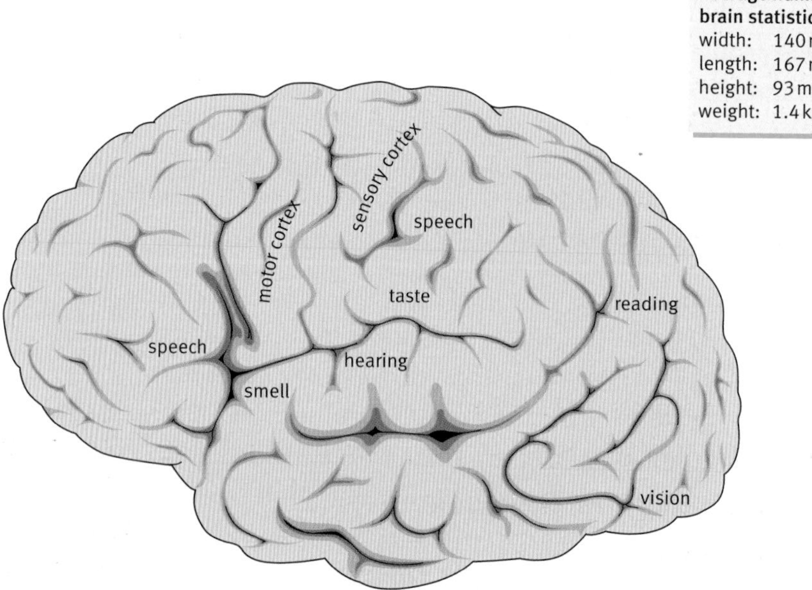

The human cerebral cortex is a highly folded region. Although it is only 5 mm thick, its total area is about 0.5 m^2. This map of the cerebral cortex shows the regions responsible for some of its functions.

The conscious mind

When you are awake, you are aware of yourself and your surroundings. This is called **consciousness**. The part of your brain where this happens is the **cerebral cortex**. This part is also responsible for intelligence, language, and memory. Brain processes to do with thoughts and feelings happen in the cerebral cortex and are what is called your 'mind'.

The cerebral cortex is very large in humans compared with other mammals. Studying what goes on in this part of the brain helps us to understand what it means to be human.

Finding out about the brain

In the 1940s a Canadian brain surgeon, Wilder Penfield, was working with patients who had epilepsy. Penfield applied electricity to the surface of their brains in order to find the problem areas. The patients were awake during the operations. There are no pain receptors in the brain so they did not feel pain.

Penfield watched for any movement the patient made as he stimulated different brain regions. From this information he was able to identify which muscles were controlled by specific regions of the motor cortex.

Injured brains

Scientists study patients whose brains are partly destroyed by injuries or diseases like strokes. Studies of injured soldiers have been important for research on how the brain functions.

Brain imaging

Modern imaging techniques such as magnetic resonance imaging (MRI) scans provide detailed information about brain structure and function without having to open up the skull. MRI can be used to show which parts of the brain are most active when a patient does different tasks. These scans are called functional MRI (fMRI) scans. The active parts of the brain have a greater flow of blood.

Key words

- consciousness
- cerebral cortex
- neuroscientist

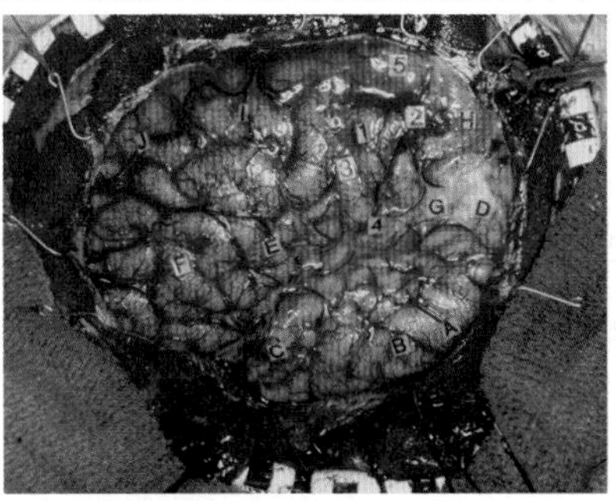

Penfield mapped the motor cortex by stimulating the exposed brain during open brain surgery. Regions of this brain have been identified and labelled in a similar way.

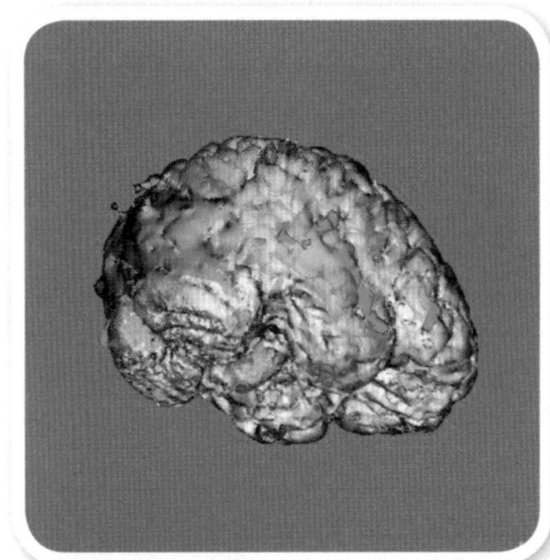

This functional MRI scan shows up areas of the brain that are active as a patient carries out a specific task. This patient was reading aloud.

Questions

1. What is your brain made up of?
2. Why is a complex brain so important for survival?
3. Which four functions of the brain happen in the cerebral cortex?
4. Explain why it is necessary for blood flow to increase to parts of the brain that are very active.
5. Compare the diagram of the brain with the functional MRI scan. How can you tell that the person was reading aloud?
6. What ethical issues should scientists consider when using injured humans to study the brain?

Learnt behaviour

Find out about

- ✓ **how conditioned reflexes can help you survive**

The lion cub below is just a few weeks old. She was born with reflexes that are helping her to stay alive. But much of her behaviour, for example, how to hunt or how to get on with other lions in the pride, she will **learn** from her mother. This is learnt behaviour.

Learnt behaviour is just as important for this lion cub's survival as reflexes.

Being able to learn new behaviour by experience is very important for survival. It means that animals can change their behaviour if their environment changes.

Pavlov's experiment

Pavlov's dog salivated when presented with food.

The food is the stimulus and salivation is the response.

Pavlov rang a bell while his dog was eating its food.

After a while the dog salivated when it heard the bell, even if no food was around.

The dog had learned to link the stimulus of ringing the bell with food. This type of learning is called **conditioning**.

Learning to link a new stimulus with a reflex action allows animals to change their behaviour. This is called a **conditioned reflex**. The final response – salivation – has no direct connection to the primary stimulus – food.

Conditioned reflexes

In 1904 the Russian scientist Ivan Pavlov won a Nobel Prize for his study on how the digestive system works. In his research Pavlov trained a dog to expect food whenever it heard a bell ring. The diagrams on the left explain what happened.

Conditioning aids survival

Conditioned reflexes can help animals survive. For example, bitter-tasting caterpillars are usually brightly coloured. A bird that tries to eat one learns that these bright colours mean that caterpillars will have a nasty taste. After a first experience the bird responds to the colours by leaving them alone. So this helps the caterpillars to survive.

If the brightly coloured insect is also poisonous, this reflex will help the bird survive as well. If other very tasty insects have similar colours and patterns, the bird does not eat them because of this conditioned response. You might have been caught out by this too – harmless yellow-and-black striped hoverflies sometimes alarm people who have been stung by a wasp.

'Warning' colours protect this caterpillar from predators.

Conditioning your pet

Open a can of soup in your kitchen. If you have a dog or a cat, this sound may get them very interested. But they are not hoping for soup! The animal's reflex response to food has been conditioned. It has learnt through experience that the sound of a tin being opened may be followed by food being put into its dish.

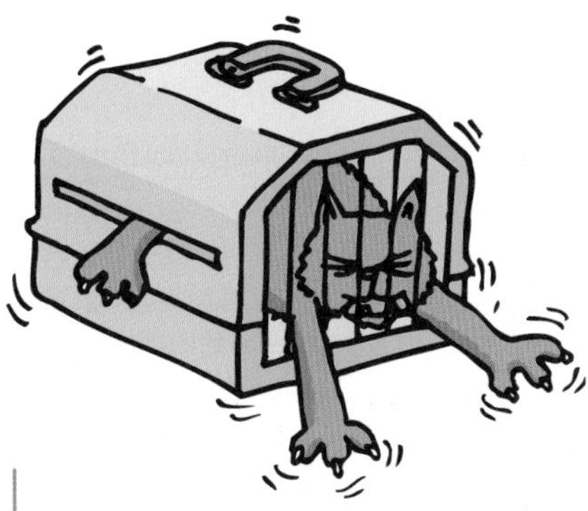

If a cat only uses its basket when you are taking it to the vet for an injection, it may become conditioned to link the basket with a frightening experience. The cat will then always be frightened by the stimulus of the basket. It will fight to keep out of the basket, even if you are only trying to take it to a new home.

Goldfish become conditioned to expect food when they see you in the room. They swim to the front of the bowl when you appear. The goldfish are linking the stimulus of seeing you with the original stimulus – food in the water.

Questions

1 Draw a flow diagram to explain how a cat or other pet can become conditioned to expect food when it hears a bathroom shower being run.

Use the key words from this section in your answer.

2 Adverts often have glamorous, funny, or exciting images and catchy tunes. Write down a list of photos and tunes from adverts that remind you of things you could buy. How is conditioning involved in making us more likely to buy these products?

Key words

- learn
- conditioning
- conditioned reflex

Hundreds of neurons interact to coordinate the responses you make when you are receiving this many stimuli.

Conscious control of reflexes

Most human reflex arcs are coordinated by the spinal cord. Only reflexes with receptors on the head are coordinated by the brain. A reflex arc only has simple connections between a sensory neuron and a motor neuron.

You do not think about reflex responses – your brain does not have to make a decision. They happen automatically, because they are designed to help you survive.

But sometimes a reflex may not be what you want to happen. Some reflexes can be modified by **conscious** control. Imagine picking up a hot plate. Your pain reflex makes you drop it. But if your dinner is on the plate, you can overcome this reflex and hold onto the plate until you put it down safely. The conscious control of your brain overcomes the reflex response. The diagram below explains how this happens.

Hundreds of complex pathways in the brain have to be used to succeed in this fast-moving sport.

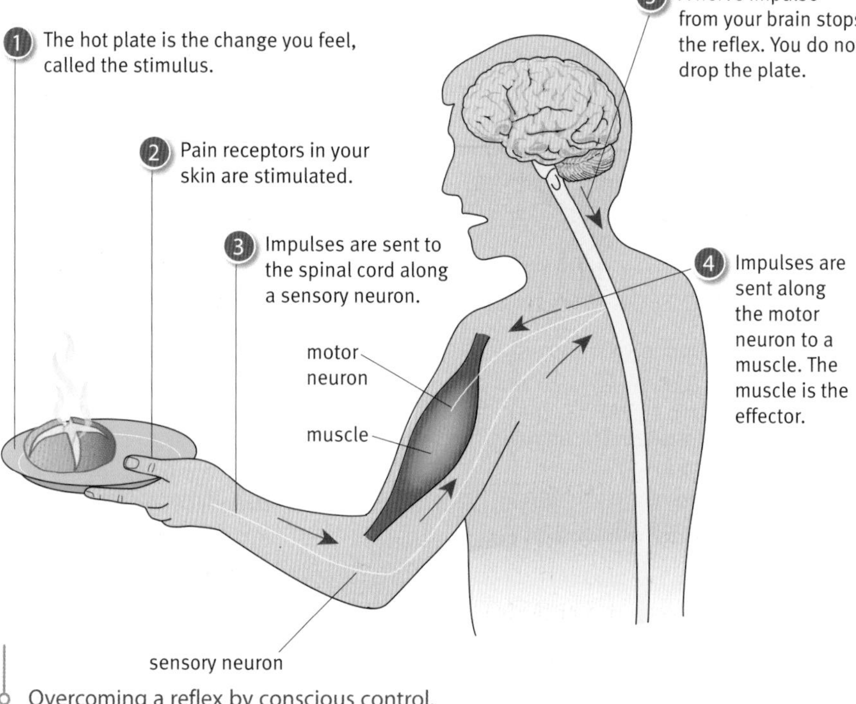

Overcoming a reflex by conscious control.

It's all in the mind – more complex behaviour

Connections in the brain usually involve hundreds of other neurons with different connections. Using these complex pathways, your brain can process highly complicated information, such as music, smells, and moving pictures. Different parts of the brain also store information (memory) and use it to make decisions for more complicated behaviour.

This complicated behaviour involves highly complex pathways in the brains of both these animals.

Complex behaviour allows us to learn from experience. For instance, **social behaviour** is learnt as humans develop.

Early humans learnt how to make and use tools for food and protection. Their ability to learn language meant that they could communicate new ideas. This gave them a survival advantage.

Making ethical decisions

Our understanding of the brain and behaviour has mainly come from experimenting on animals and (in the past) humans. This has allowed us to improve our theories about human learning, and to develop new treatments for diseases and injuries.

Some people argue that using animals for medical research is acceptable. Other people think that tampering with a vertebrate's brain in the name of science cannot be justified.

Scientists have learnt a lot about the brain from studying humans with mental-health problems. But is it fair and right to experiment on epileptic patients, as Penfold did? Some people argue that experimenting on ill patients results in improved medical knowledge that will benefit many others.

Studies of soldiers whose brains were damaged in war have helped scientists learn how the human brain works. But new technologies like MRI now mean that it is possible to build up a clearer picture of how the brain works just by observing a healthy brain in action.

Questions

3 List two reflexes you can overcome, and two that you cannot.

4 A baby urinates whenever its bladder is full. Draw a labelled diagram to show how nerve impulses from the brain overcome this reflex when the child is older.

5 Give three examples of how early humans' ability to learn gave them a survival advantage.

6 Do you think there is any ethical difference between using a rat and using a monkey for experiments to find out how the brain works?

7 Under what circumstances do you think it could be right to conduct scientific experiments on a human with a brain disorder?

Key words

- ✓ conscious
- ✓ social behaviour

G Human learning

Find out about

- how human beings learn new things
- explanations that scientists have for how your memory works

Mammals have complex brains made up of millions of neurons. When humans and other mammals experience something new, they can develop new ways of responding. Experience changes human behaviour, and this is called learning. The way that people and animals behave towards each other socially is also learned.

The evolution of larger brains gave some early humans a better chance of survival. Intelligence, memory, consciousness, and language are complex functions carried out by the outer layer of the brain, which is called the cerebral cortex. These functions are all involved in learning.

How does learning happen?

Neurons in your brain are connected together to form complicated **pathways**. How do these pathways develop? The first time a nerve impulse travels along a particular pathway, from one neuron to another, new connections are made between the neurons. New experiences set up new neuron pathways in your brain.

If the experience is repeated, or the stimulus is particularly strong, more nerve impulses follow the same nerve pathway. Each time this happens, the connections between these neurons are strengthened. Strengthened connections make it easier for nerve impulses to travel along a pathway. As a result, the response you produce becomes easier to make.

The brains of human babies develop new nerve pathways very quickly. Your brain can develop new pathways all your life. This means you can still learn as you get older, though more slowly.

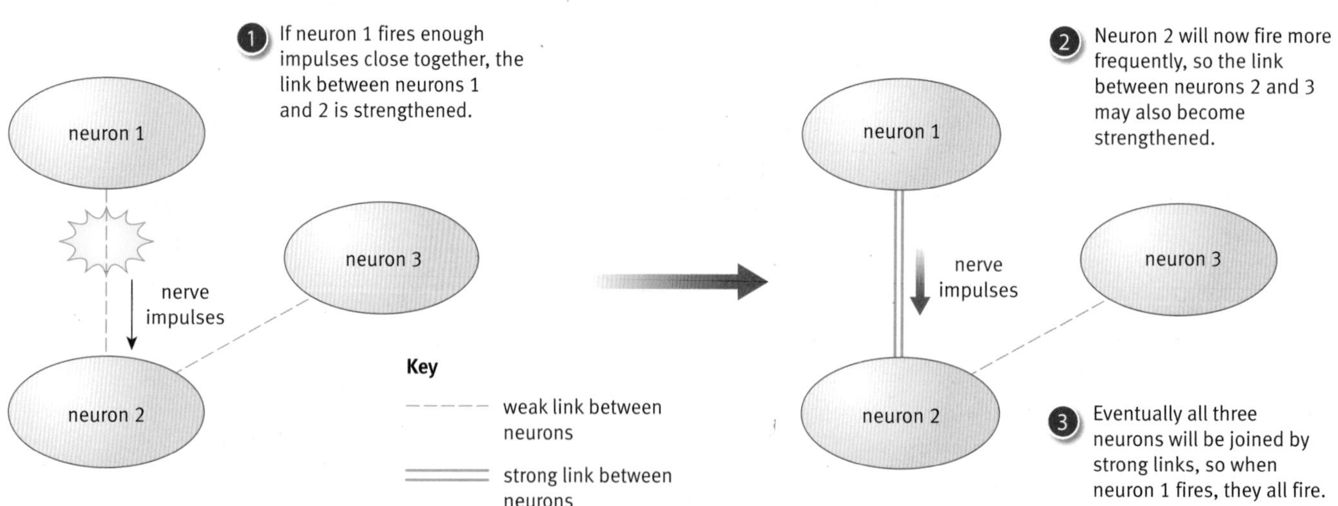

Nerve pathways form in a baby's brain as a result of a stimulus from its environment. Repeating the stimulus strengthens the pathway. The baby then responds in the same way each time it receives the stimulus. Some neurons in the brain do not take part in any pathway. Many of these unused neurons are destroyed.

Repetition

Repetition helps you learn because it strengthens the pathways the brain uses to carry out a particular skill. Perhaps you have learned to ride a bicycle, play a musical instrument, perform a new dance sequence, or touch type. To do these things you created new nerve pathways then strengthened them through repetition. This made it easy for you to respond in the way that you practised.

For example, Marie is a gymnast. When she has to learn new movements she stands still and imagines going through the motions – the position of her body and muscles being used at each stage. Visualisation works because thinking about using a muscle triggers nerve impulses to that muscle. This strengthens the pathways the impulse takes. After a period of visualisation, the actual movement is a lot easier to perform.

Marie visualises new movements to help her learn them.

Age and learning

You learn to speak through repetition because you are surrounded by people talking. Children learn language extremely easily up to the age of about eight years. Their brains easily make new neuron pathways in the language processing region. As we get older it becomes harder for this part of the brain to make new pathways.

Feral children

In 1799, in southern France, a remarkable creature crept out of the forest. He acted like an animal but looked human. He could not talk. The food he liked and the scars on his body showed he had lived wild for most of his life. He was a wild, or **feral**, child. The local people guessed that he was about twelve years old and named him Victor.

Victor was taken to Paris. He lived with a doctor who tried to tame him and teach him language. At first people thought Victor had something wrong with his tongue or voice box. He could only hiss when people tried to teach him the names of objects. He communicated in howls and grunts.

Victor never learnt to say more than a couple of words. By the time he was found, the time in his development when it was easy to learn language had passed.

Key words

- ✓ **pathways**
- ✓ **repetition**
- ✓ **feral**

Questions

1. Write a few sentences to explain how you learn by experience. Use the key words 'pathway' and 'repetition' in your answer.
2. Explain why repeating a skill helps you learn it.
3. Write a list of skills you could practise by visualisation.

What is memory?

Find out about

- short-term and long-term memory
- the multistore model of memory
- the working-memory model

Psychologists are scientists who study the human mind. They describe **memory** as your ability to store and retrieve information.

Short-term and long-term memory

Read this sentence:

- As you read this sentence you are using your **short-term memory**.

Short-term memory lasts for about 30 seconds in most people. If you have no short-term memory you will not be able to make sense of this sentence. By the time you get to the end of the sentence, you will have forgotten the beginning.

Think about a song you know the words to:

- To remember the words you use your **long-term memory**.

Verbal memory is *any information* you store about words and language. It can be divided into short-term and long-term memory. Long-term memory is a lasting store of information. There seems to be no limit to how much can be stored in long-term memory. And the stored information can last a lifetime.

Different memory stores work separately

People with advanced **Alzheimer's disease** suffer severe short-term memory loss. They cannot remember what day it is, or follow simple instructions. But they may still remember their childhood clearly.

Some people lose long-term memory because of brain damage or disease. Their short-term memory is normal. This evidence is important because it shows that long-term and short-term memory must work separately in the brain.

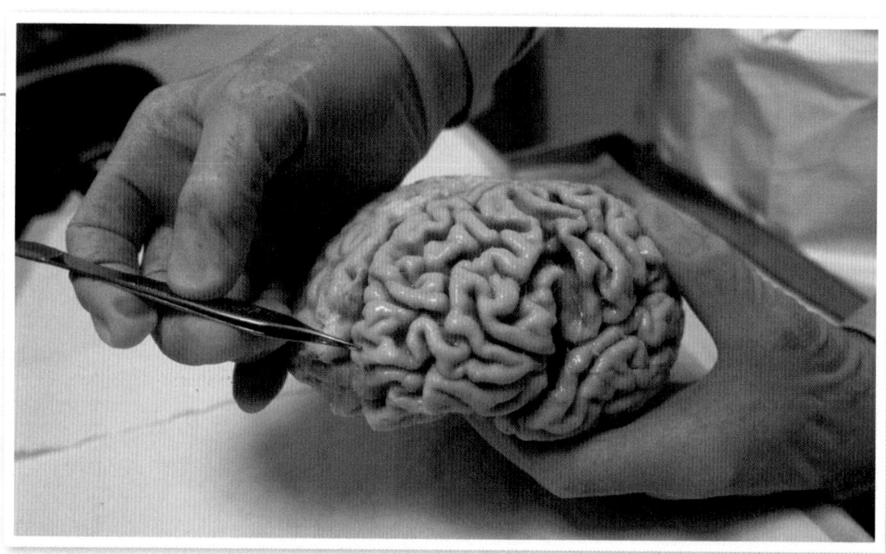

The 'Nun Study' at the University of Kentucky, USA, has had the participation of 678 School Sisters of Notre Dame. They range in age from 75 to 106 years. The sisters have allowed scientists to assess their mental and physical function every year and to examine their brains at death. The study has led to significant advances and discoveries in the area of Alzheimer's disease and other brain disorders.

Sensory memory store

You can also use a sensory memory store to store sound and visual information for a short time. When you wave a sparkler on bonfire night it leaves a trail of light. You can even write shapes in the air that other people can see. You see the trail because you store each image of the sparkler separately for a short time. The ability to store images for a short time makes the separate pictures in a film seem continuous. You can store sound temporarily in the same way.

Your sensory memory stores each image of the sparkler separately for a short time. This makes the whole shape seem continuous.

Questions

1 Write down one sentence to describe memory.

2 What is the difference between short-term and long-term memory?

3 Explain why a person with advanced Alzheimer's disease is unable to do simple things like go shopping or cook for themselves.

4 Give one piece of evidence that short-term and long-term memory are separate.

Key words

- ✔ **memory**
- ✔ **short-term memory**
- ✔ **long-term memory**
- ✔ **Alzheimer's disease**

Key words

- retrieval of information
- models of memory
- multistore model

How much can you store in your short-term memory?

Cover up the list of letters below with a piece of paper. Move the paper down so you can see just the top row. Read through the row once, then cover it up and try to write down the letter sequence. Then go down to the next row and do the same. Find out how many letters you can remember in the correct order.

N T

A N L

N F E K

B F E X A

N A Z T P L

M B T F E Q P

U N D A C X Z G

O R B V E X Z D A

R T L D C A G P V E

If you remembered more than seven letters in a row correctly, then you have excellent short-term memory. Short-term memory can only store about seven items. When you are remembering letters in a list, each letter is an 'item'. To remember more letters, chunk them into groups.

For example, the row O R B V E X Z D A has nine letters. Chunk these into groups of three: 'ORB' 'VEX' 'ZDA'.

The nine letters are easy to remember because now they are only three items. Three items doesn't overload your short-term memory.

Models of memory

The **retrieval of information**, such as word lists, is a way of testing your memory. Memory tests can tell us what memory can and cannot do. But they do not explain how the neurons in the brain work to give you memory. Explanations for how memory happens are called **models of memory**.

The multistore model: memory stores work together

Read through the list of words below once. Then cover the page and try to write down as many of the words as you can remember. They can be in any order.

dog, window, film, menu, archer, slave, lamp, coat, bottle, paper, kettle, stage, fairy, hobby, package

How many did you remember? If this type of test is carried out on large numbers of people, a pattern is seen in the words they recall. People often remember the last few words on the list and get more of them right. They also recall the first few words on the list quite well.

When you look at a list of words:

- Nerve impulses travel from your eyes to your sensory memory.
- Some sensory information is passed on to your short-term memory. Only the information you pay attention to is passed on. You will not be able to remember words you have not noticed.
- If more information arrives than the short-term memory can hold then some is lost (forgotten). You will not remember these words either.
- Some information is passed to your long-term memory. These are words you will remember – usually the first few words on the list.
- The last information your short-term memory receives will still be there when you start to write down the list. So these are also words you will remember, usually the last few words on the list.

This use of sensory, short-term and long-term memory is known as the **multistore model** of memory.

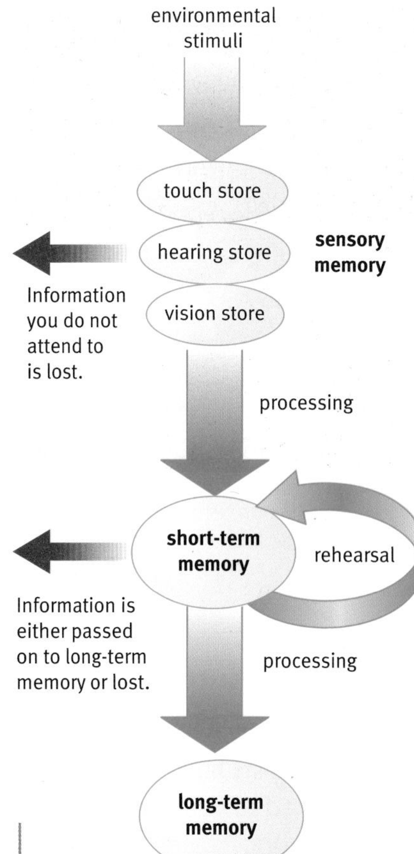

The multistore model of memory can be used to explain how some information is passed to the long-term memory store and some information is lost.

Questions

5 You read the menu on a board inside a café. When you try to tell your friend sitting outside all the choices, you forget some. Why can you not remember everything on the list?

6 'So far none of the models of memory can explain completely how your memory works.'

a What does it take for an explanation to be widely accepted by scientists?

b Why is it useful to have models that try to explain how memory works?

Rehearsal is one technique actors use to learn their lines.

Rehearsal and long-term memory

Look at this row of letters:

R T I D A C G P E V

There are too many letters in this row for you to store them separately in your short-term memory. Given time you would probably repeat the letters over and over until you remembered them. **Repetition of information** is a well-known way of memorising things. An actor can memorise a sonnet (a 14-line poem) in around 45 minutes. Psychologists think that rehearsal moves information from your short-term memory to your long-term memory store.

The working-memory model

In 1972, two psychologists, Fergus Craik and Robert Lockhart, concluded that the multistore memory model was too simple. They suggested that rehearsal is only one way to transfer information from short-term to long-term memory.

Rehearsed information is processed and stored rather than lost from short-term memory. Craik and Lockhart argued that you are more likely to remember information if you process it more deeply. They suggested that this will happen if you understand the information or it means something to you.

For example, if you can see a pattern in the information, you process it more deeply. So

AAT, BAT, CAT, DAT, EAT

is much easier to remember than

DAT, AAT, EAT, CAT, BAT

You also process information more deeply if there is a strong stimulus linked to the information, for example, colour, light, smell, or sound.

An active working memory

Short-term memory is now seen as an active '**working memory**'. Here you can hold and process information that you are consciously thinking about. Communication between long-term and working memory is in both directions. This way you can retrieve information you need, and also store information you may need later.

Putting it into practice

You can apply what the psychologists have discovered to your own school work.

- *Repetition:* If you are struggling to remember a piece of information you have read, read it several times.
- *Rehearsal:* Read sections of what you have to learn that are short enough to keep in your short-term memory. Make notes from memory to help move the information to your long-term memory.
- *Active memory:* Use highlighter pens and spider diagrams to process information for learning.

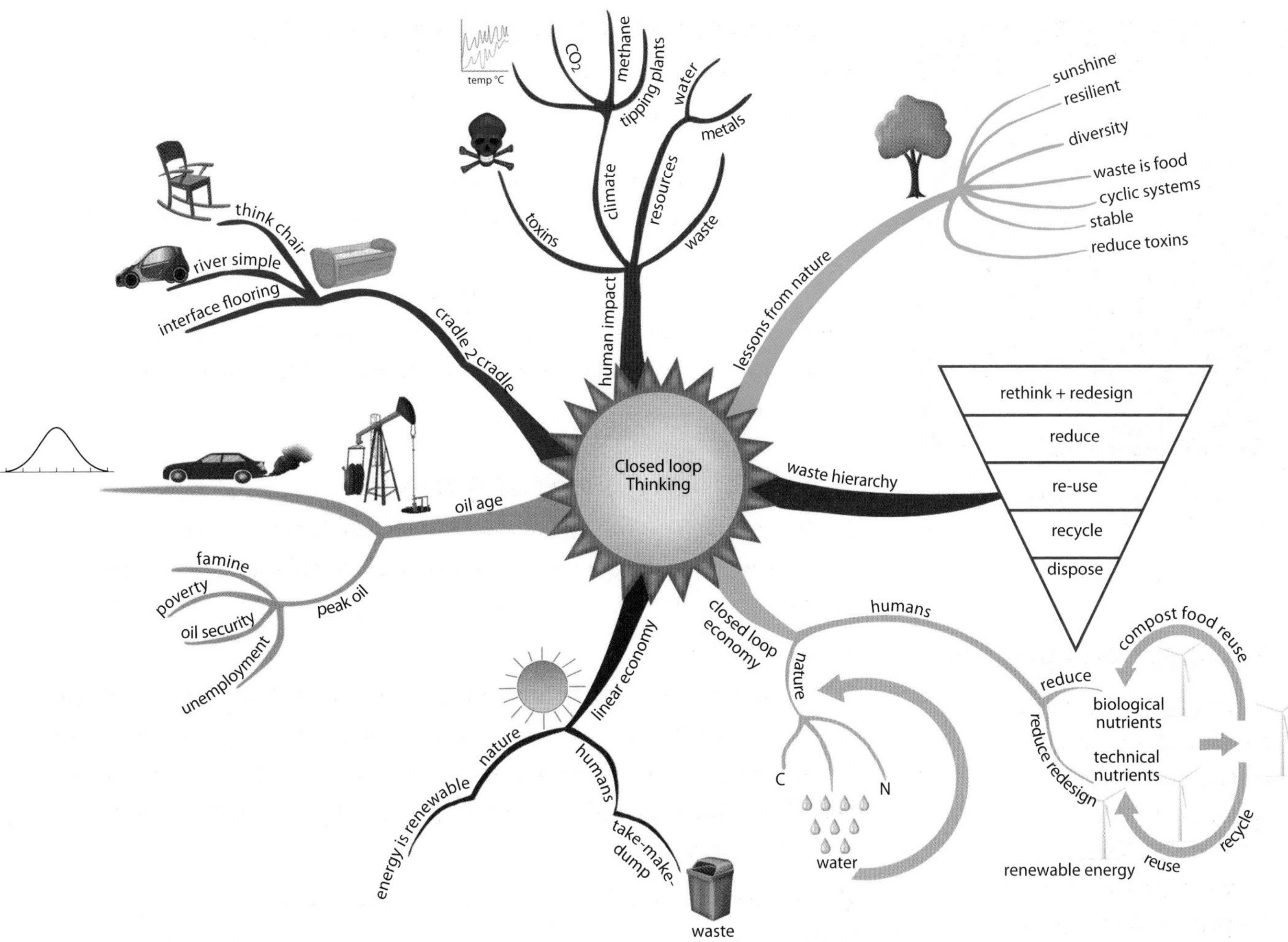

Questions

7 Make two lists of 10 different things to buy from a supermarket. Try to remember one list. Put the second list into 'families', for example tins, bakery, cleaning, and memorise it. Which list is easier to remember and why?

8 Give an example of something you can remember because a strong stimulus is linked to the information:
a a colour b a smell c a sound

9 Write down an example of something you have memorised through rehearsal, for example, the directions to the cinema, or a complicated set of moves in a computer game. How has rehearsal helped you to remember this?

10 Explain why using a highlighter pen to pick out key facts makes your revision more successful.

11 Construct a spider diagram to show the information you have learnt about memory and learning.

Key words

- **repetition of information**
- **working memory**

Science Explanations

There have been big advances in neuroscience recently, but we still do not know everything about how the human brain works. Knowing how the brain works is very important for understanding how we learn and about the mental-health diseases of old age.

You should know:

- that a stimulus is a change in the environment of an organism
- what simple reflexes are and why they are important in the role of receptors, processing centres, and effectors in body systems such as human vision
- that the body uses electrical impulses and chemical hormones for short- and long-lived responses, respectively
- the relationship between the central nervous system and peripheral nervous system (sensory and motor neurons) in humans and other vertebrates
- that transmitter substances carry nerve impulses from one nerve to the next at the synapse and how some toxins and drugs affect the transmission
- that nerve impulses travel through relay neurons in the spinal cord and connect sensory and motor neurons, allowing automatic, rapid responses
- how scientists map the regions of the cerebral cortex to particular functions
- how the evolution of a larger brain gave some early humans a better chance of survival
- the role of short-term memory and long-term memory in the storage and retrieval of information
- what helps humans to learn and recall information
- how the 'multistore model' of memory provides a working model for short-term memory, long-term memory, repetition, storage, retrieval, and forgetting
- how simple models like the multistore model develop into more complex models such as the working-memory model.

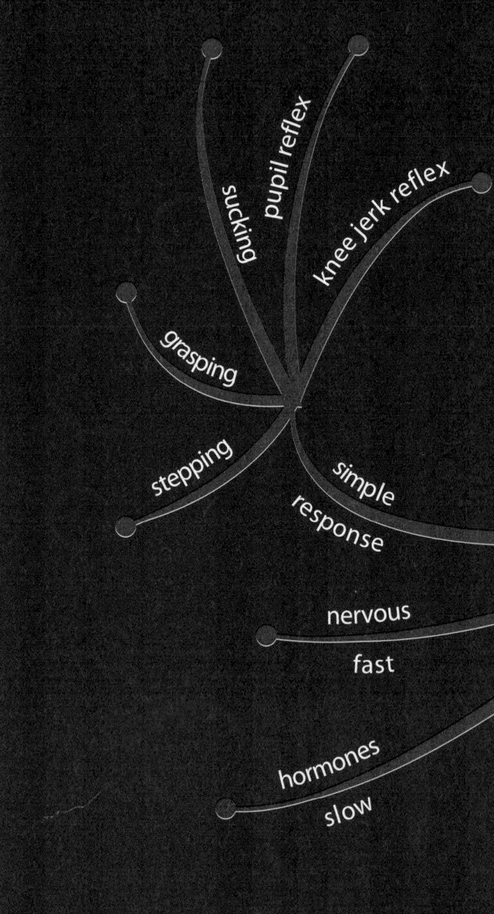

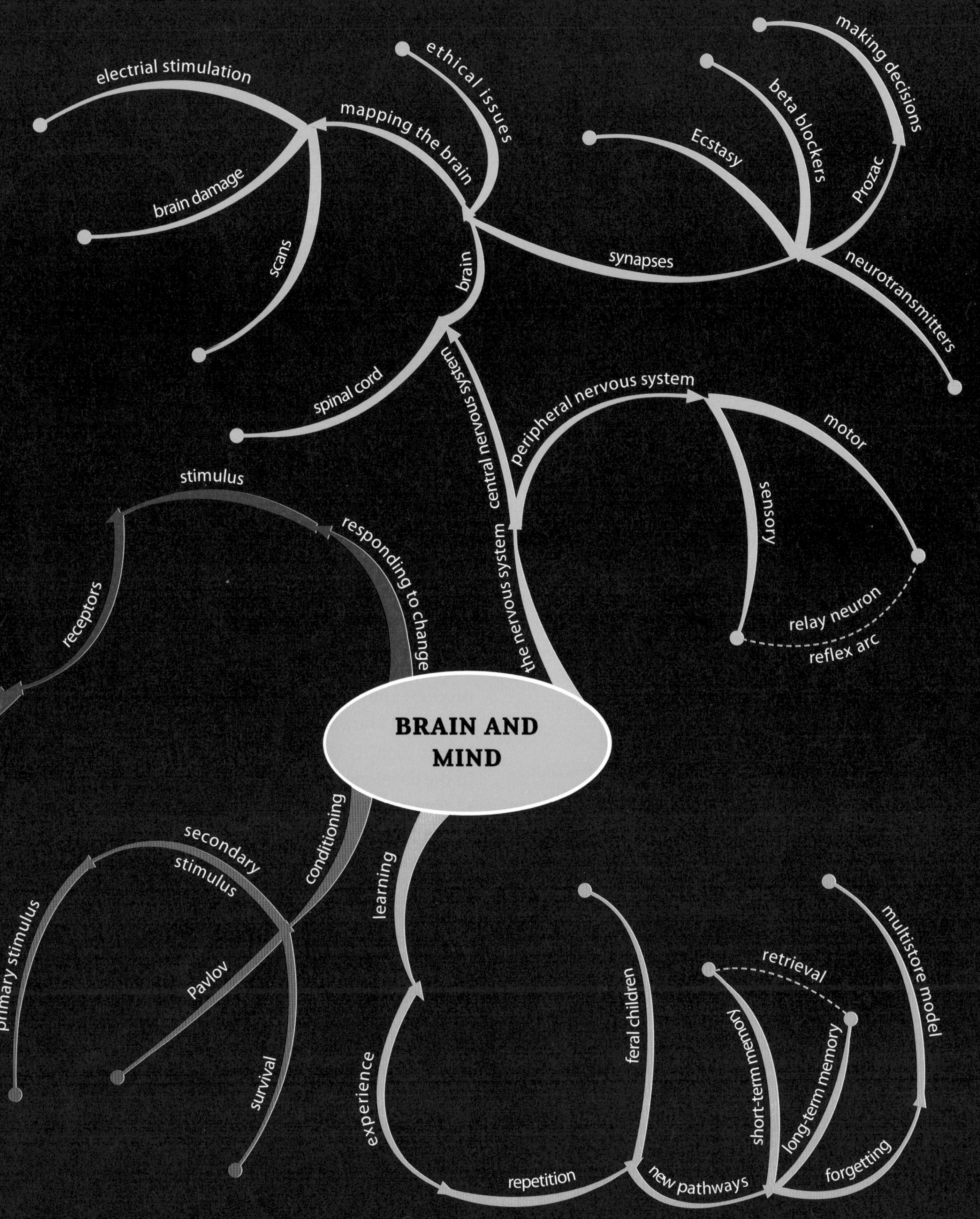
BRAIN AND MIND
the nervous system
central nervous system
brain
mapping the brain
electrial stimulation
brain damage
scans
ethical issues
spinal cord
synapses
Ecstasy
beta blockers
making decisions
Prozac
neurotransmitters
peripheral nervous system
motor
sensory
relay neuron
reflex arc
responding to change
stimulus
receptors
conditioning
secondary stimulus
primary stimulus
Pavlov
survival
learning
experience
repetition
feral children
new pathways
short-term memory
long-term memory
retrieval
forgetting
multistore model

Ideas about Science

Most knowledge about the brain has come from experiments on animals and humans. This has allowed scientists to refine theories about human learning and to develop new treatments for diseases and injuries.

You need to identify and develop arguments around ethical issues in scientific work, and summarise the different views that may be held.

- Some people argue that using animals for medical research is acceptable if there are benefits for humans. Other people think that tampering with a vertebrate's brain cannot be justified.
- Scientists have learnt a lot about the brain from studying humans with mental-health problems. But is it fair and right to experiment on people who have an illness?
- Studies of soldiers whose brains were damaged in war have helped scientists learn how the human brain works.

It is sometimes difficult to decide what is right and what is wrong. Some people think that the right decision is one that leads to the best outcome for the greatest number of people.

Other people think that certain actions are either right or wrong whatever the consequences. You need to be able to identify and develop arguments that are based on these two different ideas.

- With new technologies such as MRI it is possible to build up a picture of how the brain works by observing a healthy brain in action. This reduces the need for controversial experiments on animals and ill people.

This module examines some models that were developed to explain how humans remember and learn things.

- The simple multistore model is useful for only some of the observations or data on learning and memory. Fergus Craik and Robert Lockhart concluded in 1972 that the multistore memory model was too simple.

Craik and Lockhart used creative thought to produce another explanation to explain more completely how people remember. You should be able to identify where creative thinking is involved in the development of an explanation.

- The working-memory model provides a way of explaining a wider range of how people remember things.

When looking at science explanations you need to identify the better of two given scientific explanations for a phenomenon, and give reasons for your choice.

Review Questions

1 The gaps between sensory and motor neurons are called synapses. When an impulse is transmitted a series of events take place at the synapse.
These statements are in the wrong order.
One statement is incorrect.

A Chemicals are released into the synapse.
B The receptor molecules produce chemicals.
C Chemicals bind with receptor molecules on the motor neuron membrane.
D Chemicals diffuse across the synapse.
E The impulse travels along a motor neuron.
F An impulse reaches the end of a sensory neuron.

Select the five correct statements and write them in the correct order.

2 Pavlov used salivation in dogs to study conditioned reflexes. Pavlov used a series of steps over a period of time to produce a conditioned reflex.

a At each step a **different** stimulus was provided.

A Dog hears bell ringing.
B Dog shown food.
C Dog shown food and hears bell ringing.

Copy and complete the table below by writing A, B, or C in the blank cells.

step 1: initial reflex		
	dog salivates	dog given food

step 2: repeated many times		
	dog salivates	dog given food

step 3: conditioned reflex		
	dog salivates	dog given food

b Conclusions can be made following this investigation.

Decide which of these conclusions are true and which are false.

The bell was used as a primary stimulus.

The conditioned reflex response had a direct connection to the primary stimulus.

The dog learned to associate the secondary stimulus with the primary stimulus.

3 Pip is a young puppy. His brain contains billions of neurons.
Explain what will happen to the neuron pathways in Pip's brain as he develops.

4 Scientists can gather useful information about how the brain works by studying people with brain damage.
Write down the ethical issues that are involved with this research and two different views that may be held.

5 Explain how recreational drugs such as Ecstasy can affect the transmission of impulses across the synapses between neurons in the brain.

6 Describe one model that can be used to understand how the brain stores and retrieves information as memories.

C6 Chemical synthesis

Why study chemical synthesis?

We use chemicals to preserve food, treat disease, and decorate our homes. Many of these chemicals do not occur naturally: they are synthetic. Developing new products, such as drugs to treat disease, depends on chemists who synthesise and test new chemicals.

What you already know

- Atoms are rearranged during chemical reactions.
- The number of atoms of each element stays the same in a chemical reaction.
- Raw materials can be used to make synthetic materials.
- Alkalis neutralise acids to make salts.
- Chemical reactions can be represented by word equations and balanced symbol equations.
- Some substances are made up of electrically charged particles called ions.
- Data is more reliable if it can be repeated.

Find out about

- the importance of the chemical industry
- a theory to explain acids and alkalis
- reactions that give out and take in energy
- techniques for controlling the rate of chemical change
- the steps involved in the synthesis of a new chemical
- ways to measure the efficiency of chemical synthesis.

The Science

Chemists who synthesise new chemicals need practical skills and an understanding of science explanations. They must control reactions so that they are neither too slow nor too fast. They must calculate how much of the reactants to use to make the amount of product required. Chemists also take into account any energy changes. Acids are important reactants in synthesis. Ionic theory explains the characteristic behaviours of these chemicals.

Ideas about Science

Chemists make sure they use the right grade of chemical for a reaction. Technical chemists test chemicals from suppliers to check the purity. They take measurements and make sure the data they collect is as accurate and reliable as possible. They can then make the best estimate of the true value of the purity.

A The chemical industry

Find out about

- the chemical industry
- bulk and fine chemicals
- the importance of chemical synthesis

The chemical industry converts raw materials into pure chemicals, which are then used in synthesis to make a wide range of products.

Industrial chemists work in the plant and the laboratory.

Key words

- chemical industry
- bulk chemicals
- fine chemicals
- plant
- pilot plant
- scale up

The **chemical industry** converts raw materials, such as crude oil, natural gas, minerals, air and water, into useful products. The products include chemicals for use as food additives, fertilisers, pigments, dyes, paints, and pharmaceutical drugs.

The industry makes **bulk chemicals** on a scale of thousands or even millions of tonnes per year. Examples are ammonia, sulfuric acid, sodium hydroxide, chlorine, and ethene.

On a much smaller scale, the industry makes **fine chemicals** such as drugs, herbicides, and pesticides. It also makes small quantities of speciality chemicals needed by other manufacturers for particular purposes. These include such things as flame retardants, food additives, and the liquid crystals for flat-screen televisions and computer displays.

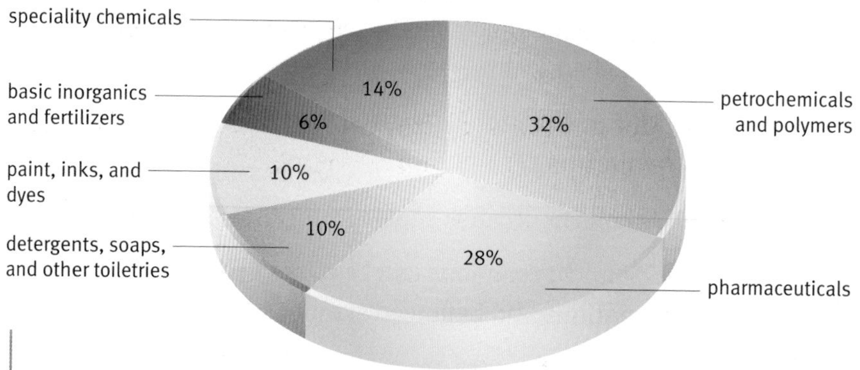

The range of products made by the chemical industry in the UK by value of sales.

The part of a chemical works that produces a chemical is called a **plant**. Some of the chemical reactions occur at a high temperature, so a source of energy is needed. Also, a lot of electrical power is needed for pumps to move reactants and products from one part of the plant to another. Sometimes the energy to produce this power can be supplied by chemical reactions that give out energy.

Sensors monitor the conditions, such as temperature and pressure, in all areas of the plant. The data is fed to computers in the control centre, where the technical team controls the plant.

People in the chemical industry

People with many different skills are needed in the chemical industry. Research chemists work in laboratories to find new processes and develop new products.

The industry needs new processes so that it can be more competitive and more sustainable. The aim is to use smaller amounts of raw materials and energy while creating less waste.

People devising new products have to work closely with people in the marketing and sales department. They are able to say if the novel product is wanted. If the new product is promising, it may first be tried out by making small amounts of it in a **pilot plant**.

As part of the market research, possible new products are given to customers for trial. At the same time, financial experts estimate the value of the new product in the market. They then compare this with the cost of making the product to check that the new process will be profitable.

Chemical engineers have to **scale up** the process and design a full-scale plant. This can cost hundreds of millions of pounds.

Some chemicals from the industry go directly on sale to the public, but most of them are used to make other products. Transport workers carry the chemicals to the industry's customers.

Every chemical plant needs managers and administrators to control the whole operation. There are also people in service departments who look after the needs of the people working in the plant. These include medical and catering staff, and training and safety officers.

Questions

1. Classify the following as raw materials or products of the chemical industry: air, ammonia, aspirin, water, crude oil, polythene.
2. Give the name and chemical formula of a bulk chemical.
3. List these chemicals under two headings: 'bulk chemical' and 'fine chemical'.
   - the drug aspirin
   - the hydrocarbon ethene
   - the perfume chemical citral
   - the acid sulfuric acid
   - the herbicide glyphosate
   - the alkali sodium hydroxide
   - the food dye carotene

Plant operators monitor the processes from a control room.

Maintenance workers help to keep the plant running.

B Acids and alkalis

Find out about

- acids and alkalis
- the pH scale
- reactions of acids

Key words

- acid
- alkali

Acids

The word **acid** sounds dangerous. Nitric, sulfuric, and hydrochloric acids are very dangerous when they are concentrated. You must handle them with great care. These acids are less of a hazard when diluted with water. Dilute hydrochloric acid, for example, does not hurt the skin if you wash it away quickly, but it stings in a cut and rots clothing. It is in fact present in our stomachs where it helps to break down food and kill bacteria. Our stomachs are lined with a protective layer of mucus.

Not all acids are dangerous to life. Many acids are part of life itself. Biochemists have discovered the citric acid cycle. This is a series of reactions in all cells. The cycle harnesses the energy from respiration for movement and growth in living things.

Organic acids

Organic acids are molecular. They are made of groups of atoms. Their molecules consist of carbon, hydrogen, and oxygen atoms. The acidity of these acids arises from the hydrogen in the —COOH group of atoms.

Citric and tartaric acids are examples of solid organic acids. Ethanoic acid is a liquid organic acid.

Acetic acid (chemical name: ethanoic acid) is a liquid. It is the acid in vinegar. Most white vinegar is just a dilute solution of acetic acid. Brown vinegars have other chemicals in the solution that give the vinegar its colour and flavour. Most microorganisms cannot survive in acid, so vinegar is used as a preservative in pickles (E260).

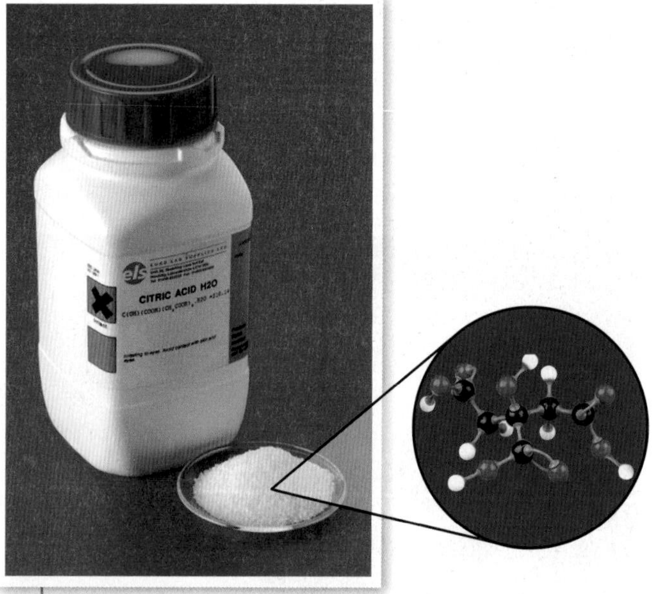

Citric acid (a solid acid) is found in citrus fruits like oranges and lemons. The human body processes about 2 kg of citric acid a day during respiration. Citric acid and its salts are added to food to prevent them reacting with oxygen in the air, and to give a tart taste to drinks and sweets.

Mineral acids

Sulfuric, hydrochloric, and nitric acids come from inorganic or mineral sources. The pure acids are all molecular. Sulfuric acid and nitric acid are liquids at room temperature. Hydrogen chloride is a gas, and becomes hydrochloric acid when it dissolves in water.

Alkalis

Pharmacists sell antacids in tablets to control heartburn and indigestion. The chemicals in these medicines are the chemical opposites of acids. They are designed to neutralise excess acid produced in the stomach – hence the name 'antacids'.

Antacids in medicines are usually insoluble in water. Other chemical antacids are soluble in water and give a solution with a pH above 7. Chemists call them **alkalis**. Common alkalis are sodium hydroxide, NaOH; potassium hydroxide, KOH; and calcium hydroxide, $Ca(OH)_2$.

The traditional name for sodium hydroxide is caustic soda. The word caustic means that the chemical attacks living tissue, including skin. Alkalis can do more damage to delicate tissues than dilute acids. Caustic alkalis are used in the strongest oven and drain cleaners. They have to be used with great care.

Sulfuric acid, H_2SO_4, is manufactured from sulfur, oxygen, and water. The pure, concentrated acid is an oily liquid. The chemical industry in the UK makes about 2 million tonnes of the acid each year. The acid is essential for the manufacture of other chemicals, including detergents, pigments, dyes, plastics, and fertilisers.

Hydrogen chloride forms when concentrated sulfuric acid is added to salt (sodium chloride) crystals. Hydrogen chloride, HCl, is a gas that fumes in moist air and is very soluble in water.

Oven cleaners often contain caustic alkalis.

Questions

1 From the pictures of molecules work out the formulae of:
   a acetic acid
   b citric acid.

2 What are the formulae of these antacids?
   a Magnesium hydroxide made up of magnesium ions, Mg^{2+}, and hydroxide ions, OH^-.
   b Aluminium hydroxide made up of aluminium ions, Al^{3+}, and hydroxide ions, OH^-.

pH

14 — dilute sodium hydroxide
13
12 — limewater
11
alkaline
10 — some brands of toothpaste
9
8
7 — blood
neutral
— pure water
— fresh cows' milk
6 — distilled water
5
4
3 — vinegar
acidic
2 — lemon juice
1 — digestive fluids in the stomach
0 — dilute hydrochloric acid

The pH scale.

A pH meter can be used to measure pH values.

Using a feather to brush away hydrogen bubbles while etching a metal plate with acid.

Indicators and the pH scale

Indicators change colour to show whether a solution is acidic or alkaline. Blue litmus turns red in acid solution and red litmus turns blue in alkalis. Special mixed indicators, such as universal indicator, show a range of colours and can be used to estimate pH values.

pH values can also be measured electronically using a pH meter with an electrode that dips into the solution. The meter can be read directly from the display or it may be connected to a datalogger or computer.

The term pH appears on many cosmetic, shampoo, and food labels. It is a measure of acidity. The **pH scale** is a number scale that shows the acidity or alkalinity of a solution in water. Most laboratory solutions have a pH in the range 1–14.

Hydrangea flowers contain natural indicators – they are blue if grown on acid soil and pink on alkaline soil. Note that this is the opposite of the litmus colours.

Reaction of acids

Acids with metals

Acids react with **metals** to produce **salts**. The other product is hydrogen gas.

$$\text{acid} + \text{metal} \longrightarrow \text{salt} + \text{hydrogen}$$

For example: $2HCl(aq) + Mg(s) \longrightarrow MgCl_2(aq) + H_2(g)$

Not all metals will react in this way. You may remember the list of metals in order of reactivity in C5, G. Metals below lead in the list do not react with acids, and even with lead it is hard to detect any change in a short time.

Acids with metal oxides or hydroxides

An acid reacts with a **metal oxide** or **hydroxide** to form a salt and water. No gas forms.

$$\text{acid} + \text{metal oxide (or hydroxide)} \longrightarrow \text{salt} + \text{water}$$

For example: $2HCl(aq) + MgO(s) \longrightarrow MgCl_2(aq) + H_2O(l)$

The reaction between an acid and a metal oxide is often a vital step in making useful chemicals from ores.

Acids with carbonates

Acids react with **carbonates** to form a salt, water, and bubbles of carbon dioxide gas.

$$\text{acid} + \text{metal carbonate} \longrightarrow \text{salt} + \text{water} + \text{carbon dioxide}$$

Geologists can test for carbonates by dripping hydrochloric acid onto rocks. If they see any fizzing, the rocks contain a carbonate. This is likely to be calcium carbonate or magnesium carbonate.

The word equation is:

$$\text{hydrochloric acid} + \text{calcium carbonate} \longrightarrow \text{calcium chloride} + \text{water} + \text{carbon dioxide}$$

The balanced equation is:

$$2HCl(aq) + CaCO_3(s) \longrightarrow CaCl_2(aq) + H_2O(l) + CO_2(g)$$

This is a foolproof test for the carbonate ion. So the term 'the acid test' has come to be used to describe any way of providing definite proof.

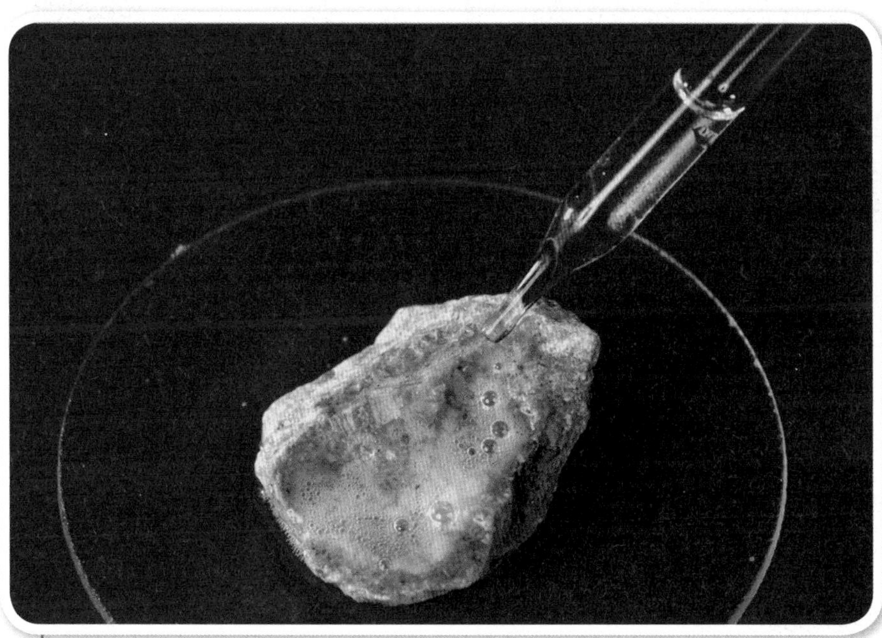

Testing for carbonate using hydrochloric acid.

Key words

- pH scale
- indicators
- metals
- salts
- metal oxide
- metal hydroxide
- carbonates

Questions

3 A pattern can be etched onto a zinc plate using hydrochloric acid to react with the zinc, forming soluble zinc chloride, $ZnCl_2$. Write a word equation and a balanced symbol equation for the reaction.

4 Magnesium hydroxide, $Mg(OH)_2$, is an antacid used to neutralise excess stomach acid, HCl. Write a word equation and a balanced symbol equation for the reaction.

5 There is a volcano in Tanzania, Africa, whose lava contains sodium carbonate, Na_2CO_3. The cooled lava fizzes with hydrochloric acid. Write a word equation and a balanced symbol equation for the reaction.

6 Limescale forms in kettles where hard water is heated. Limescale consists of calcium carbonate. Three acids are often used to remove limescale: citric acid, acetic acid (in vinegar), and dilute hydrochloric acid. Which acid would you use to de-scale an electric kettle and why?

Find out about

- an ionic explanation for neutralisation reactions
- salts and their formulae

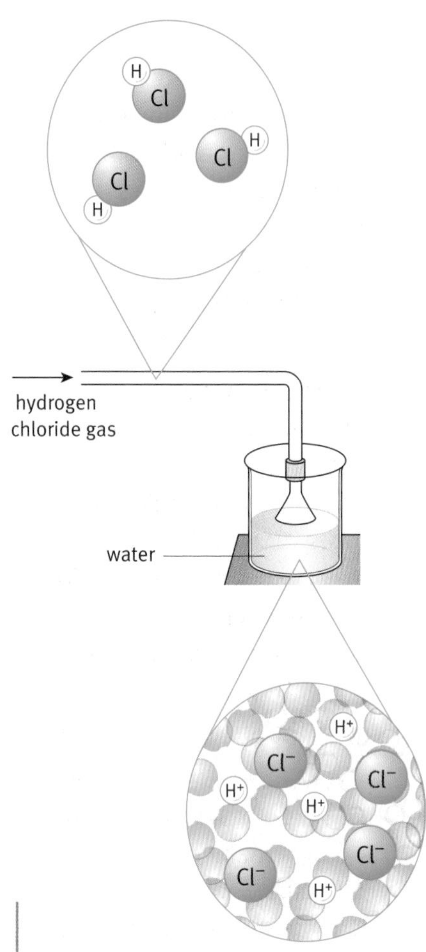

Hydrogen chloride dissolves in water to make hydrochloric acid. The HCl molecules react with water to form ions.

Key words

- hydrogen ions
- hydroxide ions
- neutralisation reaction

What makes an acid an acid?

Chemists have a theory to explain why all the different compounds that are acids behave in a similar way when they react with indicators, metals, carbonates, metal oxides, and metal hydroxides.

It turns out that acids do not simply mix with water when they dissolve. They react, and when they react with water they produce hydrogen ions (H^+). For example, hydrochloric acid is a solution of hydrogen chloride in water. The HCl molecules react with the water to produce **hydrogen ions** and chloride ions.

$$HCl(g) \xrightarrow{\text{water}} H^+(aq) + Cl^-(aq)$$

The theory of acids is an ionic theory. Any compound is an acid if it produces hydrogen ions when it dissolves in water.

All acids contain hydrogen in their formula. Nitric acid, HNO_3, and phosphoric acid, H_3PO_4, both contain hydrogen. But not all chemicals that contain hydrogen are acids. Ethane, C_2H_6, and ethanol, C_2H_5OH, are not acids.

In an organic acid it is only the hydrogen atom in the —COOH group that can ionise when the acid dissolves in water.

What makes a solution alkaline?

Alkalis such as the soluble metal hydroxides are ionic compounds. They consist of metal ions and **hydroxide ions** (OH^-). When they dissolve, they add hydroxide ions to water. It is these ions that make the solution alkaline.

$$NaOH(s) \xrightarrow{\text{water}} Na^+(aq) + OH^-(aq)$$

Neutralisation

Sodium hydroxide and hydrochloric acid react to produce a salt (sodium chloride) and water.

$$Na^+(aq) + OH^-(aq) + H^+(aq) + Cl^-(aq) \longrightarrow Na^+(aq) + Cl^-(aq) + H_2O(l)$$

During a **neutralisation reaction** the hydrogen ions from an acid react with hydroxide ions from the alkali to make water.

$$H^+(aq) + OH^-(aq) \longrightarrow H_2O(l)$$

The remaining ions in the solution make a salt.

Salts

Salts form when a metal oxide, or hydroxide, neutralises an acid. So every salt can be thought of as having two parents. Salts are related to a parent metal oxide or hydroxide and to a parent acid.

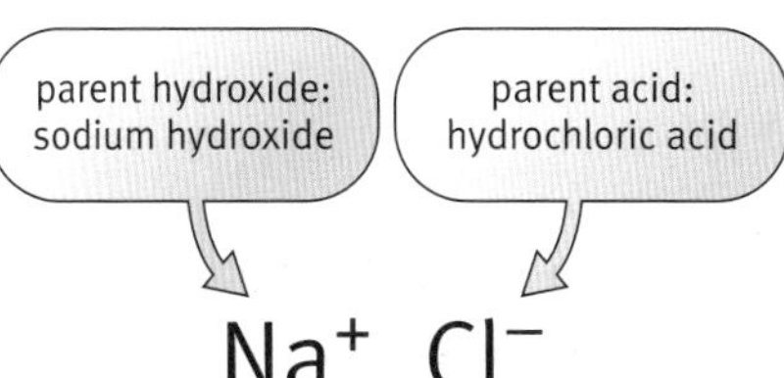

Salts are ionic (see C4, J: Ionic theory). Most salts consist of a positive metal ion combined with a negative non-metal ion. The metal ion comes from the parent metal oxide or hydroxide. The non-metal ion comes from the parent acid.

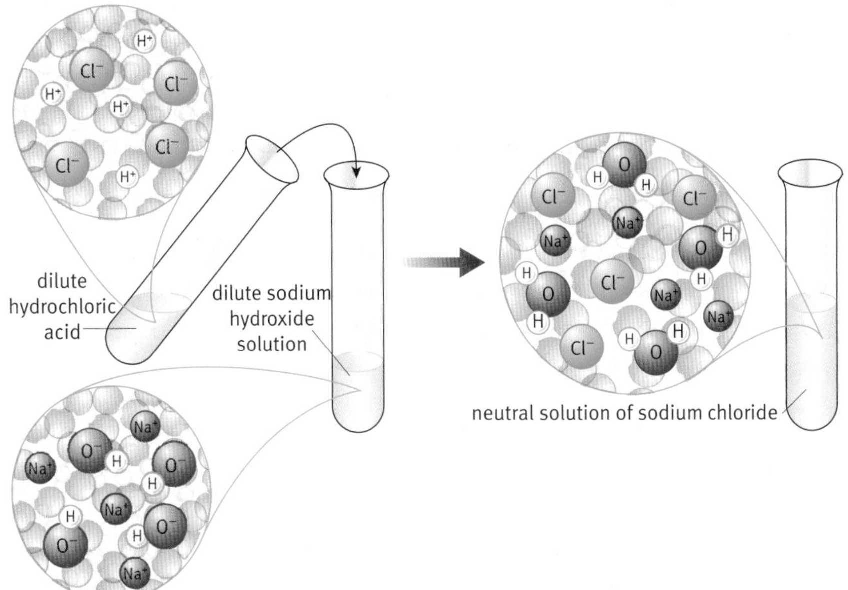

Dilute sodium hydroxide solution neutralises dilute hydrochloric acid, forming a neutral solution of sodium chloride. Water molecules not involved in the reaction are shown in a paler colour.

It is possible to work out the formulae of salts knowing the charges on the ions. Remember that all compounds are overall electrically neutral (see C4, K: Ionic theory and atomic structure). Some non-metal ions consist of more than one atom. The table below includes some examples. In the formula for magnesium nitrate, $Mg(NO_3)_2$, the brackets around the NO_3 show that two complete nitrate ions appear in the formula.

Non-metal ions that consist of more than one atom	Symbols
carbonate	CO_3^{2-}
hydroxide	OH^-
nitrate	NO_3^-
sulfate	SO_4^{2-}

Questions

1. Write equations to show what happens when these compounds dissolve in water:
   a. nitric acid
   b. sulfuric acid
   c. calcium hydroxide.
2. Write down the name of the salt produced in the following reactions:
   a. lithium hydroxide with hydrochloric acid
   b. calcium carbonate with nitric acid
   c. magnesium oxide with sulfuric acid.
3. Use the tables of ions in C4 Section K and on this page to write down the formulae of these salts:
   a. potassium nitrate
   b. magnesium carbonate
   c. sodium sulfate
   d. calcium nitrate.
4. Use the tables of ions in C4 Section K and on this page to write down the charge on the metal ion in each of these salts:
   a. $CuCO_3$
   b. $PbBr_2$
   c. Fe_2O_3.

D Purity of chemicals

Find out about

- purity
- titrations for testing purity

CALCIUM CARBONATE PRECIPITATED CP
QTY: 1kg BNO: C1042/R6 - 708717

Assay	99%
Chloride (Cl)	0.005%
sulphate (SO_4)	0.05%
Iron (Fe)	0.002%
Lead (Pb)	0.002%

Label on a bottle of laboratory grade calcium carbonate. The term 'assay' tells you how pure the chemical is. The calcium carbonate is 99% pure with the small amounts of impurities shown.

Key words

- titration
- burette
- end point

Questions

1 Uses of sodium chloride (salt) include:
  i flavouring food
  ii melting ice on roads
  iii saline drips in hospitals.
  Put these in order of the grade of sodium chloride required, with the purest first.

2 From 'steps involved in a titration' in which step is:
  a a solution made?
  b a pipette used, and what is it used for?
  c the end point reached, and how does the technician know?

Grades of purity

The reactions of acids with metals, oxides, hydroxides, and carbonates can be used to make valuable salts. For uses such as food or medicines, these salts have to be made pure so that they are safe to swallow.

Chemicals do not always have to be pure. Calcium carbonate, for example, is used in a blast furnace to extract iron from its ores. The iron industry can use limestone straight from a quarry. Limestone has some impurities but they do not stop it from doing its job in a blast furnace.

Suppliers of chemicals offer a range of grades of chemicals. In a school laboratory you might use one of these grades: technical, general laboratory, and analytical. The purest grade is the analytical grade.

Purifying a chemical is done in stages. Each stage takes time and money, and becomes more difficult. So the higher the purity, the more expensive the chemical. Manufacturers therefore buy the grade most suitable for their purpose.

When deciding what grade of chemical to use for a particular purpose, it is important to know:

- the amount of impurities
- what the impurities are
- how they can affect the process
- whether they will end up in the product, and whether it matters if they do.

Testing purity

Medicines contain an active ingredient. Other ingredients are included to make the medicine pleasant to taste and easy to take. This means that the pharmaceutical companies that make medicines need sweeteners, food flavours, and other additives.

The companies buy in many of their ingredients. Technical chemists working for the companies have to make sure that the suppliers are delivering the right grade of chemical.

Citric acid is often added to syrups, such as cough medicines, to control their pH. Technicians can check the purity of the acid using a procedure called a **titration**, which measures the volume of alkali that it can neutralise. The technician has to know the accurate concentration of the alkali.

Steps involved in a titration

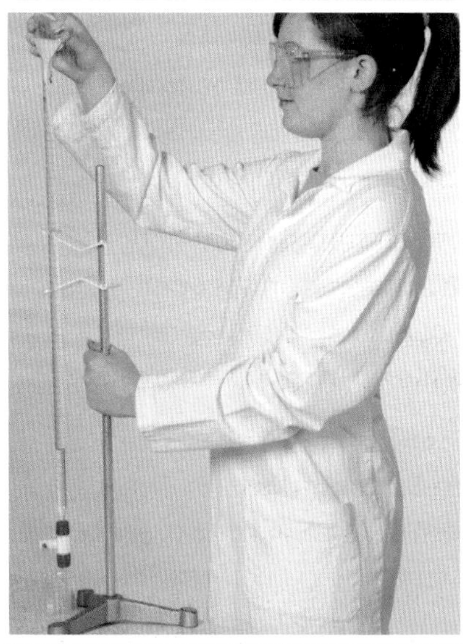

1 The technician fills a **burette** with a solution of sodium hydroxide. She knows the concentration of the alkali.

2 The technician weighs out a sample of citric acid accurately.

3 The technician dissolves the acid in pure water. Then she adds a few drops of phenolphthalein indicator. The indicator is colourless in the acid solution.

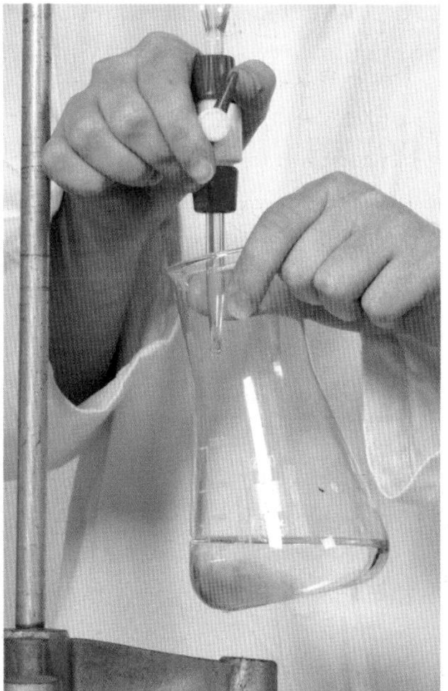

4 The technician adds alkali from the burette. She swirls the contents of the flask as the alkali runs in. Near the end she adds the alkali drop by drop. At the **end point** all the citric acid is just neutralised. The indicator is now permanently pink.

The technician will repeat the titration several times. If there are any results that differ greatly from the rest, and there is reason to doubt their accuracy, they will be discarded. The remaining values will be averaged to find the mean.

Questions

3 A 1.35 g sample of impure citric acid was dissolved in water and titrated with sodium hydroxide solution of concentration 40 g/dm^3. The average titre was found to be 20.6 cm^3.

- a Use the following equation to find the mass of citric acid in the sample: mass of citric acid (g) = average titre (cm^3) × 0.064.
- b Use the following equation to calculate the percentage purity of the sample:
  percentage purity = $\frac{\text{mass of citric acid}}{\text{mass of sample}} \times 100\ \%$

4 A technical chemist measures the purity of tartaric acid by titration, and obtains these results: 98.7%, 99.0%. 105.4%, 80.0%, 98.8%, 98.5%

- a Suggest reasons why the results are not exactly the same.
- b Which two values should be checked?
- c Calculate the mean value of purity after discarding the two outlying values.
- d Suggest the limits between which the true value is likely to lie.

E Energy changes in chemical reactions

Find out about

- ✓ reactions that give out energy and reactions that take in (absorb) energy

Both exothermic and endothermic reactions have practical uses. An exothermic reaction provides the energy for welding. A cold pack absorbs energy from an injured muscle using an endothermic reaction.

Exothermic and endothermic reactions

Most chemical reactions need a supply of energy to get them started. Some also need energy to keep them going. That is why Bunsen burners and electric heating mantles are so common in laboratories.

Most reactions give out energy once they are going – combustion (burning), and neutralisation of acids to make salts are common examples. Sulfuric acid is an important chemical in industry. One of the key reactions in making this acid has to be carefully controlled because it gives out a great deal of energy.

Reactions that give out energy to the surroundings (the surroundings get hotter) are called **exothermic**. There are also reactions that absorb energy from the surroundings (the surroundings get cooler). These are called **endothermic**.

It is possible to tell whether a chemical reaction is exothermic or endothermic by measuring the temperature of the reactants before the reaction starts, and the temperature of the products after it has finished. If the temperature has risen, the reaction is exothermic. If it has dropped, the reaction is endothermic.

Energy changes in the chemical industry

Scientists working in the chemical industry need to know whether a reaction is exothermic or endothermic when it is part of a synthesis. There are several reasons for this:

- It takes fuel, which costs money, to provide the energy input for endothermic reactions.
- The energy given out by exothermic reactions can be used elsewhere in the plant, to produce electricity, for example.
- A temperature increase makes chemical reactions go faster (see Section F) – a reaction that gives out heat energy will tend to get faster and faster, and may 'run away', possibly causing an explosion.

Such a 'runaway' reaction may have been the cause of a major disaster at a chemical plant in Bhopal, India, in 1984. Between 2000 and 15 000 people may have died and over 25 years later, the health of many people is still being affected by the incident. Proper understanding and control of the energy changes in the reactions concerned might have prevented the disaster.

Energy-level diagrams

All chemical reactions give out or take in energy. Understanding energy changes helps chemists to control reactions.

- In an exothermic reaction (where energy is given out), the products must have less energy than the reactants.
- In an endothermic reaction (where energy is absorbed), the products must have more energy than the reactants.

Chemists keep track of the changes in energy in chemical reactions using **energy-level diagrams**. These show the energies of the reactants and products. Notice that energy is plotted vertically up the diagram. Reactants are on the left and products on the right. These are usually shown using a balanced equation.

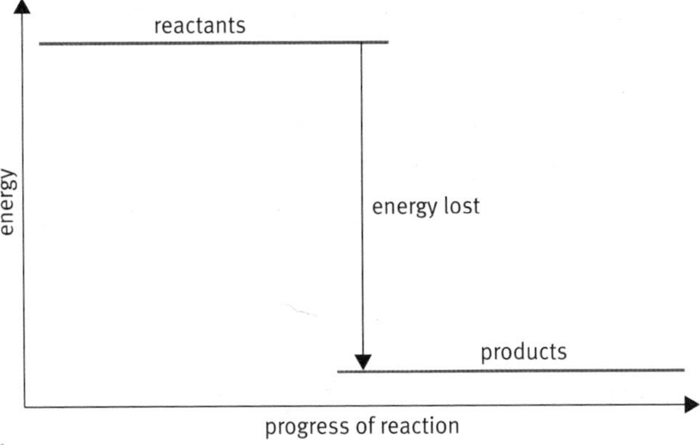

An exothermic reaction.

The energy-level diagram for the reaction of magnesium and hydrochloric acid, which is exothermic. Energy is given out in this reaction, so the products have less energy than the reactants.

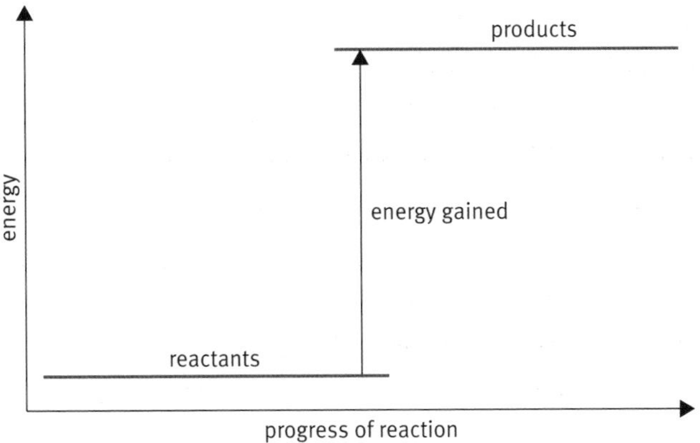

An endothermic reaction.

Energy-level diagram for the reaction of citric acid and sodium hydrogencarbonate, which is endothermic. Energy is absorbed in this reaction, so the products have more energy than the reactants.

Key words

- exothermic
- endothermic
- energy-level diagram

An electric heating mantle, used to heat up reactions without a naked flame.

Question

1 If you have ever had a plaster cast for a broken bone, apart from the pain of the injury, you might remember the pleasant feeling of warmth as the wet plaster begins to set. If you put a sherbet sweet on your tongue you will feel your mouth getting cold. Both of these are chemical reactions.

  a Which reaction is exothermic and which is endothermic?

  b Draw an energy-level diagram for each reaction. Use 'reactants' and 'products' for the names of the chemicals involved.

Rates of reaction

Find out about

- measuring rates of reaction
- factors affecting rates of reaction
- catalysts in industry
- collision theory

An explosion is an example of a very fast chemical reaction.

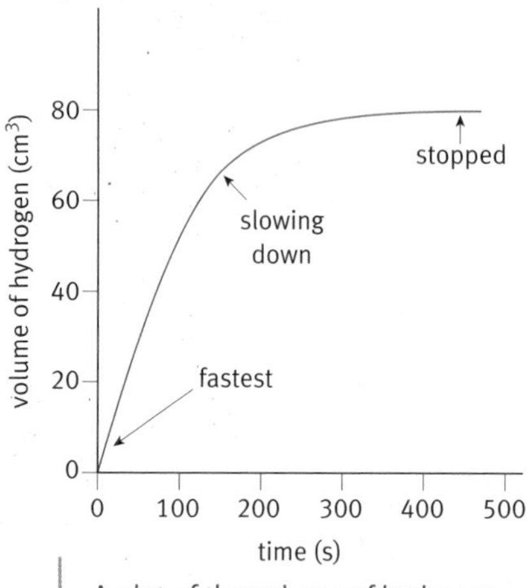

A plot of the volume of hydrogen formed against time for a reaction of magnesium with hydrochloric acid.

Key word

- rate of reaction

Controlling reaction rates

Some chemical reactions seem to happen in an instant. An explosion is an example of a very fast reaction.

Other reactions take time – seconds, minutes, hours, or even years. Rusting is a slow reaction and so is the rotting of food.

It is the chemist's job to work out the most efficient way to synthesise a chemical. It is important that the chosen reactions happen at a convenient speed. A reaction that occurs too quickly can be hazardous. A reaction that takes several days to complete is not practical because it ties up equipment and people's time for too long, which costs money.

Measuring rates of reaction

Your pulse rate is the number of times your heart beats every minute. The production rate in a factory is a measure of how many articles are made in a particular time. Similar ideas apply to chemical reactions.

Chemists measure the **rate of a reaction** by finding the quantity of product produced or the quantity of reactant used up in a fixed time.

For the reaction

$$Mg(s) + 2HCl(aq) \longrightarrow MgCl_2(aq) + H_2(g)$$

the rate can be measured quite easily by collecting and measuring the hydrogen gas produced.

$$\text{average rate} = \frac{\text{change in the volume of hydrogen}}{\text{time for the change to happen}}$$

In most chemical reactions, the rate changes with time. The graph on the left is a plot of the volume of hydrogen formed against time for the reaction of magnesium with acid. The graph is steepest at the start, showing that the rate of reaction was greatest at that point. As the reaction continues the rate decreases until the reaction finally stops. The steepness of the line is a measure of the rate of reaction.

Question

1 For each of these reactions, pick a method from the opposite page that could be used to measure the rate of reaction: (Hint: Look at the states of the reactants and products.)

a $CaCO_3(s) + 2HCl(aq) \longrightarrow CaCl_2(aq) + CO_2(g) + H_2O(l)$

b $Zn(s) + H_2SO_4(aq) \longrightarrow ZnSO_4(aq) + H_2(g)$

c $Na_2S_2O_3(aq) + 2HCl(aq) \longrightarrow 2NaCl(aq) + SO_2(g) + S(s) + H_2O(l)$

Methods of measuring rates of reaction

Collecting and measuring a gas product

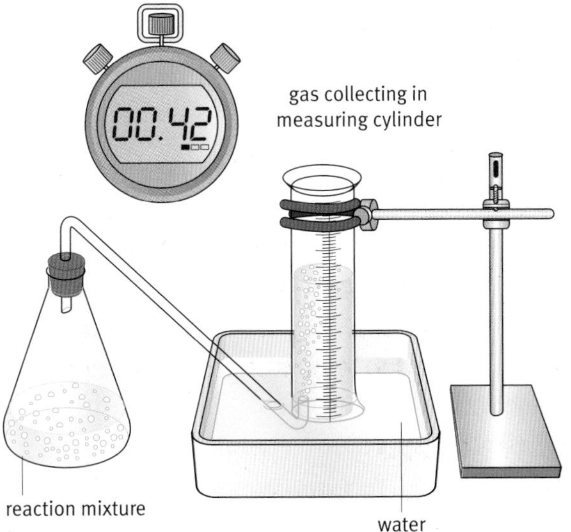

Record the volume at regular intervals, such as every 30 or 60 seconds. A gas syringe could be used to collect the gas, instead of a measuring cylinder – see the next page.

Measuring the loss of mass as a gas forms

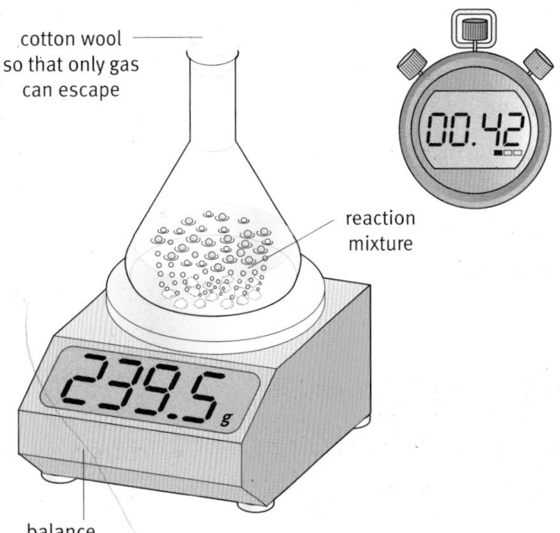

Record the mass at regular intervals such as every 30 or 60 seconds.

Timing how long it takes for a small amount of solid reactant to disappear

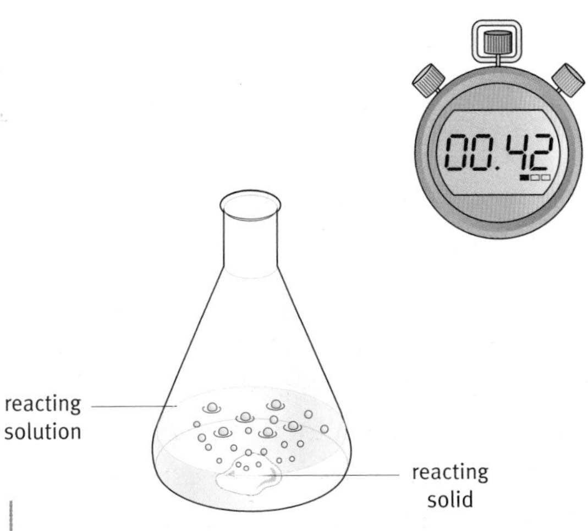

Mix the solid and solution in the flask and start the timer. Stop it when you can no longer see any solid.

Timing how long it takes for a solution to turn cloudy

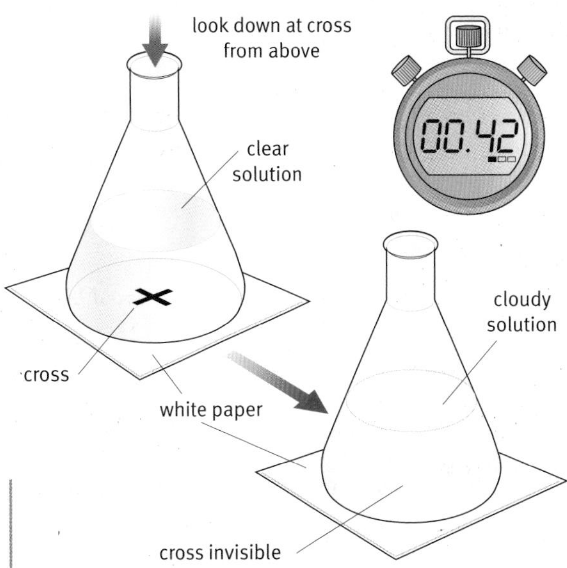

This is for reactions that produce an insoluble solid. Mix the solutions in the flask and start the timer. Stop it when you can no longer see the cross on the paper through the solution.

Manual timing of reactions is only suitable for reactions that are over in a few minutes (or more). Modern methods using lasers and dataloggers allow chemists to measure the rates of reactions that are over in less than one thousand billion billionths of a second (one femtosecond).

Key words

- concentration
- surface area
- catalyst

Factors affecting reaction rates

Powdered Alka-Seltzer reacts faster in water than a tablet. Milk standing in a warm kitchen goes sour more quickly than milk kept in a refrigerator. Changing the conditions alters the rate of these processes and many others.

Factors that affect the rate of chemical reactions are:

- the *concentration* of reactants in solution – the higher the concentration, the faster the reaction
- the *surface area* of solids – powdering a solid increases the surface area in contact with a liquid, solution, or gas, and so speeds up the reaction.
- the *temperature* – typically a 10 °C rise in temperature can roughly double the rate of many reactions
- *catalysts* – these are chemicals that speed up a chemical reaction without being used up in the process.

The factors in action

The apparatus in the diagram was used in an investigation into the effect of changing the conditions on the reaction of zinc metal with sulfuric acid. The graph below shows the results.

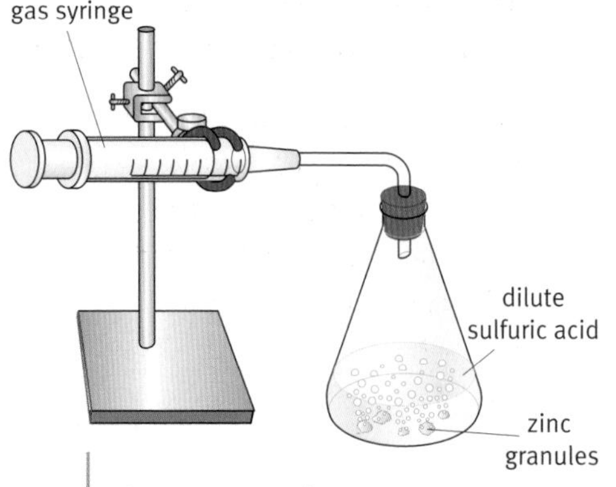

Apparatus used to investigate the factors affecting the rate of reaction of zinc with sulfuric acid.

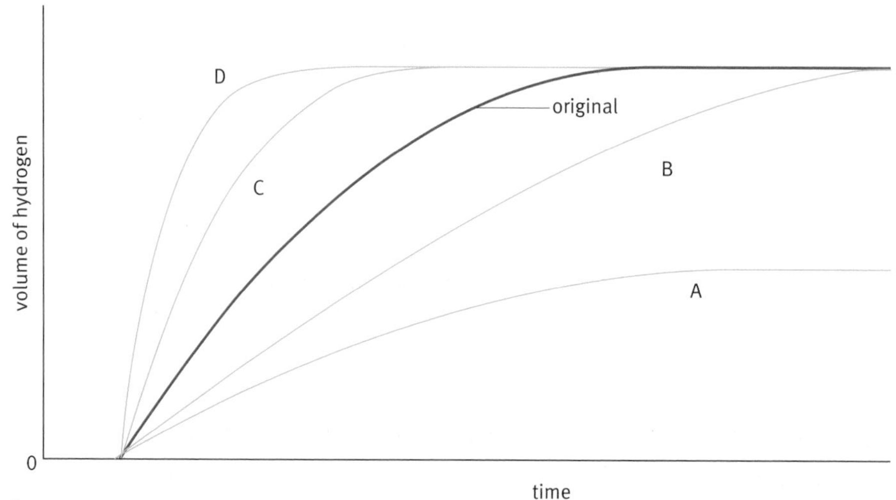

The volume of hydrogen formed over time during an investigation of the factors affecting the rate of reaction of zinc with sulfuric acid. The investigator used the same mass of zinc each time. There was more than enough metal to react with all the acid.

$$Zn(s) + H_2SO_4(aq) \longrightarrow ZnSO_4(aq) + H_2(g)$$

The red line on the graph plots the volume of hydrogen gas against time using zinc granules and 50 cm^3 of dilute sulfuric acid at 20 °C. The reaction gradually slows down and stops because the acid concentration falls to zero. There is more than enough metal to react with all the acid. The zinc is in excess.

The effect of concentration

Line A on the graph shows the result of using acid that was half as concentrated while leaving all the other conditions the same as in the original set up.

The investigator added 50 cm^3 of this more dilute acid. Halving the acid **concentration** lowers the rate at the start. The final volume of gas is cut by half because there was only half as much acid in the 50 cm^3 of solution to start with.

The effect of surface area

Line B on the graph shows the result of using the same excess of zinc metal in fewer larger pieces. All other conditions were the same as in the original setup. Fewer larger lumps of metal have a smaller total **surface area** so the reaction starts more slowly. The amount of acid is unchanged and the metal is still in excess so that the final volume of hydrogen is the same.

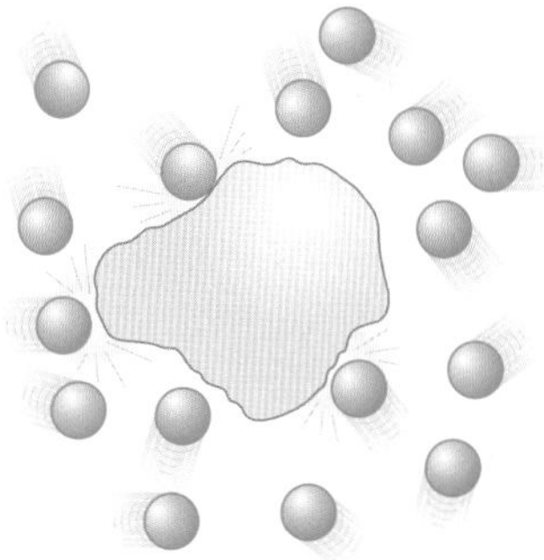

one big lump (slow reaction)

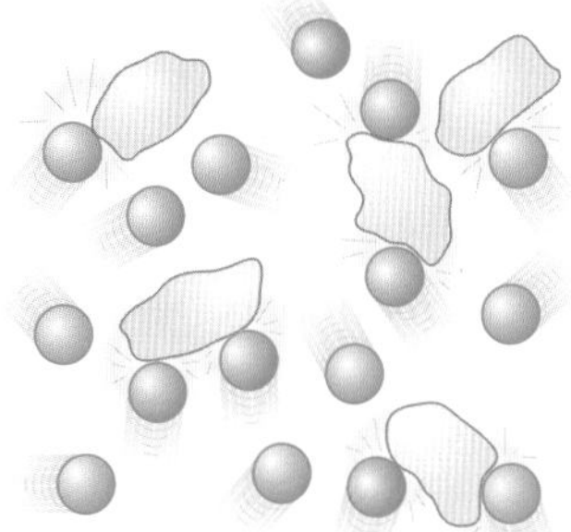

several small lumps (fast reaction)

Breaking up a solid into smaller pieces increases the total surface area. This increases the amount of contact between the solid and the solution, making it possible for the reaction to go faster.

The effect of temperature

Line C on the graph shows the result of carrying out the reaction at 30 °C while leaving all the other conditions the same as in the original set up. This more or less doubles the rate at the start. The quantities of chemicals are the same so the final volume of gas collected is the same as it was originally.

The effect of adding a catalyst

Line D on the graph shows what happens when the investigation is repeated with everything the same as in the original set up but with a **catalyst** added. The reaction starts more quickly. Catalysts do not change the final amount of product, so the volume of gas at the end is the same as before.

Questions

2 When a lump of calcium carbonate is placed into a beaker of acid it reacts and carbon dioxide is given off. Suggest three ways to speed up the reaction.

3 How is it possible to control conditions to speed up these changes:
   - a the setting of an epoxy glue
   - b the cooking of an egg
   - c the conversion of oxides of nitrogen in car exhausts to nitrogen.

4 When investigating the effect of temperature on a chemical reaction, why is it important to keep all other conditions the same?

5 The effect of concentration on the rate of reaction between zinc and sulfuric acid was investigated. The results were plotted on a graph.

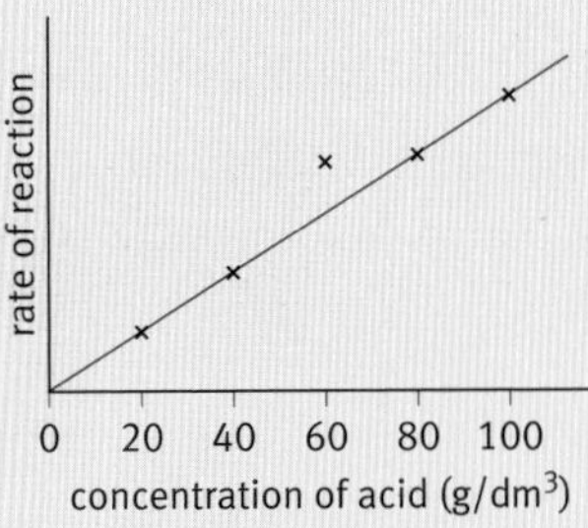

   - a Is there a correlation? If so, describe it.
   - b Which result is an outlier? Suggest a reason why this result is different from the expected value.

Catalysts in industry

What is a catalyst?

A catalyst is a chemical that speeds up a chemical reaction. It takes part in the reaction, but is not used up.

Modern catalysts can be highly selective. This is important when reactants can undergo more than one chemical reaction to give a mixture of products. With a suitable catalyst it can be possible to speed up the reaction that gives the required product, but not speed up other possible reactions that create unwanted by-products.

The manufacture of ethanoic acid from methanol and carbon monoxide uses a catalyst to speed up the reaction.

methanol	+	carbon monoxide	$\longrightarrow$	ethanoic acid
$CH_3OH(g)$	+	$CO(g)$	$\longrightarrow$	$CH_3COOH(g)$

Better catalysts

Catalysts are essential in many industrial processes. They make many processes economically viable. This means that chemical products can be made at a reasonable cost and sold at affordable prices.

Research into new catalysts is an important area of scientific work. This is shown by the industrial manufacture of ethanoic acid (see Section B) from methanol and carbon monoxide. This process was first developed by the company BASF in 1960 using a cobalt compound as the catalyst at 300 °C and at a pressure 700 times atmospheric pressure.

About six years later the company Monsanto developed a process using the same reaction, but a new catalyst system based on rhodium compounds. This ran under much milder conditions: 200 °C and 30–60 times atmospheric pressure.

In 1986, the petrochemical company BP bought the technology for making ethanoic acid from Monsanto. It has since devised a new catalyst based on compounds of iridium. This process is faster and more efficient. Iridium is cheaper, and less of the catalyst is needed. Iridium is even more selective so the amount of ethanoic acid produced (yield) is greater and there are fewer by-products. This makes it easier and cheaper to make pure ethanoic acid and there is less waste.

Collision theory

Chemists have a theory to explain how the various factors affect reaction rates.

The basic idea is that particles, such as molecules, atoms, and ions can only react if they bump into each other. Imagining these particles colliding with each other leads to a theory that can account for the effects of concentration, temperature, and catalysts on reaction rates.

According to **collision theory**, when molecules collide some bonds between atoms can break while new bonds form. This creates new molecules.

Molecules are in constant motion in gases, liquids, and solutions. There are millions upon millions of collisions every second. Most reactions would be explosive if every collision led to a reaction. It turns out that only a very small proportion of all the collisions are successful and actually lead to a reaction. These are the collisions in which the molecules are moving with enough energy to break bonds between atoms.

Any change that increases the number of successful collisions per second has the effect of increasing the rate of reaction. Increasing the concentration of solutions of dissolved chemicals increases the frequency of collisions. The same small proportion of these collisions will be successful, but now there are more of them. This means that there will also be more successful collisions.

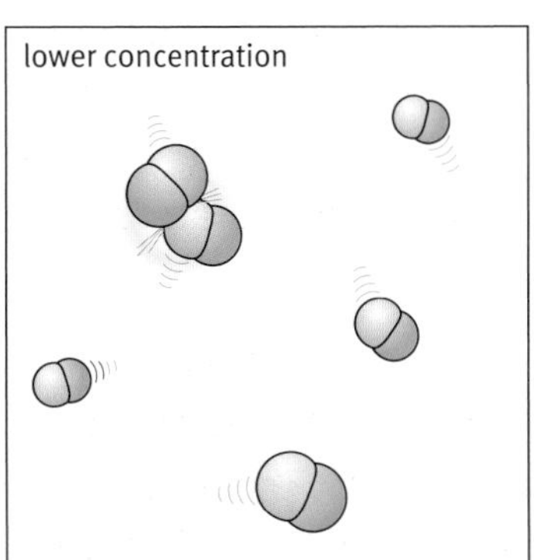

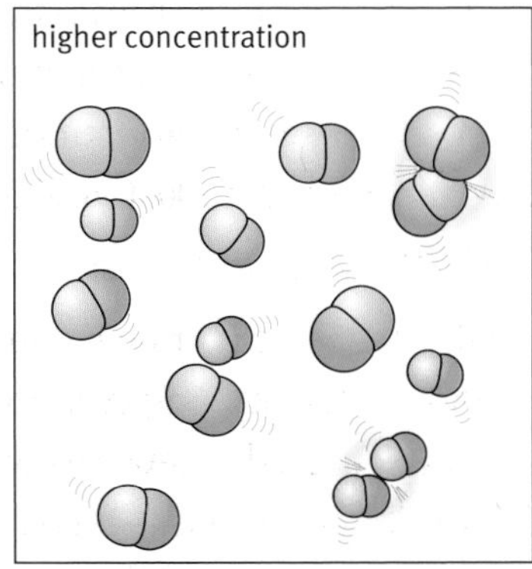

Molecules have a greater chance of colliding in a more concentrated solution. More frequent collisions means – faster reactions. Reactions get faster if the reactants are more concentrated.

Breaking up a solid into smaller pieces increases its surface area. Increased surface area means that there are more atoms, molecules, or ions of the solid available to react. This speeds up the reaction by increasing the frequency of successful collisions with particles in liquid, solution, or gas.

Key word

- collision theory

Questions

6 a Where do cobalt, rhodium and iridium appear in the periodic table?
  b Why is not surprising that these three metals can be used to make catalysts for the same process?

7 Suggest reasons why it is important to develop industrial processes that:
  a run at lower temperatures and pressures
  b produce less waste.

Stages in chemical synthesis

Find out about

- steps in synthesis
- making a soluble salt

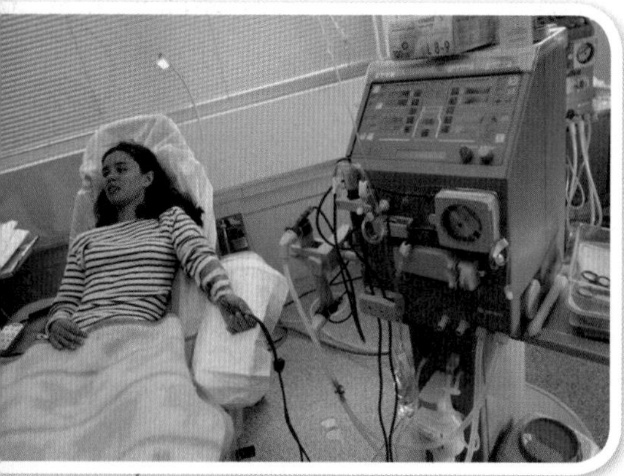

Kidney dialysis.

Making calcium chloride for dialysis

Chemical synthesis is a way of making new compounds. Synthesis puts things together to make something new. It is the opposite of analysis, which takes things apart to see what they are made of.

The kidneys remove toxic chemicals from the blood. In cases of kidney failure, patients are put on dialysis machines that do the job outside the body while they await a transplant. Blood passes out of the body through a tube into the dialysis machine.

Inside the machine, the blood flows past a special membrane. On the other side of the membrane is a solution containing a mixture of salts at the same concentrations as the same salts in the blood. The toxic chemicals pass from the blood through the membrane into the solution and are carried away. It is also possible for useful salts to pass back into the blood.

One of the salts in the dialysis solution is calcium chloride. It is a particularly important salt as the level of calcium in the blood has to be maintained at a particular level. Just a little bit too much or too little and the patient could become very ill indeed. The calcium chloride therefore has to be very pure, and the quantity added to the solution has to be measured accurately.

The process for making calcium chloride is shown in the flow chart.

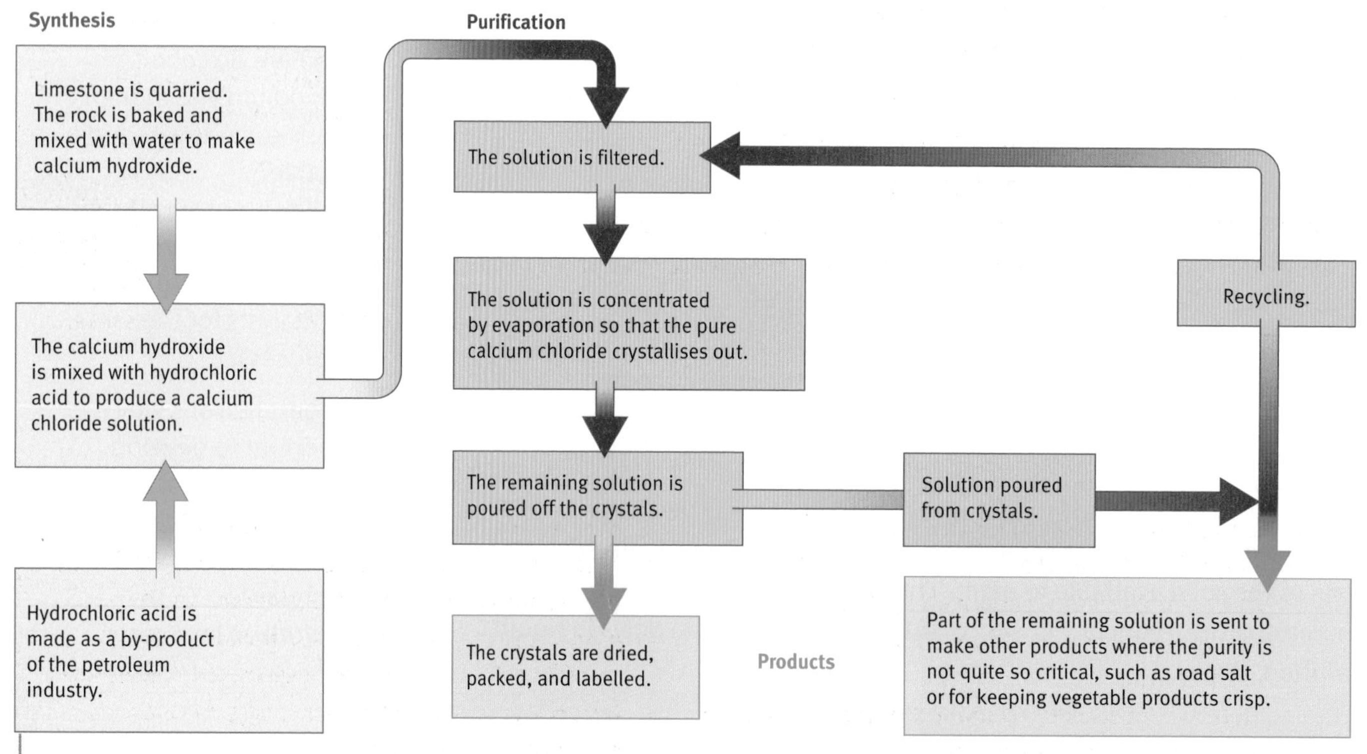

The process for making very pure calcium chloride for dialysis.

Making a sample of magnesium sulfate

Magnesium sulfate is another salt produced by the chemical industry. It has many uses including as a nutrient of plants.

The process of making magnesium sulfate (or any other soluble salt) on a laboratory scale illustrates the stages in a chemical synthesis.

In the following method an excess of solid is added to make sure that all the acid is used up. This method is only suitable if the solid added to the acid is either insoluble in water or does not react with water.

Choosing the reaction

Any of the characteristic reactions of acids can all be used to make salts:

- acid + metal ⟶ salt + hydrogen
- acid + metal oxide or hydroxide ⟶ salt + water
- acid + metal carbonate ⟶ salt + carbon dioxide + water

Magnesium metal is relatively expensive because it has to be extracted from one of its compounds. So it makes sense to use either magnesium oxide or carbonate as the starting point for making magnesium sulfate from sulfuric acid.

Carrying out a risk assessment

It is always important to minimise exposures to risk. You should take care to identify hazardous chemicals. You should also look for hazards arising from equipment or procedures. This is a **risk assessment**.

In this preparation the magnesium compounds are not hazardous. The dilute sulfuric acid is an irritant, which means that you should keep it off your skin and especially protect your eyes. You should always wear eye protection when handling chemicals, for example.

An operator emptying magnesium sulfate into the tank of a sprayer on a farm. He is wearing protective clothing because magnesium sulfate is harmful. As well as being a micronutrient needed for healthy plant growth, the salt is a:

- raw material in soaps and detergents
- laxative in medicine
- refreshing additive in bath water
- raw material in the manufacture of other magnesium compounds
- supplement in feed for poultry and cattle
- coagulant in the manufacture of some plastics.

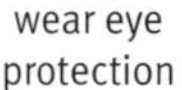

wear eye protection

harmful

Questions

1 Refer to the flow diagram for the process of making calcium chloride.
   a Write the word and balanced symbol equations for the reaction used to make the salt.
   b Identify steps taken to make the yield of the pure salt as large as possible.

2 Epsom salts consist of magnesium sulfate. Magnesium sulfate is soluble in water. Produce a flow diagram to show how you could remove impurities that are insoluble in water from a sample of Epsom salts. Processes you might use include: crystallisation, dissolving, drying, evaporation, filtration.

Questions

3 Write the balanced equation for the reaction of magnesium carbonate with sulfuric acid.

4 Why does the mixture of magnesium carbonate and sulfuric acid froth up?

5 What is the advantage of:
  a using powdered magnesium carbonate?
  b warming when most of the acid has been used up?
  c adding a slight excess of the solid to the acid?

6 Why is it impossible to use this method to make a pure metal sulfate by the reaction of dilute sulfuric acid with:
  a lithium metal?
  b sodium hydroxide?
  c potassium carbonate?

7 Look at the procedure on this and the following page for making magnesium sulfate, and identify the step when risks might arise from:
  a chemicals that react vigorously and spill over
  b chemicals that might spit or splash on heating
  c hot apparatus that might cause burns
  d apparatus that might crack and form sharp edges.

Working out the quantities to use

Reacting masses can be used to work out the amount of reactants needed to produce a particular amount of product (see Section H). In this procedure the solid is added in excess. This means that the amount of product is determined by the volume and concentration of the sulfuric acid. The concentration of dilute sulfuric acid is 98 g/litre, and a volume of 50 cm^3 dilute sulfuric acid is used. This contains 4.9 g of the acid.

Carrying out the reaction in suitable apparatus under the right conditions

The reaction is fast enough at room temperature, especially if the magnesium carbonate is supplied as a fine powder.

This reaction can be safely carried out in a beaker. Stirring with a glass rod makes sure that the magnesium oxide or carbonate and acid mix well. Stirring also helps to prevent the mixture frothing up and out of the beaker.

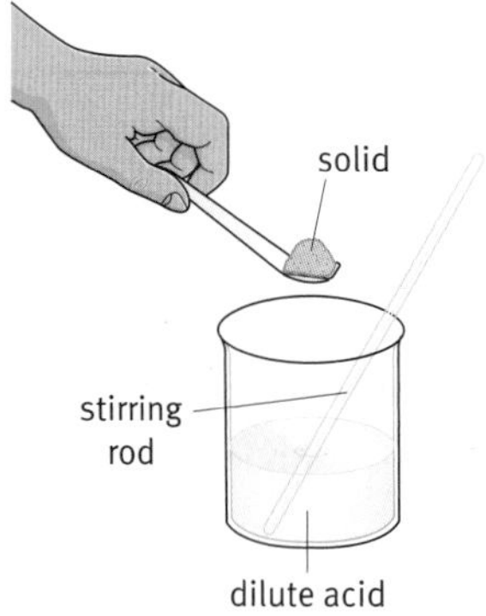

1 Measure the required volume of acid into a beaker. Add the metal oxide or carbonate bit by bit until no more dissolves in the acid. Warm gently until all of the acid has been used up. Make sure that there is a slight excess of solid before moving on to the next stage.

Separating the product from the reaction mixture

Filtering is a quick and easy way of separating the solution of the product from the excess solid. The mixture filters more quickly if the mixture is warm.

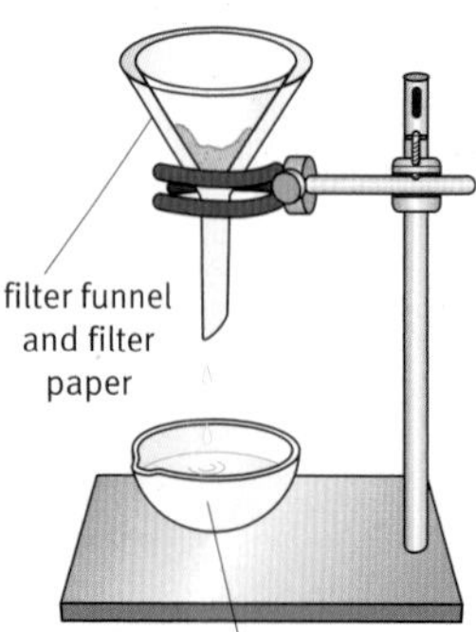

2 Filter off the excess solid, collecting the solution of the salt in an evaporating basin. The residue on the filter paper is the excess solid.

Key word

✓ **risk assessment**

Purifying the product

After the mixture has been filtered, the filtrate contains the pure salt dissolved in water. Evaporating much of the water speeds up crystallisation. This is conveniently carried out in an evaporating basin. The concentrated solution can then be left to cool and crystallise. The drying can be completed in a warm oven. The crystals can then be transferred to a desiccator. This is a closed container that contains a solid that absorbs water strongly.

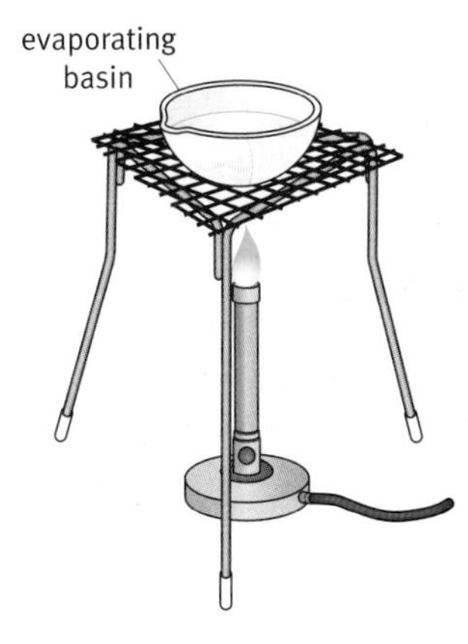

3 Heat gently to evaporate some of the water. Evaporate until crystals form when a droplet of solution picked up on a glass rod crystallises on cooling.

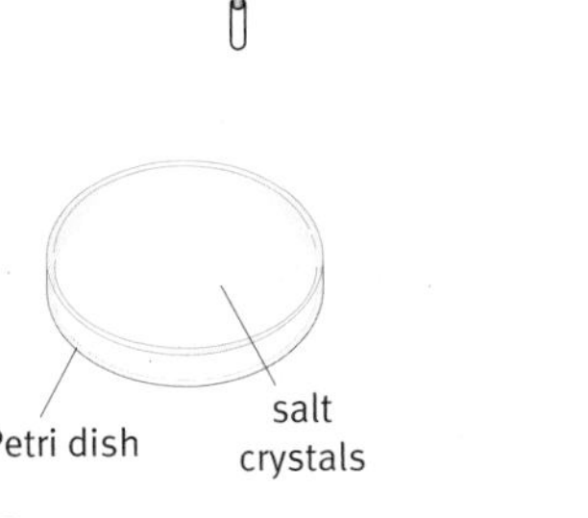

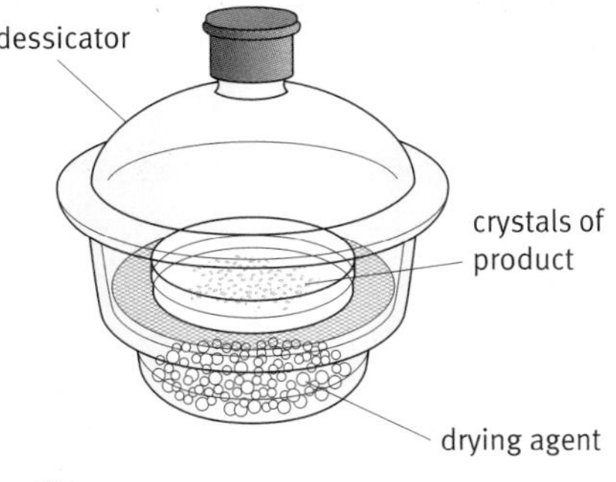

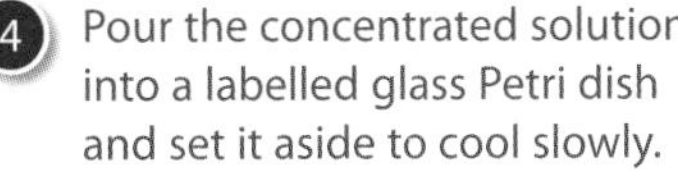

4 Pour the concentrated solution into a labelled glass Petri dish and set it aside to cool slowly.

5 Complete the drying in an oven and then store in a dessicator.

Crystals of pure magnesium sulfate seen through a Polaroid filter (×60).

Measuring the yield and checking the purity of the product

The final step is to transfer the dry crystals to a weighed sample tube and re-weigh it to find the actual yield of crystals. Often it is important to carry out tests to check that the product is pure.

The appearance of the crystals can give a clue to the purity of the product. A microscope can help if the crystals are small. The crystals of a pure product are often well-formed and even in shape.

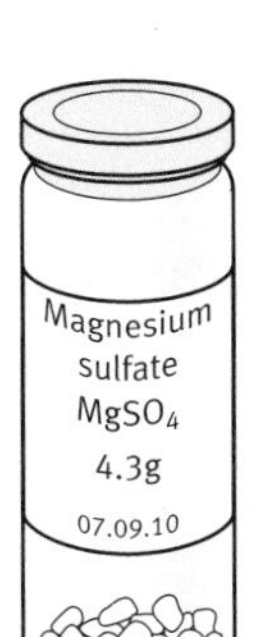

6 The weighed sample of product showing the name and formula of the chemical, the mass of product, and the date it was made.

Questions

8 Identify the impurities removed during the purification stages of the magnesium sulfate synthesis.

9 Why is it important that the magnesium carbonate is added in excess to the sulfuric acid?

H Chemical quantities

Find out about

- reacting masses
- yields from chemical reactions

Key words

- relative formula mass
- reacting mass
- actual yield
- theoretical yield
- percentage yield

Question

1 What mass of:
   a HCl in hydrochloric acid reacts with 100 g calcium carbonate?
   b HNO_3 in dilute nitric acid neutralises 56 g of potassium hydroxide?
   c copper(II) oxide (CuO) would you need to react with sulfuric acid (H_2SO_4) to make 319 g of copper sulfate ($CuSO_4$)?

Chemists wanting to make a certain quantity of product need to work out how much of the starting materials to order. Getting the sums right matters – especially in industry, where a higher yield for a lower price can mean better profits.

The trick is to turn the symbols in the balanced chemical equation into masses in grams or tonnes. This is possible given the relative masses of the atoms in the periodic table.

Reacting masses

Adding up the relative atomic masses for all the atoms in the formula of a compound gives the **relative formula mass** of chemicals . Given the relative formula masses, it is then possible to work out the masses of reactants and products in a balanced equation. These are the **reacting masses**.

RULES FOR WORKING OUT REACTING MASSES

STEP 1 Write down the balanced symbol equation.

STEP 2 Work out the relative formula mass of each reactant and product.

STEP 3 Write the relative reacting masses under the balanced equation, taking into account the numbers used to balance the equation.

STEP 4 Convert to reacting masses by adding the units (g, kg, or tonnes).

STEP 5 Scale the quantities to amounts actually used in the synthesis or experiment.

Worked example

What are the masses of reactants and products when sulfuric acid reacts with sodium hydroxide?

Step 1 $2NaOH + H_2SO_4 \longrightarrow Na_2SO_4 + 2H_2O$

Step 2 relative formula mass of NaOH = 23 + 16 + 1 = 40

relative formula mass of $H_2SO_4 = (2 \times 1) + 32 + (4 \times 16) = 98$

relative formula mass of $Na_2SO_4 = (2 \times 23) + 32 + (4 \times 16) = 142$

relative formula mass of $H_2O = (2 \times 1) + 16 = 18$

Steps 3 & 4	2NaOH	+	H_2SO_4	$\longrightarrow$	Na_2SO_4	+	$2H_2O$
	$2 \times 40 = 80$		98		142		$2 \times 18 = 36$
	80 g		98 g		142 g		36 g

Yields

The yield of any synthesis is the quantity of product obtained from known amounts of starting materials. The **actual yield** is the mass of product after it is separated from the mixture, purified, and dried.

Theoretical yield

The **theoretical yield** is the mass of product expected if the reaction goes exactly as shown in the balanced equation. This is what could be obtained in theory if there are no by-products and no losses while chemicals are transferred from one container to another. The actual yield is always less than the theoretical yield.

Worked example

What is the theoretical yield of ethanoic acid made from 8 tonnes of methanol?

Step 1 *Write down the balanced equation*

methanol + carbon monoxide ⟶ ethanoic acid

$CH_3OH(g) + CO(g) \longrightarrow CH_3COOH(g)$

32 → 60

Step 2 *Work out the relative formula masses*

methanol: $12 + 4 + 16 = 32$

ethanoic acid: $24 + 4 + 32 = 60$

Steps 3 & 4 *Write down the relative reacting masses and convert to reacting masses by adding the units*

Theoretically, 32 tonnes of methanol should give 60 tonnes of ethanoic acid.

Step 5 *Scale to the quantities actually used*

If the theoretical yield of ethanoic acid = *x* tonnes, then

$$\frac{\text{mass of ethanoic acid}}{\text{mass of methanol}} = \frac{60\text{ tonnes}}{32\text{ tonnes}} = \frac{x\text{ tonnes}}{8\text{ tonnes}}$$

So, the yield of ethanoic acid from 8 tonnes of methanol should be

$$8\text{ tonnes} \times \frac{60\text{ tonnes}}{32\text{ tonnes}} = 15\text{ tonnes}$$

Percentage yield

The **percentage yield** is the percentage of the theoretical yield that is actually obtained. It is always less than 100%.

Questions

2 What is the mass of salt that forms in solution when:

- a hydrochloric acid neutralises 4 g sodium hydroxide?
- b 12.5 g zinc carbonate, $ZnCO_3$, reacts with excess sulfuric acid?

3 A preparation of sodium sulfate began with 8.0 g of sodium hydroxide.

- a Calculate the theoretical yield of sodium sulfate from 8.0 g sodium hydroxide.
- b Calculate the percentage yield, given that the actual yield was 12.0 g.

Worked example

What is the percentage yield if 8 tonnes of methanol produces 14.7 tonnes of ethanoic acid?

From the previous example:

theoretical yield = 15 tonnes
actual yield = 14.7 tonnes

$$\text{percentage yield} = \frac{\text{actual yield}}{\text{theoretical yield}} \times 100$$

$$= \frac{14.7\text{ tonnes}}{15\text{ tonnes}} \times 100$$

$$= 98\%$$

Science Explanations

Chemists use their knowledge of chemical reactions to plan and carry out the synthesis of new compounds.

You should know:

- that the chemical industry provides useful products such as food additives, fertilisers, dyestuffs, paints, pigments, and pharmaceuticals
- how chemists use indicators and pH meters to detect acids and alkalis and to measure pH
- that common alkalis include the hydroxides of sodium, potassium, and calcium
- that there are characteristic reactions of acids with metals, metal oxides, metal hydroxides, and metal carbonates that produce salts
- how chemists use ionic theory to explain why acids have similar properties, and how alkalis neutralise acids to form salts
- that safety precautions are important when working with hazardous chemicals such as corrosive acids and alkalis
- how to follow the rate of a change by measuring the disappearance of a reactant or the formation of a product and then to analyse the results graphically
- that the concentrations of reactants, the particle size of solid reactants, the temperature, and the presence of catalysts are factors that affect the rates of reaction
- how collision theory can explain why changing the concentration of reactants, or the particle size of solids, affects the rate of a reaction
- that reactions are exothermic if they give out energy, and endothermic if they take in energy
- that a chemical synthesis involves a number of stages and a range of practical techniques, which are important to achieving a good yield of a pure product in a safe way
- that a titration is a procedure that can be used to check the purity of chemicals used in synthesis
- how to use the balanced equation for a reaction to work out the quantities of chemicals to use in a synthesis, and to calculate the theoretical yield.

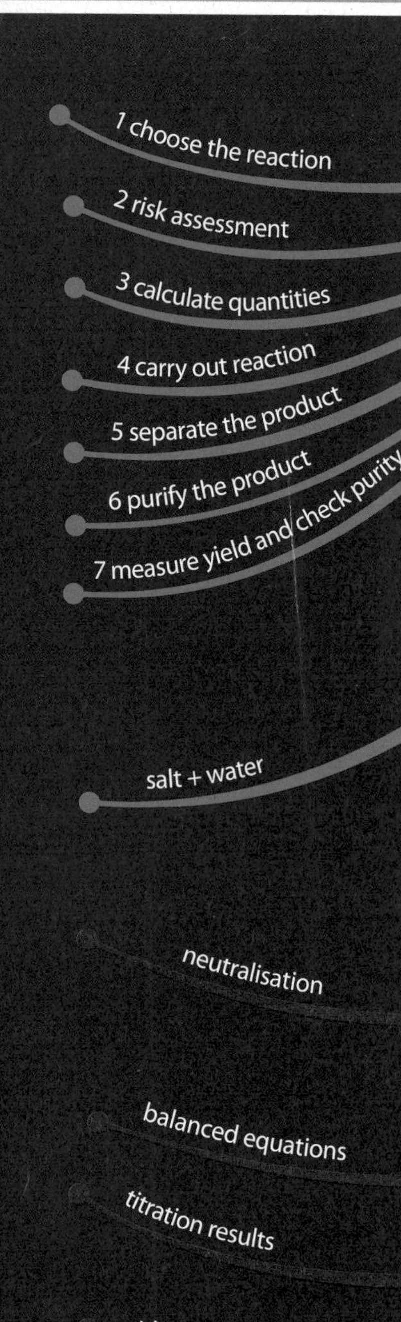

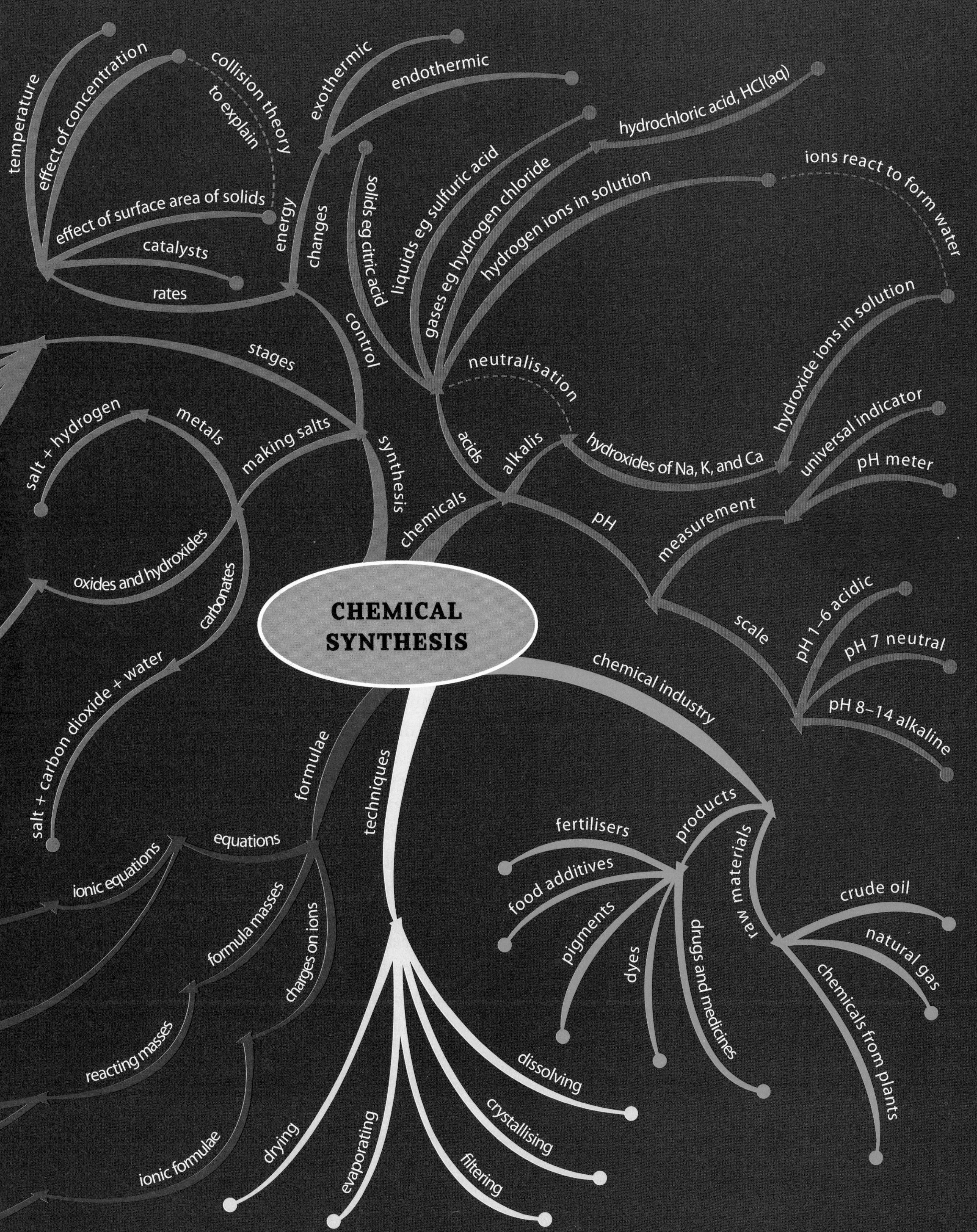
CHEMICAL SYNTHESIS
temperature
effect of concentration
collision theory
to explain
effect of surface area of solids
catalysts
rates
energy
changes
exothermic
endothermic
control
stages
making salts
metals
salt + hydrogen
oxides and hydroxides
carbonates
salt + carbon dioxide + water
synthesis
chemicals
acids
solids eg citric acid
liquids eg sulfuric acid
gases eg hydrogen chloride
hydrogen ions in solution
hydrochloric acid, HCl(aq)
ions react to form water
neutralisation
alkalis
hydroxides of Na, K, and Ca
hydroxide ions in solution
pH
measurement
universal indicator
pH meter
scale
pH 1–6 acidic
pH 7 neutral
pH 8–14 alkaline
chemical industry
products
fertilisers
food additives
pigments
dyes
drugs and medicines
raw materials
crude oil
natural gas
chemicals from plants
formulae
equations
ionic equations
formula masses
charges on ions
reacting masses
ionic formulae
techniques
drying
evaporating
filtering
crystallising
dissolving

Ideas about Science

Scientists can never be sure that a measurement tells them the true value of the quantity being measured. Data is more reliable if it can be repeated. If you take several measurements of the same quantity, for example, in a titration, the results are likely to vary. This may be because:

- the quantity you are measuring is varying, for example, the purity of separate solid samples of a product may not be uniform
- there are variations in the judgement of when the indicator colour corresponds to the end-point, or of the position of a meniscus
- there are limitations in the measuring equipment, such as burettes.

Usually the best estimate of the true value of a quantity is the mean (or average) of several repeat measurements. You should:

- be able to calculate the mean from a set of repeat measurements
- know that a measurement may be an outlier if it lies well outside the range of the other values in a set of repeat measurements
- be able to give reasons to explain why an outlier should be retained as part of the data or rejected when calculating the mean.

When comparing sets of titration results you should know that:

- a difference between their means is real if their ranges do not overlap.

To investigate the relationship between a factor and an outcome, it is important to control all the other factors that might affect the outcome, such as temperature and the concentrations of other reactants. When investigating the rates of chemical reactions you should be able to:

- identify the outcome and factors that may affect it
- suggest how an outcome might alter when a factor is changed.

In a plan for an investigation of the effect of a factor on an outcome, you should:

- be able to explain why it is necessary to control all the factors that might affect the outcome
- recognize that the fact that other factors are controlled is a positive design feature, and the fact that they are not is a design flaw.

If an outcome occurs when a specific factor is present but does not when it is absent, or if an outcome variable increases (or decreases) steadily as an input variable increases, we say that there is a correlation between the two. In the context of studying reaction rates you should:

- understand that a correlation shows a link between a factor and an outcome
- be able to identify where a correlation exists when data is presented as text, as a graph, or in a table
- understand that a correlation does not always mean that the factor causes the outcome.
- identify that where there is a machanism to explain a correlation, scientists are likely to accept that the factor causes the outcome.

Review Questions

1 A student is developing a new sports pack. She needs to find two chemicals that will cool an injury when mixed inside the pack.

a Should the reaction be endothermic or exothermic?

b Which of the energy-level diagrams below represents the reaction that is likely to cool injuries more effectively?

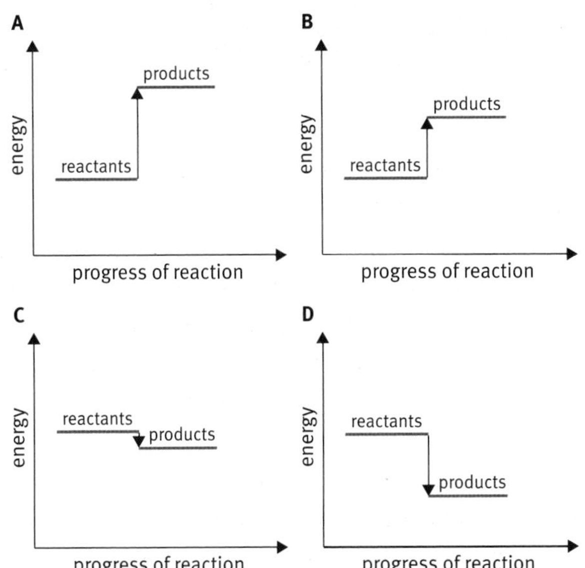

2 A chemist tested the purity of a sample of tartaric acid by titrating with dilute sodium hydroxide. The table shows his results.

	Run 1	Run 2	Run 3	Run 4
Initial burette reading (cm^3)	0.1	0.1	25.0	21.9
Final burette reading (cm^3)	25.2	25.0	50.0	46.9
Volume of acid added (cm^3)	25.1	24.9	25.0	

a Explain why the chemist repeated the titration four times.

b Calculate the volume of alkali added in run 4.

c The chemist calculates the mean of his titration results. Why does he do this?

d Suggest the range within which the true value probably lies.

3 A teacher adds 3 g of zinc granules to dilute hydrochloric acid in a flask. She uses a gas syringe to measure the volume of hydrogen gas given off.

Time (min)	Volume of gas in syringe (cm^3)
1	32
2	56
3	74
4	87
5	95
6	95

a Draw a line graph of the results.

b Calculate the average rate of reaction between three and four minutes, in cm^3 per minute.

c The teacher decides to repeat the experiment with 3 g of zinc powder. She keeps all other factors the same. Draw and label a line on the graph to show the expected results. Use collision theory to explain the effect of this change.

d Explain why it was important for the teacher to keep all other factors the same.

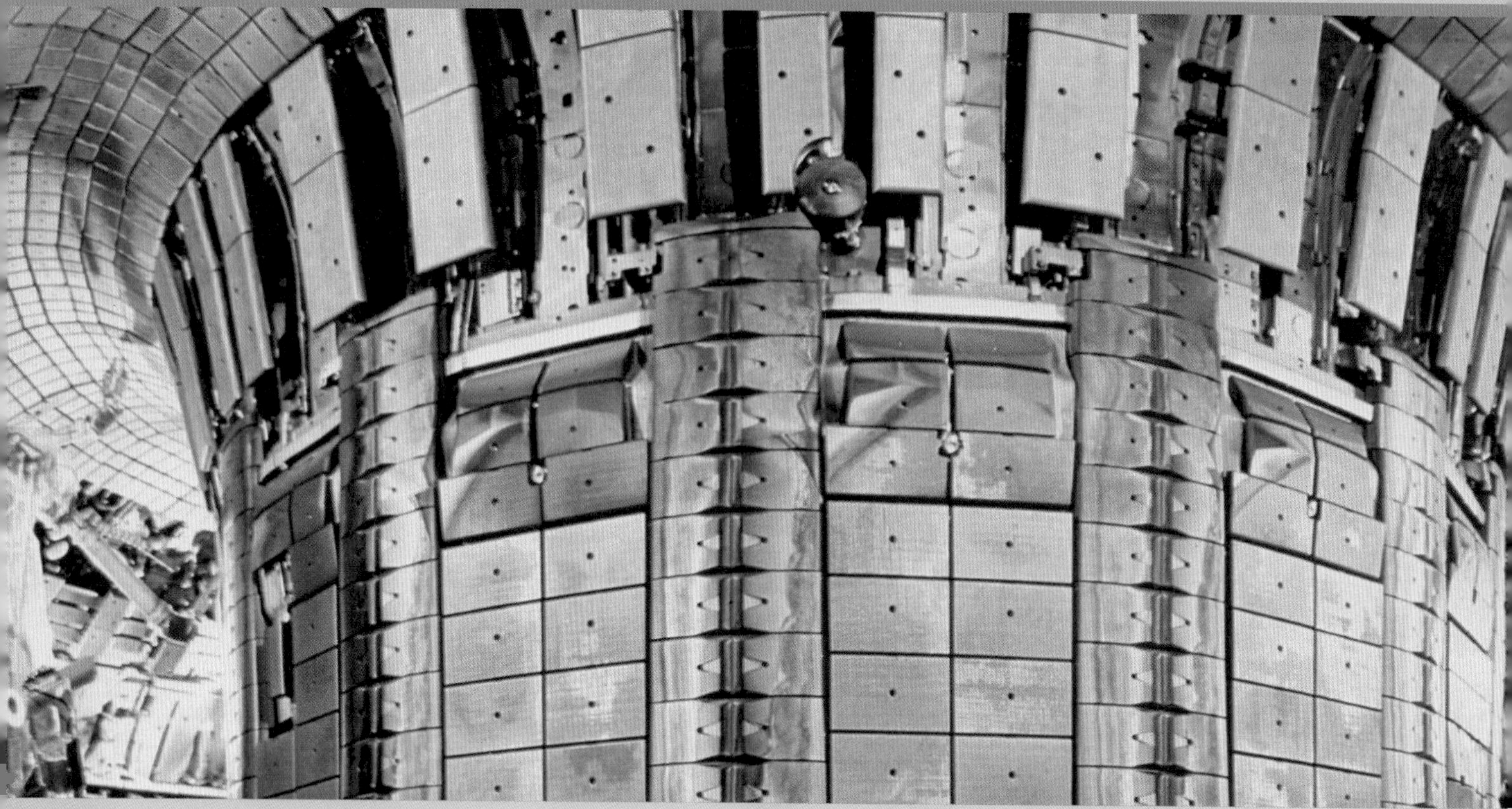

P6 Radioactive materials

Why study radioactive materials?

People make jokes about radioactivity. If you visit a nuclear power station, or if you have hospital treatment with radiation, they may say you will 'glow in the dark'. People may worry about radioactivity when they don't need to. Most of us take electricity for granted. But today's power stations are becoming old and soon will need replacement. Should nuclear power stations be built as replacements? Should we research nuclear fusion as a long-term energy solution?

What you already know

- Some materials are radioactive, and naturally emit gamma rays.
- Gamma rays are ionising radiation.
- Ionising radiation can damage living cells.
- Nuclear power stations produce radioactive waste.
- Contamination by a radioactive material is more dangerous than a short period of irradiation.

Find out about

- radioactive materials and emissions
- radioactive materials being used to treat cancer
- ways of reducing risks from radioactive materials
- nuclear power stations
- nuclear fusion research.

The Science

The discovery of radioactivity changed ideas about matter and atoms. The nuclear model of the atom helped scientists explain many observations – including radioactivity and the colour of stars. Having a model for how the atom behaves enables scientists to make and test predictions, including nuclear fission and nuclear fussion.

Ideas about Science

Radioactive materials are used in many applications. Making decisions about how they are used requires decisions to be made about the balance of risk and benefit, both to the individual but also to the wider community.

A Radioactive materials

Find out about

- radioactive decay
- what makes an atom radioactive
- types of radiation

In this research the scientist is measuring the radioactivity of the fruit from plants grown with radioactive water.

Key words

- radioactive
- ionising radiation
- radioactive decay
- nucleus
- unstable
- alpha particles
- beta particles
- gamma radiation

What do these elements have in common?

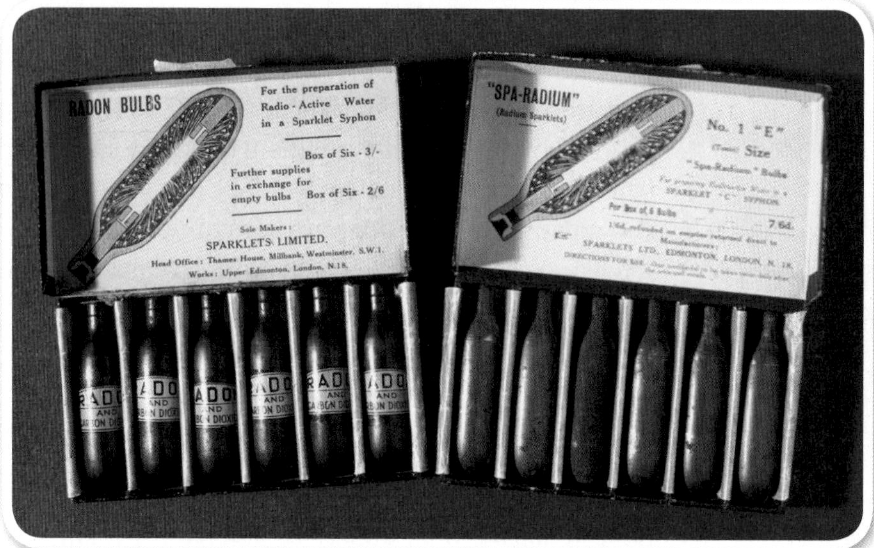

Radon is a radioactive gas. Radium is a radioactive metal. In the early 1900s, these bulbs were used to make drinking water radioactive.

Uranium ore – uranium is used as a fuel in nuclear power stations.

They are all **radioactive**. If you test them with a Geiger counter you will hear it click.

When radioactivity was first discovered, people did not know that the radiation was **ionising** and could damage or kill living cells. They thought that it was natural and healthy. Manufacturers made all kinds of products using radioactive materials. When scientists realised the danger the products were banned. Safety rules for using radioactive materials were introduced.

Radioactive elements can be naturally occurring, or they can be man-made. The man-made elements may be produced because they are useful. For example, the radioactive hydrogen in the water molecules is used to water the plants in the photograph. Or they may be a waste product, like waste from nuclear power stations.

Changes inside the atom

Many elements have more than one type of atom. For example, some carbon atoms are radioactive. In most ways they are identical to other carbon atoms. All can:

- be part of coal, diamond, or graphite
- burn to form carbon dioxide
- be a part of complex molecules.

A cut diamond sitting on a lump of coal. Each of these is made of carbon atoms. Some of the atoms will be radioactive.

Radioactive decay

The main difference is that most carbon atoms do not change. They are stable.

Radioactive carbon atoms randomly give out energetic radiation. Each atom does it only once. And what is left afterwards is not carbon, but a different element . The process is called **radioactive decay**. It is not a chemical change; it is a change *inside* the atom.

What makes an atom radioactive?

Atoms have a tiny core called the **nucleus**. In some atoms, the nucleus, is **unstable**. The atom decays to become more stable. It emits energetic radiation and the nucleus changes. This is why the word 'nuclear' appears in *nuclear reactor*, *nuclear medicine*, and *nuclear weapon*.

Three types of radiation are emitted, called alpha, beta, and gamma.

Radiation	What it is
alpha particles (α)	small, high-speed particle with + charge
beta particles (β)	smaller, higher-speed particle with − charge
gamma radiation (γ)	high-energy electromagnetic radiation

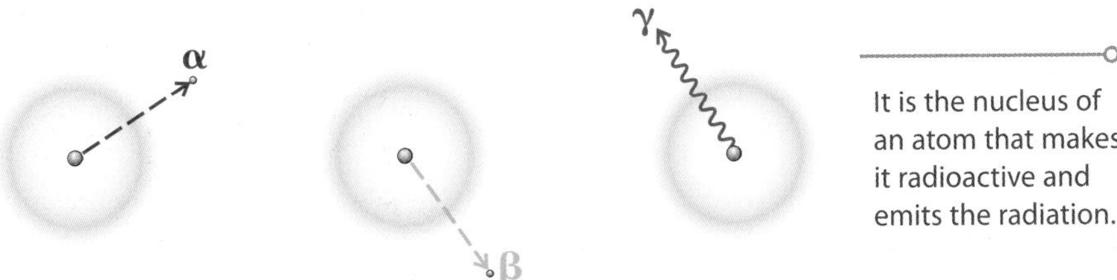

It is the nucleus of an atom that makes it radioactive and emits the radiation.

Radioactive decay happens in the nucleus of an atom. It is not affected by any physical or chemical changes that happen to the atom.

Making gold

When radioactive platinum decays it turns into a new element – gold. A good way to make money?
No. The price of gold is only half the price of platinum.

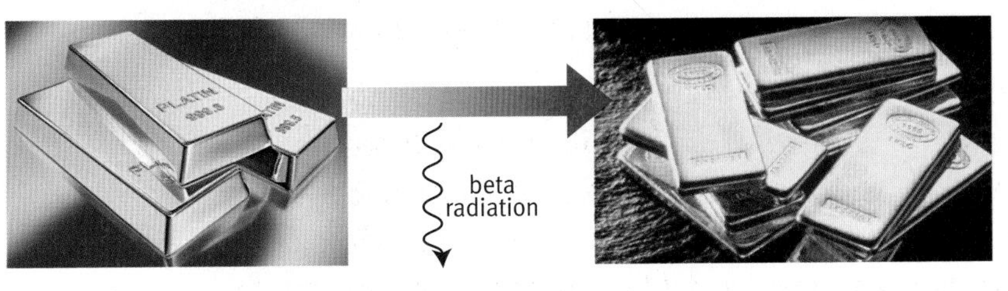

Questions

1 How can you test to see if something is radioactive?

2 Why is ionising radiation dangerous?

3 What part of the atom does the radiation come from?

4 What are the three different types of radiation from radioactive materials?

Find out about

- models of the atom
- how alpha particle scattering reveals the existence of the atomic nucleus

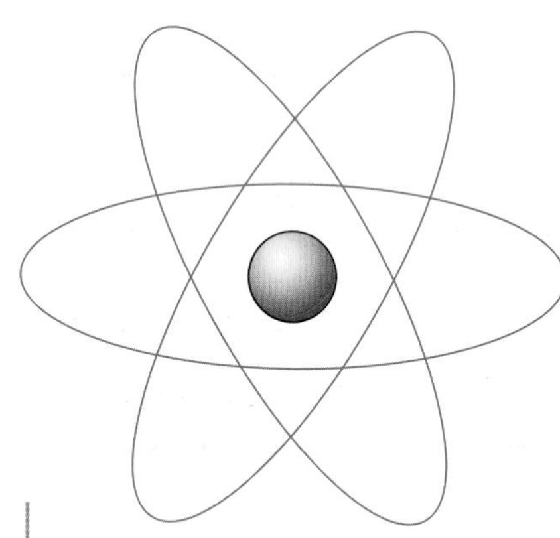

The 'solar system' model of the atom.

How do scientists know about the structure of atoms?

The diagram of an atom as a miniature 'solar system', with the nucleus at the centre and electrons whizzing round like miniature planets, has become very familiar. It is often used just to suggest that something is 'scientific'.

The 'solar system' model of the atom dates back to 1910, and an experiment thought up by Ernest Rutherford. Scientists were beginning to understand radioactivity, and were experimenting with radiation. Rutherford realised that alpha and beta particles were smaller than atoms, and so they might be useful tools for probing the structure of atoms. So he designed a suitable experiment, and it was carried out by his assistants, Hans Geiger and Ernest Marsden.

Here is how to do it:

- Start with a metal foil. Use gold, because it can be rolled out very thin, to a thickness of just a few atoms.
- Direct a source of alpha radiation at the foil. Do this in a vacuum chamber, so that the alpha particles are not absorbed by air.
- Watch for flashes of light as the alpha particles strike the detecting material on the screen at the end of the microscope.
- Work all night, counting the flashes at different angles, to see how much the alpha radiation is deflected.

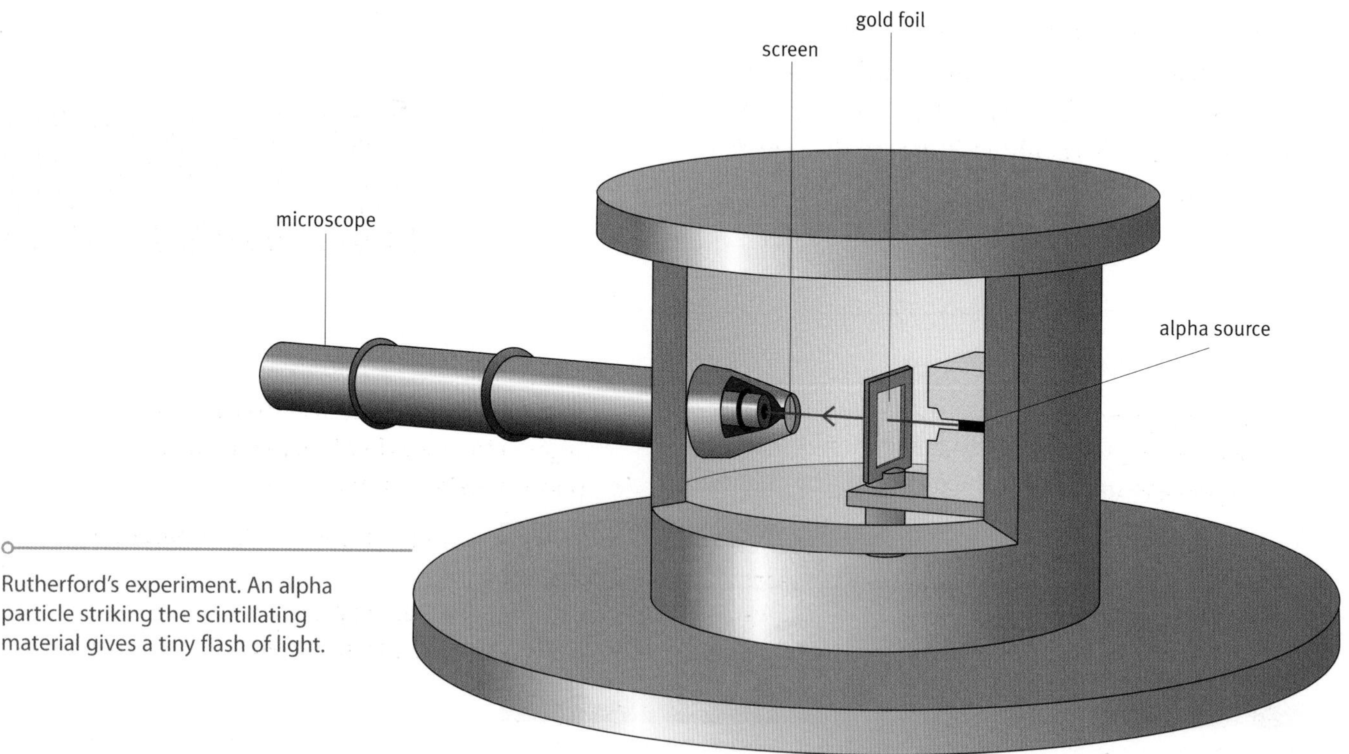

Rutherford's experiment. An alpha particle striking the scintillating material gives a tiny flash of light.

Results and interpretation

This is what Geiger and Marsden observed:

- Most of the alpha particles passed straight through the gold foil, deflected by no more than a few degrees.
- A small fraction of the alpha particles were actually reflected back towards the direction from which they had come.

And here is what Rutherford said:
'It was as if, on firing a bullet at a sheet of tissue paper, the bullet were to bounce back at you!'

In fact, less than 1 alpha particle in 8000 was back-scattered (deflected through an angle greater than 90°), but it still needed an explanation.

Rutherford realised that there must be something with positive charge that was repelling the alpha particles (which also have positive charge). And it must also have a lot of mass, or the alpha particles would just push it out of the way.

This 'something' is the nucleus of a gold atom. It contains all of the positive charge within the atom, and most of the mass. Rutherford's nuclear model is a good example of a scientist using creative thinking to develop an explanation of the data.

His analysis of his data showed that the nucleus was very tiny, because most alpha particles flew straight past without being affected by it. The diameter of the nucleus of an atom is roughly a hundred-thousandth of the diameter of the atom.

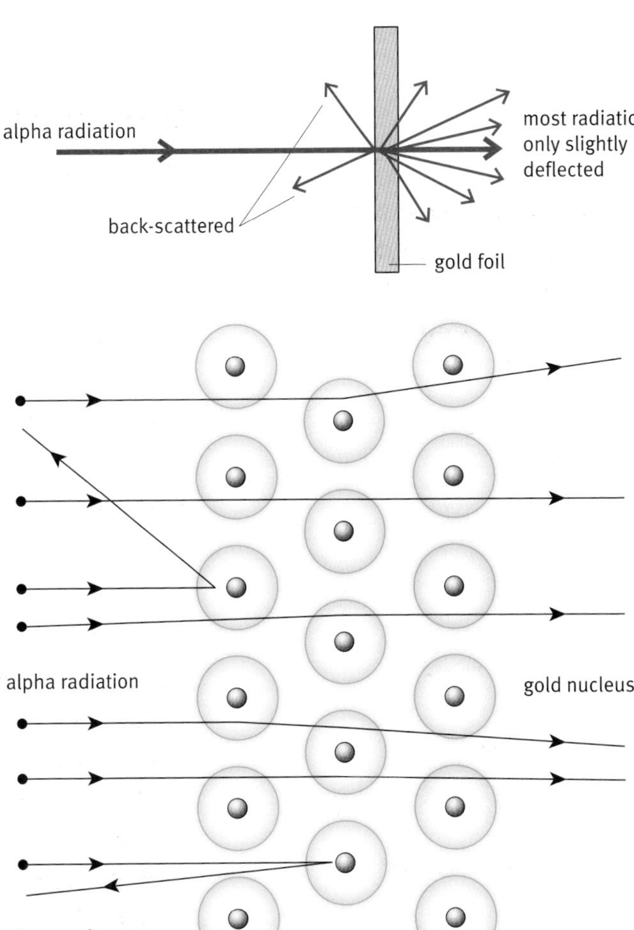

Only alpha particles passing close to a nucleus are significantly deflected.

Key word

- alpha particle scattering

Questions

1 What charge do the following have:
  a the atomic nucleus?
  b alpha radiation?
  c electrons?

2 Put these in order, from least mass to greatest: gold atom, alpha particle, gold nucleus, electron.

3 Describe and explain what happened to alpha particles that were directed:
  a straight towards a gold nucleus
  b slightly to one side of a gold nucleus
  c midway between two nuclei.

4 Suggest how the results would be different if the nucleus:
  a was positive but very large
  b was negative and very small.

C Inside the atom

Find out about

- isotopes
- protons and neutrons
- α and β particles

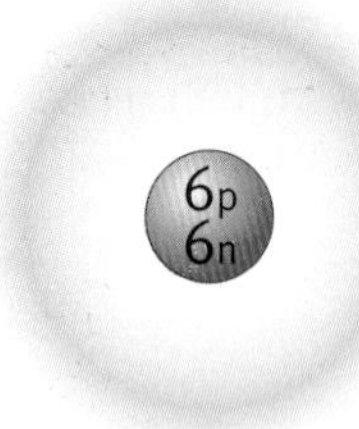

carbon-12

carbon-11

The nucleus of carbon-12 has 6 protons and 6 neutrons. Carbon-11 has 11 particles in its nucleus: 6 protons and 5 neutrons.

Atoms are small – about a ten millionth of a millimetre across. Their outer layer is made of electrons. Most of their mass is concentrated in a tiny core, called a nucleus.

Isotopes

The tiny nucleus contains two types of particle: **protons** and **neutrons**. All atoms of any element have the same number of protons. For example, carbon atoms always have six protons. But they can have different numbers of neutrons and still be carbon. The word **isotope** is used to describe different atoms of the same element. Carbon-11 and carbon-12 are different isotopes of carbon.

Carbon-11 will give out its radiation whether it is in diamond, coal, or graphite. You can burn it or vaporise it and it will still be radioactive.

Compared to the whole atom, the tiny nucleus is like a pinhead in a stadium.

Question

1 Look at these isotopes: carbon-11, boron-11, carbon-12, nitrogen-12.
   a Which two are isotopes of the same element?
   b Which ones have the same number of particles in the nucleus?
   c Do any of them have identical nuclei?
   d A nucleus of carbon-14 has i how many protons? ii how many neutrons?

Describing a nucleus

Scientists use a formula for describing an isotope.
In front of the chemical symbol are written the **proton number** and the total number of particles (protons + neutrons) in the nucleus.

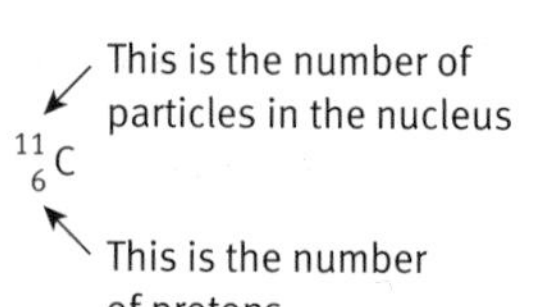

The number of neutrons can be found by taking the number of protons away from the total number of particles in the nucleus.
Neutrons = 11 − 6 = 5

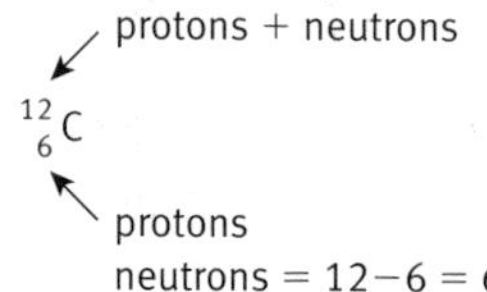

Radioactive changes

Some nuclei that are **unstable** can become more stable by emitting an alpha particle. An alpha particle is made of two protons and two neutrons – it is the same as a helium nucleus.

Other nuclei can become more stable when a neutron decays to form a proton. It does this by emitting a beta particle. A beta particle is the same as an electron, but it has come from a neutron in the nucleus. It is not one of the atom's orbital electrons.

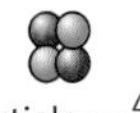

α particle = $^{4}_{2}He$

An alpha particle has two protons and two neutrons.

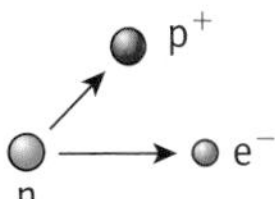

β particle

A neutron decays to a proton and an electron. The electron is a beta particle.

Counting the particles

Scientists use nuclear equations to work out what happens during radioactive decay.

Alpha decay

When plutonium-240 decays uranium-236 is formed.

$$^{240}_{94}Pu \longrightarrow {}^{336}_{92}U + {}^{4}_{2}\alpha$$

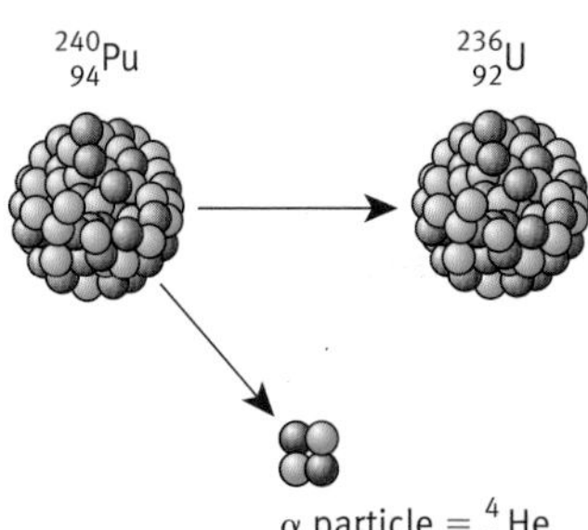

α particle = $^{4}_{2}He$

During alpha decay the nucleus loses two protons and two neutrons.

Beta decay

When carbon-14 decays nitrogen-14 is formed. Nitrogen always has 7 protons in the nucleus.

$$^{14}_{6}C \longrightarrow {}^{14}_{7}N + {}^{0}_{-1}\beta$$

The emission of either an alpha or a beta particle from an unstable nucleus produces an atom of a different element, called a 'daughter product' or 'decay product'. The daughter product may itself be unstable. There may be a series of changes, but eventually a stable end-element is formed.

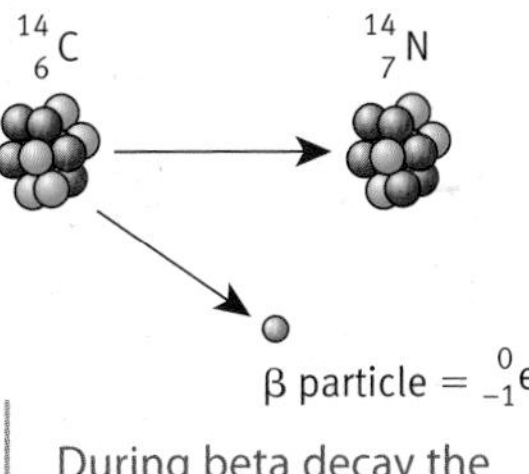

β particle = $^{0}_{-1}e$

During beta decay the proton number increases by one.

Isotope	Number of protons	Radioactive?
Hydrogen-1	1	
Hydrogen-2	1	
Hydrogen-3	1	✓
Helium-3	2	✓
Helium-4	2	
Uranium-235	92	✓
Thorium-231	90	✓

Questions

2 Copy the table of isotopes and add three extra columns for:
- the total number of particles in the nucleus
- the number of neutrons in the isotope
- the chemical symbol for the isotope.

Add the information for each of these columns to complete the table.

3 Hydrogen-3 decays to helium-3 by beta decay. Write a nuclear equation for the decay.

4 Uranium-235 decays to thorium-231 by alpha decay. Write a nuclear equation for the decay.

5 Nitrogen-16 decays by beta decay. Use your answer to question three to write a nuclear equation for the decay.

Key words

- ✓ isotope
- ✓ proton
- ✓ neutron
- ✓ unstable
- ✓ proton number

D Using radioactive isotopes

Find out about

- alpha, beta, and gamma radiation
- sterilisation using ionising radiation

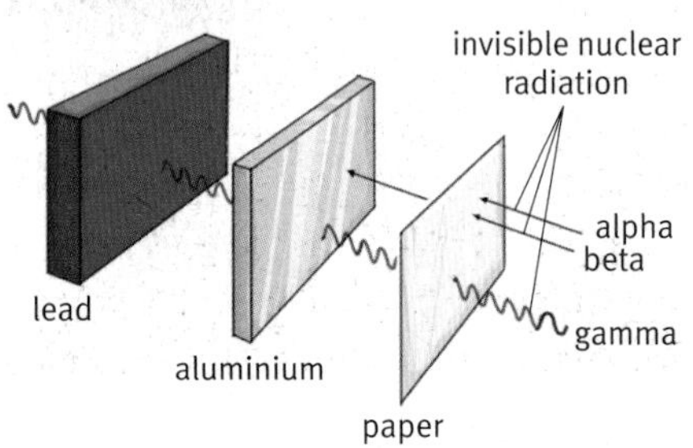

Radioactive isotopes have many uses, but they are quite rare in nature – because most of them have decayed – so radioactive isotopes are made in nuclear reactors, or in accelerators, and are prepared for use in laboratories and hospitals around the country.

Alpha, beta, or gamma?

A beam of alpha particles may be the best choice for the job, or it might be beta particles, or gamma rays. To decide, scientists consider the properties described below.

Alpha radiation

Alpha particles are much heavier than beta particles, and they quickly collide with air molecules and slow down. They gain electrons and become atoms of helium. This means that they are the least penetrating, but also the most strongly ionising radiation. They are stopped most easily.

Beta radiation

Beta particles are fast moving electrons. They are much smaller than alpha particles so less likely to collide with other particles. This means they travel further in air and other materials and are less ionising. When they have slowed down they are just like any other electrons.

Gamma radiation

Sometimes, after a nucleus emits a beta particle, the protons and neutrons remaining in the new nucleus rearrange themselves to a lower energy state. When this happens the nucleus emits a photon of electromagnetic radiation called a gamma ray. This does not cause a change of element. The photons have more energy than most X-ray photons and rarely collide with particles, so the radiation is very penetrating. It has only a very weak ionising effect.

Radiation	Range in air	Stopped by	Ionisation	Charge
alpha	a few cm	paper / dead skin cells	strong	+
beta	10–15 cm	thin aluminium	weak	–
gamma	many metres	very thick lead or several metres of concrete	very weak	no charge

Properties of ionising radiation.

Question

1 Which type of radiation:
  - a is the most penetrating?
  - b is the most ionising?
  - c has the longest range in air?

Sterilisation

Ionising radiations can kill bacteria. Gamma radiation is used for **sterilising** surgical instruments and some hygiene products such as tampons. The products are first sealed from the air and then exposed to the radiation. This passes through the sealed packet and kills the bacteria inside.

Food can be treated in the same way. Irradiating food kills bacteria and prevents spoilage. As of 2010, **irradiation** is permitted in the UK only for herbs and spices. But the label must show that they have been treated with ionising radiation. This is a useful alternative to heating or drying, because it does not affect the taste.

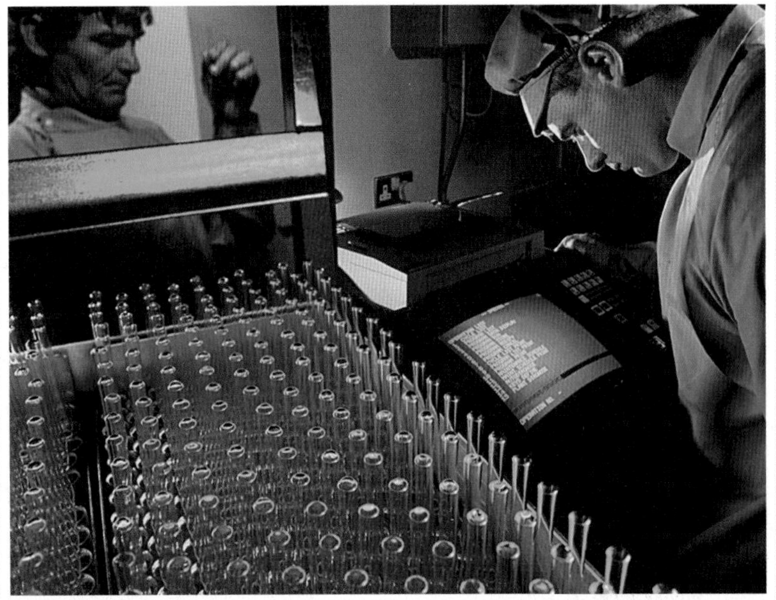

Gamma rays kill the bacteria on and inside these test tubes.

Questions

2 a Why seal the packets of surgical instruments before sterilising them?
  b Does the gamma radiation make them radioactive? Explain your answer.

3 Smoke detectors used in homes contain a source that emits alpha particles.
  a Explain why these are not dangerous in normal use.
  b What might make them dangerous?

4 A radioactive source is tested by placing sheets of material between it and a Geiger counter 2 cm away.

Sheet added	Count rate (counts per second)
None	6.8
Paper	4.9
3-mm-thick aluminium	4.7
3-cm-thick lead	0.5

Which type, or types, of radiation does the source emit? Explain your reasoning.

Food irradiation is it safe?

The logo shows that the herbs and spices have been **irradiated** with gamma radiation from Cobalt–60. Gamma rays pass through the glass and kill any bacteria in the jar. Cobalt–60 does not pass in to the jar– there is *no* **contamination.**

Uses of ionising radiation are linked to their properties.

Key words

- irradiation
- sterilisation

Radiation all around

Find out about

- background radiation
- a radioactive gas called radon
- radiation dose and risk

Radiation sources

If you switch on a Geiger counter, you will hear it click. It is picking up **background radiation**, which is all around you. Most background radiation comes from natural sources.

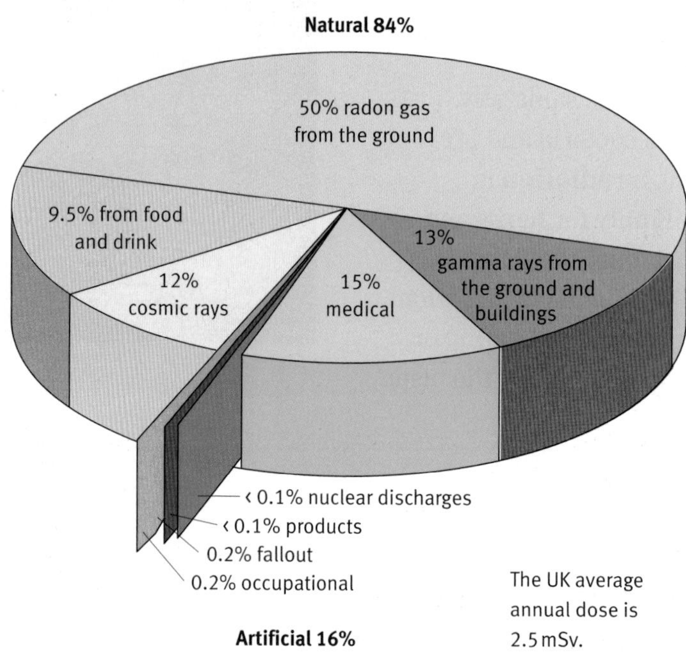

The UK average annual dose is 2.5 mSv.

How different sources contribute to the average **radiation dose** in the UK. Source HSE.

Radiation dose

Radiation dose measures the possible harm the radiation could do to the body. It is measured in millisieverts (mSv). The UK average annual dose is 2.5 mSv. For comparison, with a dose of 1000 mSv (400 times larger) three out of a hundred people, on average, develop a cancer.

Ionising radiation from outer space is called cosmic radiation. Flying to Australia gives you a dose of 0.1mSv, from cosmic rays. That's not much if you go on holiday, but it soon adds up for flight crews making repeated journeys.

What affects radiation dose?

The dose measures the potential harm done by the radiation. It depends on:

- the amount of radiation – the number of alpha particles, beta particles, or gamma photons reaching the body
- the type of radiation. Alpha is the most ionising of the three radiations. So it can cause the most damage to a cell. The same amount of alpha radiation gives a bigger dose than beta or gamma radiation. But because alpha radiation has such a short range in air, it is only a hazard if the source of radiation gets inside the body.

Questions

1. a In what units is radiation dose measured?
   b Make a reasoned estimate for the annual dose of a long-haul airline pilot.
2. On what two factors does radiation dose depend?

Radiation	Dose factor
alpha	20
beta	1
gamma	1

The damage to the body will also depend on the type of tissue affected. Lung tissue, for example, is easily damaged. Radon gas is dangerous because it emits alpha particles. If it is breathed into the lungs then the alpha radiation will be absorbed in the lung tissue.

Dose summary

Radiation dose is affected by:
- amount of radiation
- type of radiation.

Why is ionising radiation dangerous?

Ionising radiation has the energy to break molecules in the cells in the body into ions. These ions can then take part in chemical reactions that might damage the body. If the ionising radiation affects DNA molecules, this may cause the cell to be killed or to behave abnormally. Cells that behave abnormally can cause cancer.

Is there a safe dose?

There is no such thing as a safe dose. Just one radon atom might cause a cancer, just as a person might get knocked down by a bus the first time they cross a road. The chance of it happening is low, but it still exists. The lower the dose, the lower the risk. But the risk is never zero.

Irradiation and contamination

Exposure to a radiation source outside your body is called **irradiation**. Alpha irradiation presents a very low risk because alpha particles:
- only travel a few centimetres in air
- are easily absorbed.

Your clothes will stop alpha particles. So will the outer layer of dead cells on your skin.

Irradiation by beta particles is more risky as they penetrate a few centimetres into the body. Most gamma rays pass straight through the body. But they have high energy, so if they are absorbed they are dangerous.

If a radiation source enters your body, or gets on skin or clothes, it is called **contamination**. You become contaminated. If you swallow or breathe in any radioactive material, your vital organs will be exposed to continuous radiation for a long period of time. Sources that emit alpha particles are the most dangerous because alpha particles are the most ionising. Contamination by gamma sources is the least dangerous as most gamma rays will pass straight out of the body.

It is difficult to be sure about the harm that low doses of radiation can cause. In the 1970s, Alice Stewart studied the health of people working in the American nuclear industry. Her early results suggested that radiation is more harmful to children and to elderly people. She was attacked for her ideas, and the employers prevented any further access to medical records.

Key words
- background radiation
- radiation dose
- irradiation
- contamination

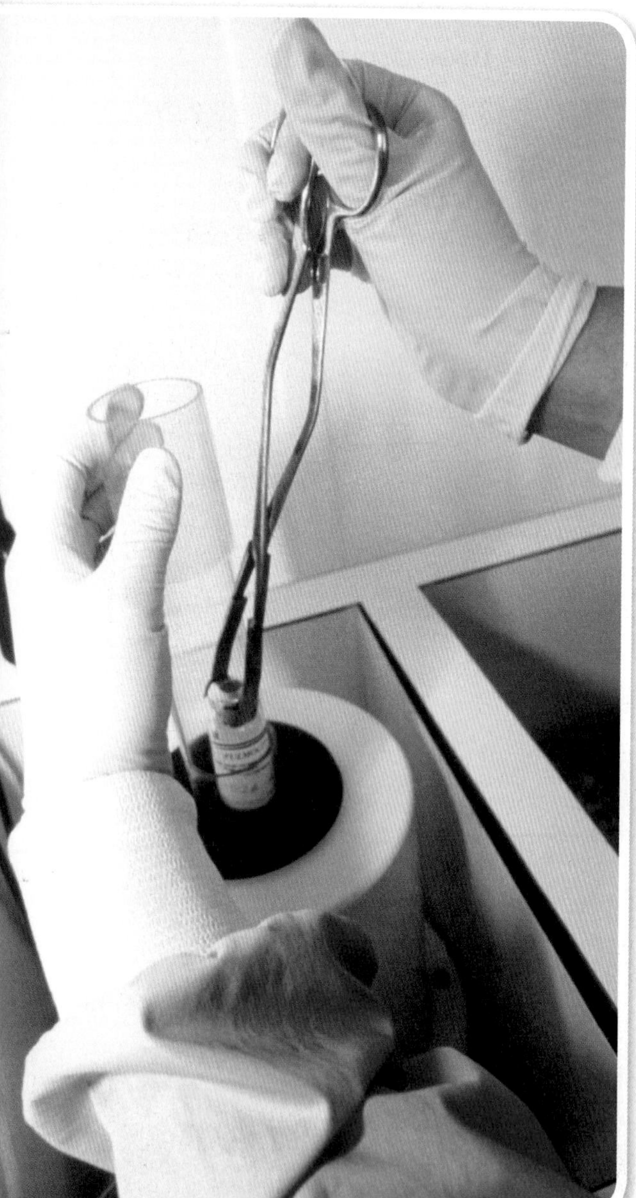
Staff handle radioactive sources with care to reduce the risk of contact with the source.

Radiation protection

Scientists study radiation hazards and give advice to protect against them. They also keep a close eye on the many people who regularly work with radioactive materials – in hospitals, industry, and nuclear installations. These people are called 'radiation workers'.

Employers must ensure that radiation workers receive a radiation dose 'as low as reasonably achievable'.

This principle applies when better equipment or procedures can reduce the risks of an activity. Any extra cost this involves must be balanced against the amount by which the risk is reduced.

To reduce their dose, medical staff take a number of precautions. They:

- use protective clothing and screens
- wear gloves and aprons
- wear special devices to monitor their dose.

The principle applies equally to hospital patients who receive radiation treatment. If doctors find that one hospital uses smaller doses but is just as effective as any other, then all hospitals are encouraged to copy them.

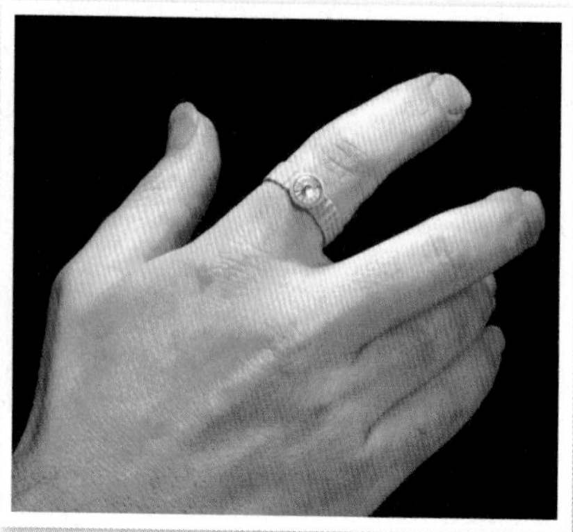
People working with radiation wear a personal radiation monitor that keeps track of their exposure to ionising radiation.

Questions

3 Explain the difference between irradiation and contamination.

4 a How big a dose of radiation do you get by catching a flight to Australia?
  b Where do cosmic rays come from?
  c Is this irradiation or contamination?

5 Imagine that alpha radiation damages a cell on the outside of your body. Why is this less risky than internal damage? Give two reasons.

6 In our food there is some potassium-40 (K-40). This is a naturally occurring radioactive isotope that sometimes decays to calcium (Ca) by emitting a beta particle. Write a nuclear equation for this decay. (Potassium has 19 protons in the nucleus.)

7 Explain how each of the precautions that medical staff take helps to keep their radiation dose as low as possible.

Living with radon

Find out about

- radon gas
- radiation dose and risk

Radon gas

Over 400 years ago, a doctor called Georgius Agricola wrote about the high death rate amongst German silver miners. He thought they were being killed by dust, and called their disease 'consumption'.

We now know that the mines contained a high concentration of radon gas. Radon gas is harmful because it is **radioactive**. It produces **ionising radiation** that can damage cells. The silver miners were dying of lung cancer.

Silver mines were contaminated with radon gas. The miners breathed it in and suffered.

Radon and lung cancer

Radon seeps into houses in some areas of the UK, as described in the leaflet on the following page. When this was realised, scientists were concerned about whether low doses over a long time could cause lung cancer. Since then they have conducted lots of studies to see if radon in homes increases the risk of lung cancer.

Scientists measure radon levels in the homes of people with lung cancer and compare them with levels in homes of people who have not got lung cancer. One study compared 413 women with lung cancer with 614 without lung cancer and showed a link between radon exposure and lung cancer. The women had lived in the same homes for 20 years. Some studies have not shown a link. This may be because they had a smaller sample size, or because it is difficult to measure radon exposure over time, especially if people move. One study got round this by analysing glass to measure radon exposure. The scientists chose glass from a mirror or picture frame that was at least 15 years old and had been with the person all the time, even if they moved house. This study did show a correlation between radon exposure and lung cancer.

The pipe runs beneath the floor of the house and a small fan sucks the radon from the building.

Questions

1 a What was the correlation that Georgius Agricola observed?

b What do we now know was i the factor and ii the outcome in what happened to the silver miners?

c Explain whether the factor *caused* the outcome by discussing the mechanism involved.

2 If you were setting up a study on radon in houses and the risk of lung cancer, explain how you would choose the samples.

A hazard at home

Radon gas builds up in enclosed spaces. In some parts of the UK, it seeps into houses.

LIVING WITH RADON

GOVERNMENT INFORMATION LEAFLET

There is radon all around you. It is radioactive and can be hazardous – especially in high doses.

Radon gives out a type of ionising radiation called **alpha radiation**. Like all ionising radiations, alpha radiation can damage cells and might start a cancerous growth.

Radon is a gas that can build up in enclosed spaces. Some homes are more likely to be contaminated with radon.

What about my home?

You and your family are at risk if you inhale radon-contaminated air. The map shows the areas where there is most contamination.

If you live in one of these areas, get your house tested for radon gas.

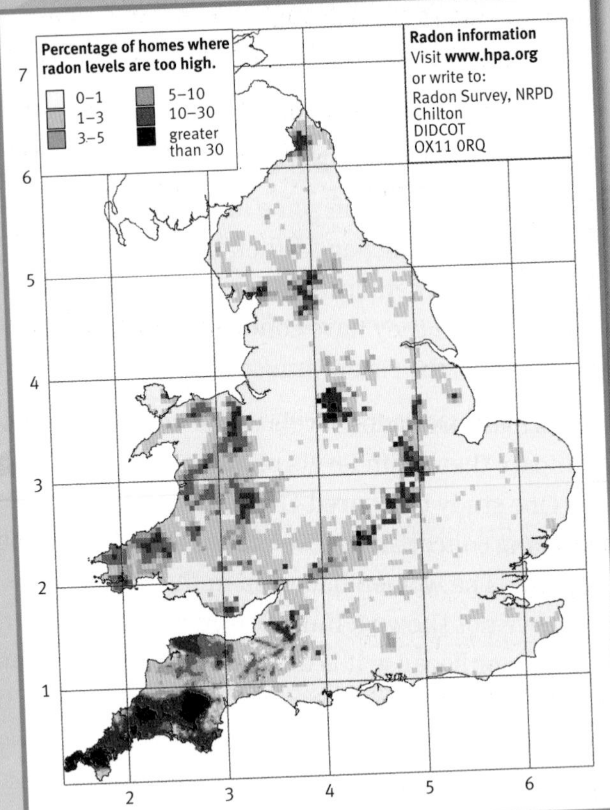

Radon-affected areas in England and Wales. Based on measurements made in over 400 000 homes.

What if the test shows radon?

Radon comes from the rocks underneath some buildings. It seeps into unprotected houses through the floorboards. If your house is contaminated, get it protected. An approved builder will put in:

- a concrete seal to keep the radon under your floorboards and
- a pump to remove it safely.

The risk is real: put in a seal.

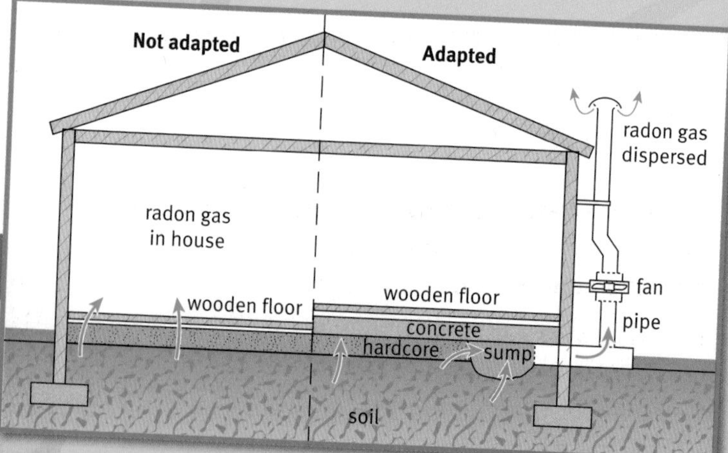

Radon gas can build up inside your home. Sealing the floor and pumping out the gas is an effective cure.

Radon and risk

The risk to miners was high because radon can build up in enclosed spaces, such as mines. In the atmosphere, the radon spreads out. It's a different story in enclosed spaces like mines. The rocks keep producing the gas and it cannot escape. So the radon concentration can be 30000 times higher than in the atmosphere.

There would be a lower concentration in a house, especially if the windows are open or it is draughty. Now that we are careful to draught proof our homes, it is more important to protect them against radon.

On average, radon makes up half the UK annual radiation dose. About 1100 people die each year from its effects, or about 1 in every 50 000 people. But radon is only one of the hazards that people face every day. There are risks associated with driving to school, sunbathing, swimming, catching a plane, and even eating.

The table shows how the risk of cancer from radon compares with some other common risks.

Many risky activities have a benefit. You need to decide whether the risk is worth taking.

Cause of death	Average number of deaths per year
cancer caused by radon	1100
cancer caused by asbestos	4000
skin cancer caused by ultraviolet radiation	1400
road deaths	2500
cancer caused by smoking	35 000
CJD	98
house fire	360
all causes	510 000

Estimated deaths per year in the UK population of 60 million (2008).

Questions

3 In the radon studies, how did analysing the glass of a mirror or picture frame improve the measurement of radon exposure?

4 There is a risk from radon gas building up in houses. Which of these are good ways to reduce the risk?
- stop breathing
- wear a special gas mask
- move house
- adapt the house

5 Choose three causes of death from the table on the left. Write down a way of reducing the risk from each chosen cause on the left.

6 Write a letter to a friend living in a high-radon area to persuade them to get their house checked for radon.

Half-life

Find out about

- the half-life of radioactive materials

Radioactive decay is random. You can never tell which nucleus will decay next. Scientists can't predict whether a particular nucleus will decay today or in a thousand years, time. But in a sample of radioactive material there are billions of atoms, so they can see a pattern in the decay.

The pattern of radioactive decay

The amount of radiation from a radioactive material is called its **activity**. This decreases with time.

- At first there are a lot of radioactive atoms.
- Each atom gives out radiation as it decays to become more stable.
- The activity of the material falls because fewer and fewer radioactive atoms remain.

When a sample of the medical tracer technetium-99m is injected into a patient at 9.00 am then on average, by 3.00 pm half of the nuclei have decayed. At 9.00 pm half of those nuclei remaining will have decayed, leaving only one quarter of the original sample. We say the **half-life** of technetium-99m is six hours. The half-life is the time it takes for the activity to drop by half. The table shows that after 24 hours (four half-lives), only one sixteenth of the original sample remains.

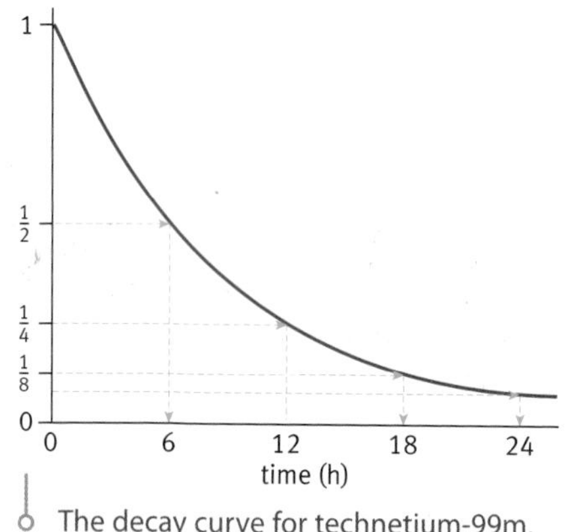

The decay curve for technetium-99m.

Time	Hours since injection	Number of half-lives	Fraction of original sample remaining
9.00 am	0	0	1
3.00 pm	6	1	$\frac{1}{2}$
9.00 pm	12	2	$\frac{1}{4}$
3.00 am	18	3	$\frac{1}{8}$
9.00 am	24	4	$\frac{1}{16}$

The radioactive decay of Technetium-99m.

Key words

- activity
- half-life

The graph shows how the activity of the sample decreases. After ten half-lives only about one thousandth of the original sample remains. For technetium-99m this is only two and a half days. The six-hour half-life makes it an ideal isotope for a tracer. It lasts long enough for doctors to get some scans of the decay, but it has almost all gone in a few days.

Different half-lives

All radioactive materials show the same pattern but they can have different half-lives. This graph shows the pattern of radioactive decay for radon. Notice that the amount of radiation halves every minute.

The **half-life** of radon-220 is one minute.

There is no way of slowing down or speeding up the rate at which radioactive materials decay. Some decay slowly over thousands of millions of years. Others decay in milliseconds – less than the blink of an eye.

The shorter the half-life, the greater the activity for the same amount of material. Of the four radioactive isotopes listed in the table, neon-17 is the most active.

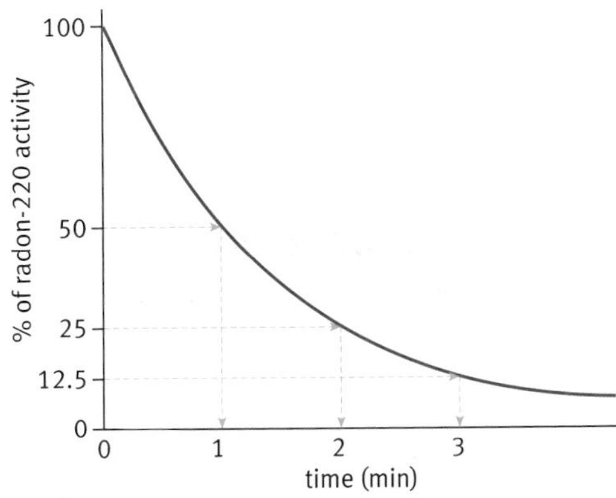

The decay curve for radon-220.

Isotope	Half-life
Iridium-192	74 days
Strontium-81	22 minutes
Uranium-235	710 million years
Neon-17	0.1 seconds

Half-lives can be short or long.

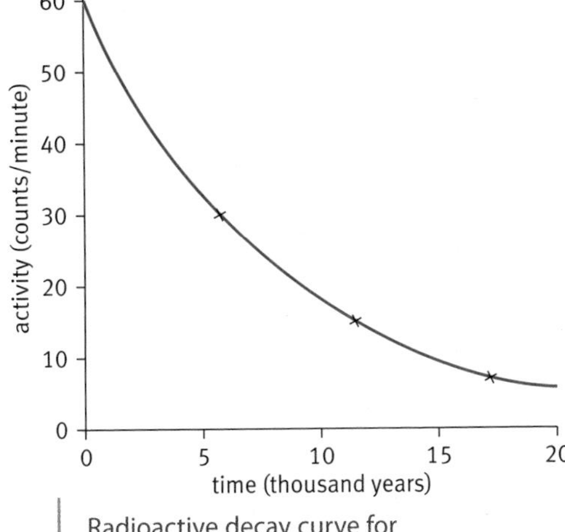

Radioactive decay curve for carbon-14.

Questions

1. Carbon-14 has a half-life of 5700 years. What fraction of its original activity will a sample have after 11 400 years?
2. Iodine-123 is used to investigate problems with the thyroid gland, which absorbs iodine. It is a gamma emitter.
   - a Explain why the element iodine is chosen.
   - b Explain why it is useful that iodine-123 gives out gamma radiation.
   - c Iodine-123 has a half-life of 13 hours. Why would it be a problem if the half-life was:
     - i a lot shorter?
     - ii a lot longer?
3. How long does it take for a sample of strontium-81 to decay to one eighth of its original size?
4. An underfloor fan is switched on to prevent any more radon-220 entering a house. The owner wants to know how long will it take for the radioactivity from the radon-220 in the house to fall to less than one thousandth of the original value.
   - a How many half-lives will this take?
   - b How long will this take?
5. The activity of a neon-17 source is 1120 decays per second. What will it be after 0.5 seconds?
6. The activity of an iridium-192 sample is 9600 decays per second. How long will it take to fall to 2400 decays per second?

H Medical imaging and treatment

Find out about

- different uses of radiation
- types of radiation
- benefits and risks of using radioactive materials
- limiting radiation dose

Radioactive materials can cause cancer. But they can also be used to diagnose and cure many health problems.

Medical imaging

Jo has been feeling unusually tired for some time. Her doctors decide to investigate whether an infection may have damaged her kidneys when she was younger.

They plan to give her an injection of DMSA. This is a chemical that is taken up by normal kidney cells.

The DMSA has been labelled as radioactive. This means its molecules contain an atom of technetium-99m (Tc-99m), which has a half-life of six hours. The kidneys cannot tell the difference between normal DMSA and labelled DMSA. They absorb both types.

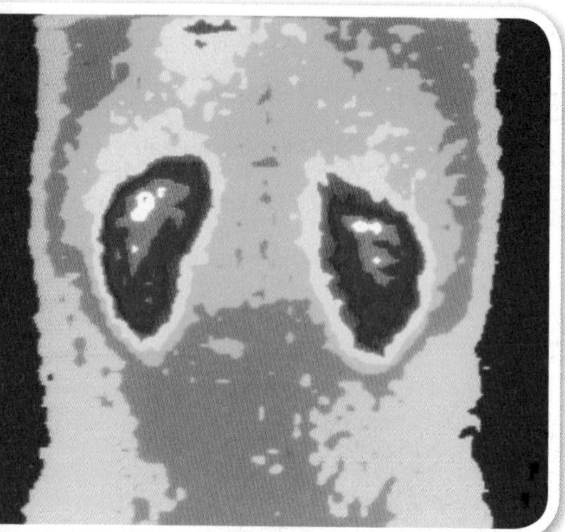

This gamma scan shows correctly functioning kidneys – the top two white areas.

The Tc-99m gives out its gamma radiation from within the kidneys. Gamma radiation is very penetrating, so nearly all of it escapes from Jo's body and is picked up by a gamma camera. The gamma camera **traces** where the technetium goes in Jo's body. Parts of the kidney, which are working normally, will appear to glow. Any dark or blank areas show where the kidney isn't working properly.

Jo's scan shows that she has only a small area of damage. The doctors will take no further action.

Glowing in the dark

Jo was temporarily contaminated by the radioactive Tc-99m. For the next few hours, until her body got rid of the technetium, she was told to:

- flush the toilet a few times after using it
- wash her hands thoroughly
- avoid close physical contact with friends and family.

Is it worth it?

There was a small chance that some gamma radiation would damage Jo's healthy cells. Before the treatment, her mum had to sign a consent form, and the doctors checked that Jo was not pregnant.

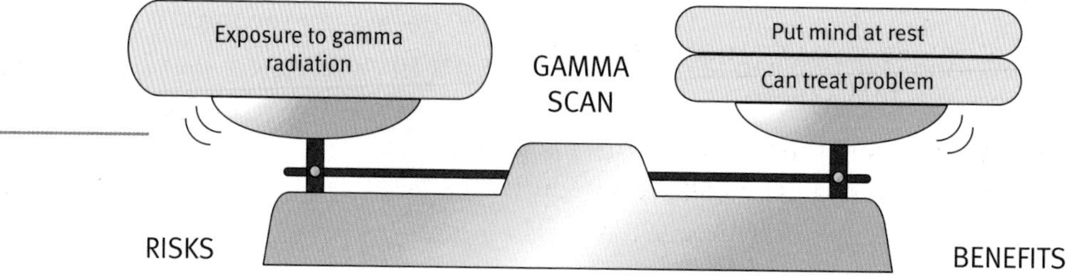

Jo's mum weighed the risk against the benefit.

Jo's mum said 'We felt the risk was very small. It was worth it to find out what was wrong. Even with ordinary medicines, there can be risks. You have to weigh these things up. Nothing is completely safe.'

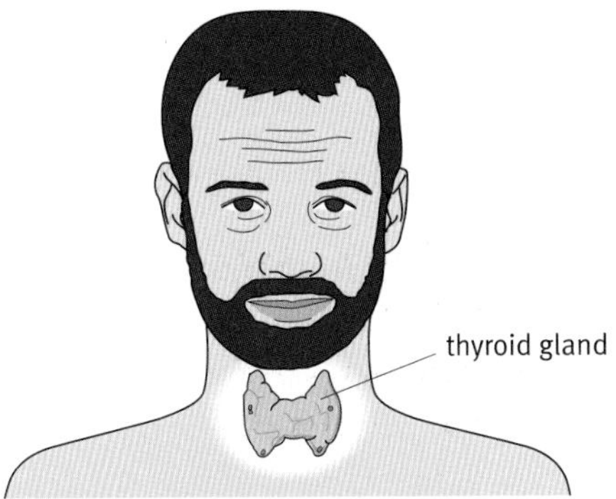

The thyroid gland is located in the front of the neck, below the voice box.

Treatment for thyroid cancer

Alf has thyroid cancer. First he will have surgery, to remove the tumour. Then he must have **radiotherapy**, to kill any cancer cells that may remain.

A hospital leaflet describes what will happen.

Radioiodine treatment

You will have to come in to hospital for a few days. You will stay in a single room.

You will be given a capsule to swallow, which contains iodine-131. This form of iodine is radioactive. You cannot eat or drink anything else for a couple of hours.

- The radioiodine is absorbed in your body.
- Radioiodine naturally collects in your thyroid, because this gland uses iodine to make its hormone.
- The radioiodine gives out beta radiation, which is absorbed in the thyroid.
- Any remaining cancer cells should be killed by the radiation.

You will have to stay in your room and take some precautions for the safety of visitors and staff. You will remain in hospital for a few days, until the amount of radioactivity in your body has fallen sufficiently.

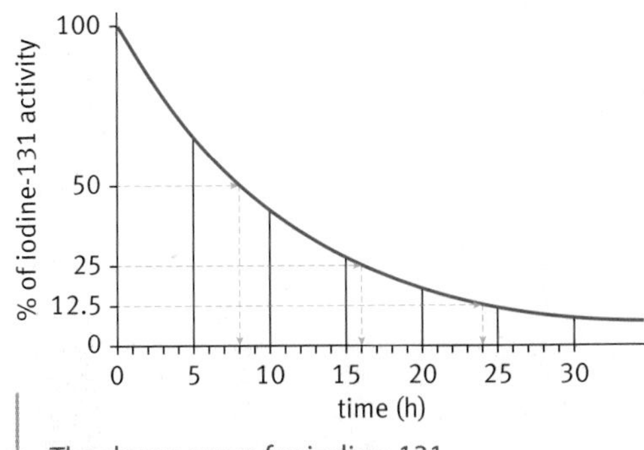

The decay curve for iodine-131.

Key words
- radiotherapy
- traces

Iodine-131 decays to an isotope of xenon.

Questions

1. Look at the precautions that Jo has to take after the scan. Write a few sentences explaining to Jo why she has to do each of them.
2. It would be safe to stand next to Jo but not to kiss her. Use the words 'irradiation' and 'contamination' to explain why.
3. What are the risks and the benefits to Jo of having the treatment?
4. Read the information leaflet about radioiodine. Describe how the risk to Alf's family and other patients is kept as low as possible.
5. Radioiodine has a half-life of eight days. Explain why a half-life of eight days is more suitable than:
   a. eight minutes
   b. eight years.
6. Alf has a check up after 40 days.
   a. How many half-lives of iodine 131 is 40 days?
   b. What fraction of the radiation remains after 40 days?

Find out about

- energy from nuclear fission
- nuclear power stations

Nuclear fission

Radioactive atoms have an unstable nucleus. Some nuclei can be made so unstable that they split in two. This process is called **nuclear fission**.

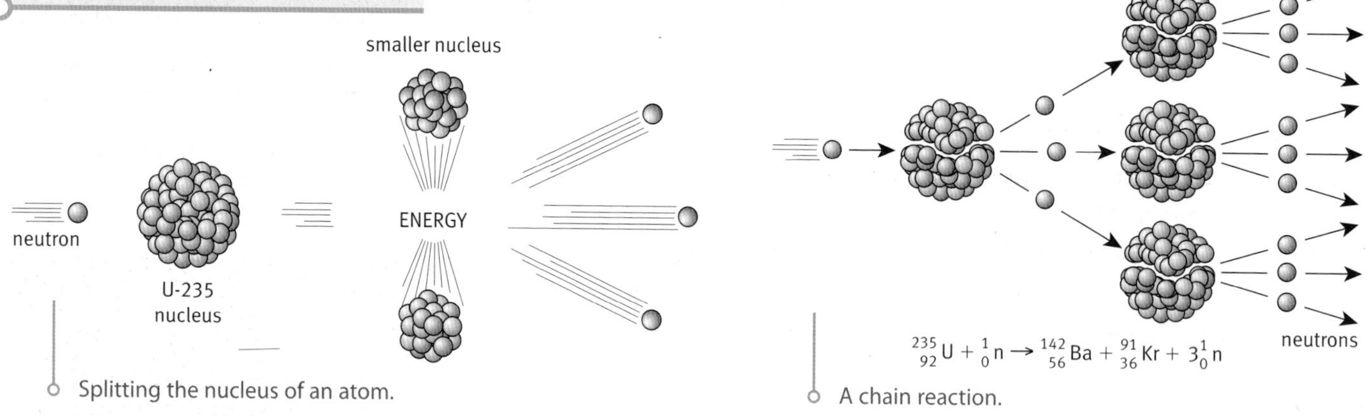

Splitting the nucleus of an atom.

A chain reaction.

Key words

- nuclear fission
- chain reaction
- Einstein's equation

For example, the nucleus of a uranium-235 atom breaks apart when it absorbs a neutron.

When this happens a small amount of the mass of the nucleus is converted to a huge amount of energy. The energy released can be calculated from **Einstein's equation**:

$$E = mc^2$$

where E = energy, m = change in mass and c = speed of light.

The energy is transferred to the fission products, so they have a lot of kinetic energy. Each fission reaction produces roughly a million times more energy than when a molecule changes during a chemical reaction.

The fission of one atom can set off several more, because each fission reaction releases a few neutrons. If there are enough U-235 atoms close together, there will be a **chain reaction**, involving more and more atoms.

The devastating power of a nuclear weapon.

Nuclear weapons

In World War 2 there was a race to 'split the atom' and harness the energy in the form of a bomb.

On 16 July 1945, in the deserts of New Mexico, a group of scientists waited tensely as they tested 'the gadget'. Some thought it would be a flop. Others worried that it might destroy the atmosphere.

At 5.29 a.m., it was detonated and filled the skies with light. The bomb vaporized the metal tower supporting it. All desert sand within a distance of 700 m was turned into glass.

Some of the scientists were worried about the power of the bomb and wanted the project stopped. A few weeks later, the Americans dropped two nuclear bombs on Japan, at Hiroshima and Nagasaki.

Controlling the chain

At the heart of a nuclear power station is a reactor. It is designed to release the energy of uranium at a slow and steady rate, by controlling a chain reaction.

- The fission takes place in **fuel rods** that contain uranium-235. This makes them extremely hot.
- **Control rods**, which contain the element boron, absorb neutrons. Moving control rods in or out of the reactor decreases or increases the reaction rate.

The Nuclear Installations Inspectorate monitors the design and operation of nuclear reactors. Reactor cores are sealed and shielded. Very little radiation gets out.

Generating electricity

A fluid, called a **coolant**, is pumped through the reactor. The hot fuel rods heat the coolant to around 500 °C. It then flows through a heat exchanger in the boiler, turning water into steam. The steam is used in the same way as in a coal-fired power station. One reason for building nuclear power stations is to reduce the need for fossil fuels.

Plutonium

The element plutonium is produced in nuclear reactors. It can also undergo nuclear fission so it can be used for nuclear weapons or fuel. Countries sometimes build nuclear reactors to obtain plutonium for weapons. Nuclear weapons inspectors try to ensure that nuclear power stations are very secure, account for all their waste, and are not operated in unstable countries.

Once fuel rods are in service, decay products build up in them. They become more radioactive.

Questions

1 Why do nuclear reactors use coolants and not circulate water to make steam directly?

2 a Complete the nuclear equation for the absorption of a neutron by Uranium-235:

$${}^{235}_{92}\text{U} + {}^{0}_{1}\text{n} \longrightarrow {}^{::}_{::}\text{U}$$

b Describe what happens to this unstable nucleus of uranium and how this can lead to a chain reaction.

c If the chain reaction runs out of control we have an atomic bomb. Explain how it is controlled in a nuclear reactor.

3 Suggest how gamma radiation from a nuclear reactor is contained, so that living things are not irradiated.

4 Write down two risks and two benefits of living in a country with nuclear power stations.

Key words

- ✓ **fuel rod**
- ✓ **control rod**
- ✓ **coolant**

Nuclear waste

Find out about

- the UK's nuclear legacy
- the half-life of radioactive materials
- possible methods of disposal

A legacy of nuclear waste

The Nuclear Decommissioning Agency (NDA) is responsible for cleaning up hazardous nuclear waste at 36 sites around the UK. These include power stations, research sites, Ministry of Defence sites and healthcare sites. Most of the radioactive waste comes from power stations. The rest comes from medical uses, industry, and scientific research. In addition to 'everyday' waste, when power stations are too old to be used anymore they must be safely dismantled. The waste radioactive materials are separated out and taken away to be stored. Nuclear waste is a cocktail made of different isotopes. They call it the UK's 'nuclear legacy'.

A long-term hazard

Radioactive waste has very little effect on the UK's average background radiation. But it is still hazardous. This is because of contamination. Imagine that some waste leaks into the water supply. This could be taken up by food, which you eat. The radioactive material is now in your stomach, where it can irradiate your internal organs.

Some radioactive materials have half-lives of thousands of years. The NDA must dispose of nuclear waste in ways that are safe and secure for many, many generations.

High-level radioactive waste is hot, so it is stored underwater.

Types of waste

The nuclear industry deals with three types of nuclear waste.

- **High-level waste** (HLW). This is 'spent' fuel rods. HLW gets hot because it is so radioactive. It has to be stored carefully but it doesn't last long. And there isn't very much of it. All the UK's HLW is kept in a pool of water at Sellafield.
- **Intermediate-level waste** (ILW). This is less radioactive than HLW. But the amount of ILW is increasing, as HLW decays to become ILW.
- **Low-level waste** (LLW). Protective clothing and medical equipment can be slightly radioactive. It is packed in drums and dumped in a landfill site that has been lined to prevent leaks.

The control room at a nuclear waste storage plant enables people to monitor the waste continuously.

Type of waste	Volume (m^3)	Radioactivity
LLW	196 000	weak
ILW	92 500	strong
HLW	1730	extremely strong

The amount of nuclear waste in store (2007). The problem of what to do with it remains unsolved.

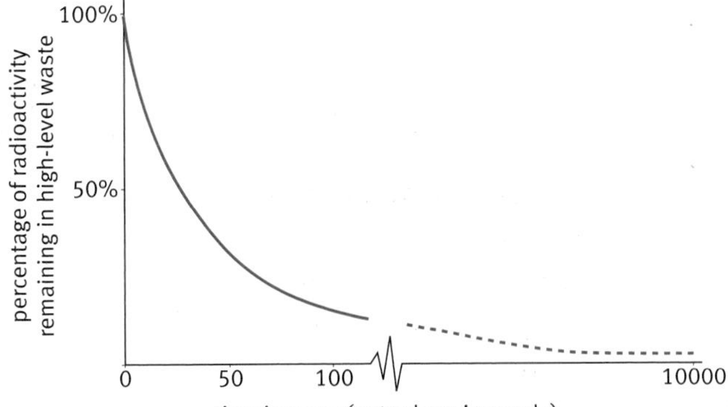

high-level waste decays quickly at first. When its activity falls, it becomes intermediate-level waste. ILW stays radioactive for thousands of years.

Sellafield

A government-owned company runs the biggest nuclear site in the UK – Sellafield, in Cumbria. Thousands of workers – professional, skilled, and unskilled – contribute to its important work. Sellafield reprocesses nuclear waste produced in the UK and abroad. It also prepares and stores nuclear waste for permanent disposal.

Risk management is a major concern at Sellafield. They must plan in advance how to maintain production and safety in the event of any possible problem.

ILW presents the biggest technical challenge, because it is very long-lived. Currently it is chopped up, mixed with concrete, and stored in thousands of large stainless-steel containers. This is secure but not permanent. The long-term solution has to be secure and permanent.

The work of the NDA

Managing waste is very expensive. In 2010 the NDA spent £28 billion on work at 19 sites including Sellafield. One priority is to complete the reprocessing of the HLW at Sellafield by 2016. In 2007 an inventory was carried out of all the radioactive material stored at the 36 sites. There have been a number of public consultations about what to do with the waste. At the time of writing these are still going on. The preferred plan at the moment is to store it until a safe geological site can be found and then to bury it.

When will it be 'safe'?

We are exposed to some radiation all the time (background radiation). When the nuclear waste only emits very low levels of radiation, similar to the background radiation, it poses little risk. The longer the half-life of the radioactive material, the longer it will take to become 'safe'.

Key words

- high-level waste
- intermediate-level waste
- low-level waste

Questions

1 Disposing of ILW needs to be both secure and permanent. Explain why both criteria are important.

2 The NDA wants to know the public's views on waste storage. Write a letter explaining what you think should be done.

3 What are the advantages and disadvantages of keeping all the waste together above ground rather than burying it in a deep shaft and sealing it?

4 A small amount of nuclear fuel produces a lot of energy, so in the 1950s scientists thought this would be a cheap way of generating electricity. Explain why the real cost is much greater than realised at the time.

5 HLW contains plutonium-239, which has a half-life of 24 100 years. Explain some of the difficulties of keeping it safe for 10 half-lives.

Nuclear fusion

Find out about

- nuclear fusion
- the attractive and repulsive forces between nuclear particles
- the iter project

A balance of forces

The nucleus of an atom is made up of protons and neutrons. This tells us something important – protons and neutrons are happy to stick together. There must be an attractive force that holds protons and neutrons together, and that can even hold two protons together despite the fact that they repel each other because of their positive charges. This force is called the **strong nuclear force**.

The strong nuclear force has a short range. It only acts when two nucleons (protons or neutrons) are very close together. In a nucleus, the particles are separated by just the right distance so that the strong nuclear force is balanced by the electrical (electrostatic) force.

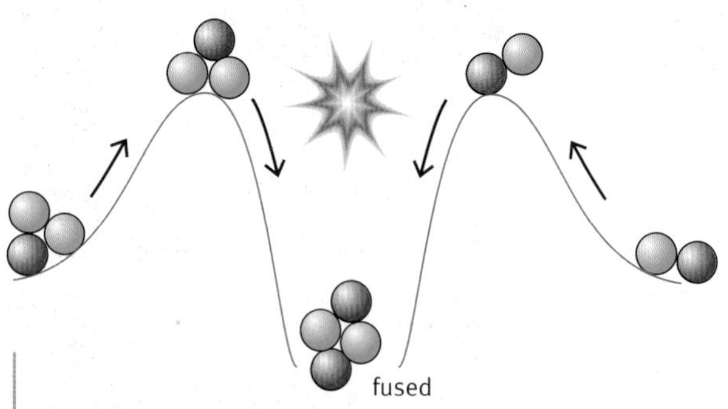

Pushing two hydrogen nuclei together. The 'hills' represent the repulsive force between them. The deep 'valley' represents the stable state they reach when they fuse together.

$${}^{2}_{1}\text{H} + {}^{3}_{1}\text{H} \rightarrow {}^{4}_{2}\text{He} + {}^{1}_{0}\text{n}$$

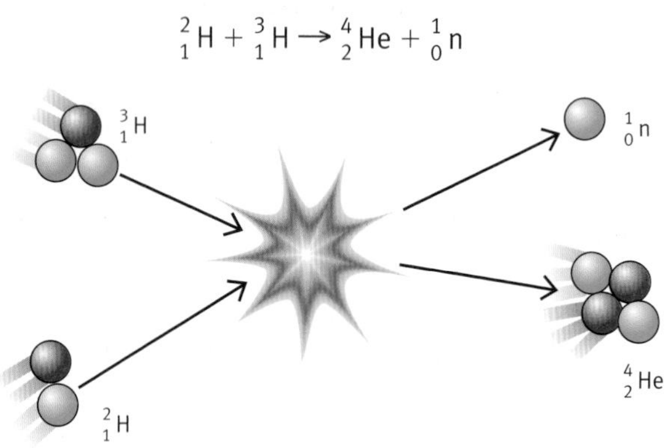

Fusion of two hydrogen isotopes gives helium.

Joining up

In the process of **nuclear fusion**, the nuclei of two hydrogen atoms join together and energy is released. The diagram on the right shows one way in which this happens. Note that the nuclei are of two different isotopes of hydrogen – both are hydrogen, because they have just one proton in the nucleus.

Picture bringing two atoms close together: their nuclei repel each other, because of the electrical (electrostatic) force. They will not fuse together. Push hard enough: they come close enough for the attractive force to take over, and the nuclei fuse. Energy is released.

You would have to do a lot of work to push two nuclei together, but you would get a lot more energy out when they fused.

Energy is released by a small amount of mass being converted to energy. The energy can be calculated from Einstein's equation $E = mc^2$.

It may seem strange that you can get energy from fission and fusion. It happens because the iron nucleus is the most stable, so nuclei lighter than iron can be joined to release energy and heavier nuclei can be split to release energy.

The quest for fusion power stations

If we could fuse the hydrogen nuclei from water to give nuclei of helium, which is an inert gas, we would have plenty of fuel (water) and the process would not produce as much nuclear waste.

Over the past 70 years there has been a lot of research. Scientists can give hydrogen nuclei enough energy to overcome the repulsive force. The problem is how to control the reaction and keeping it going?

When hydrogen isotopes are heated enough they lose their electrons and form a **plasma**. This is kept from touching the sides of the container by using magnetic fields. The JET project at Culham in the UK has researched fusion for many years. As yet no reactor has produced more energy than it used.

The H bomb

Hydrogen bombs, which fuse hydrogen, release hundreds of times more energy than atomic (fission) bombs. They are triggered using an atomic bomb to compress the hydrogen so that it fuses.

The ITER project

This is a joint project between China, the European Atomic Energy Community, India, Japan, Korea, Russia, and the U.S.A. ITER means 'the way' in Latin. Fusion research is very expensive so these countries have joined together to build a research reactor in France. Construction has begun. It will take 10 years to build, and be used for research for 20 years. ITER will investigate how plasmas behave during the hydrogen fusion reaction at 150 million °C. The goal is to be able to build a nuclear fusion power station.

You can think of fusion as the opposite of nuclear fission, the process used in nuclear power stations.

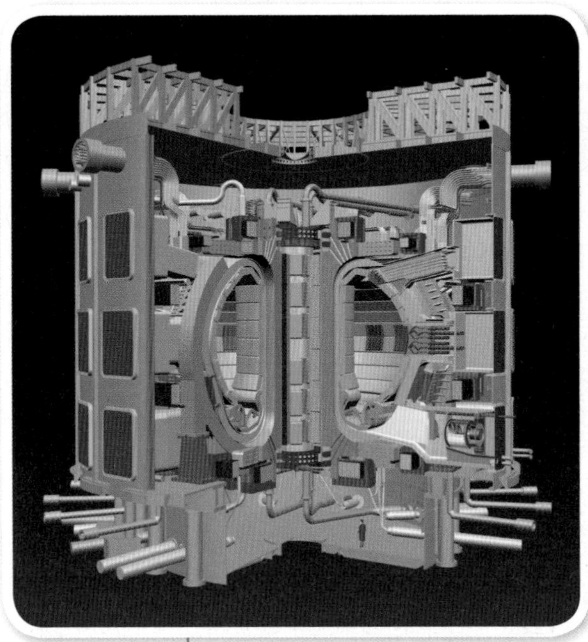

The planned ITER fusion reactor. Fusion will take place in the doughnut-shaped hole.

Questions

1 a What attractive force acts between particles in the nucleus? What repulsive force?
  b Which of these two forces has the greater range?

2 The Sun contains 1% oxygen nuclei. There are eight protons in an oxygen nucleus. Explain why these are less likely to fuse together than hydrogen nuclei.

3 Two hydrogen-2 nuclei can fuse to give a helium-3 nucleus.
  a Write a nuclear equation for the reaction.
  b What is the particle that is left over?

4 What are the advantages of countries working together on the ITER project?

5 ITER is very expensive. Write a letter to persuade the Government either:
  a to stay part of ITER or
  b to leave ITER to save money.

Key words

- ✓ **nuclear fusion**
- ✓ **strong nuclear force**

Science Explanations

Our understanding of radioactivity and the structure of the atom has enabled many applications, such as nuclear power stations and cancer treatment, to be developed. Knowledge of the way ionising radiation behaves is essential for working safely and making good risk assessments.

You should know:

- why some materials are radioactive and emit ionising radiation all the time
- how ionising radiation can damage living cells
- that atoms have shells of electrons and a nucleus made of protons and neutrons
- about the alpha scattering experiment and how it showed that the atom has a small, massive, positively charged nucleus
- that all the atoms of an element have the same number of protons
- that isotopes are atoms of the same element with different numbers of neutrons
- how the nucleus changes in radioactive decay
- how to complete nuclear equations for radioactive decay
- what alpha and beta particles and gamma radiation are, and their different properties
- that there is background radiation all around us, mostly from natural sources
- what radiation dose measures, and what factors affect it
- the difference between contamination and irradiation
- how to interpret data on risk related to radiation dose
- that radioactive materials randomly emit ionising radiation all the time and that the rate of decay cannot be changed by physical or chemical changes
- that the activity of a radioactive source decreases over time
- what is meant by the half-life of a radioactive isotope
- that radioactive isotopes have a wide range of half-life values
- how to do calculations involving half-life
- about uses of ionising radiation from radioactive materials
- about nuclear fission and energy released from the nucleus
- how nuclear power stations use the fission process to produce energy, including how the chain reaction is controlled, and about the nuclear waste produced
- about the three categories of radioactive waste, and the different methods of disposal
- that protons and neutrons in a nucleus are held together by the strong force, which acts against the electrical repulsive force between protons
- that hydrogen nuclei can fuse together to form helium if they are brought close enough together and this releases energy
- how to use Einstein's equation, $E=mc^2$, to calculate the energy released.

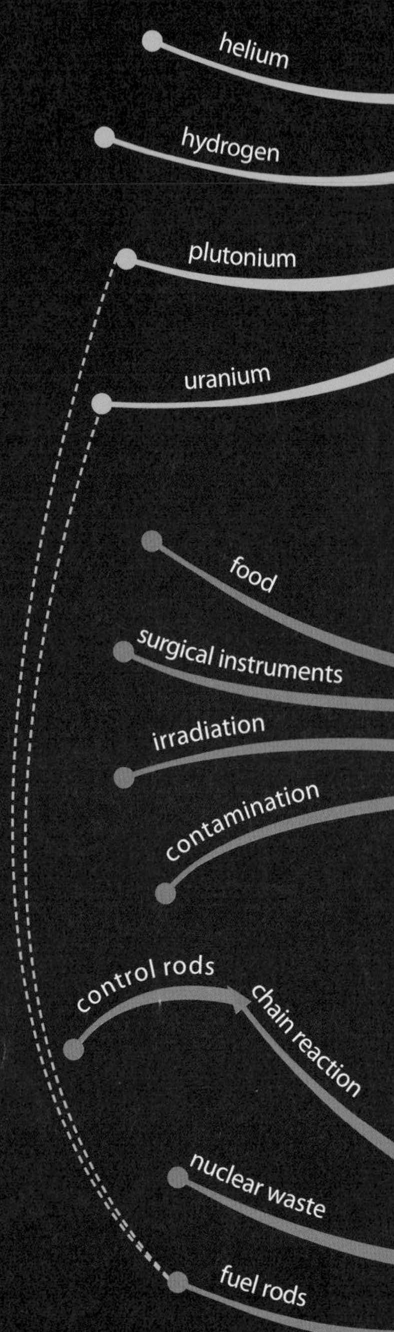

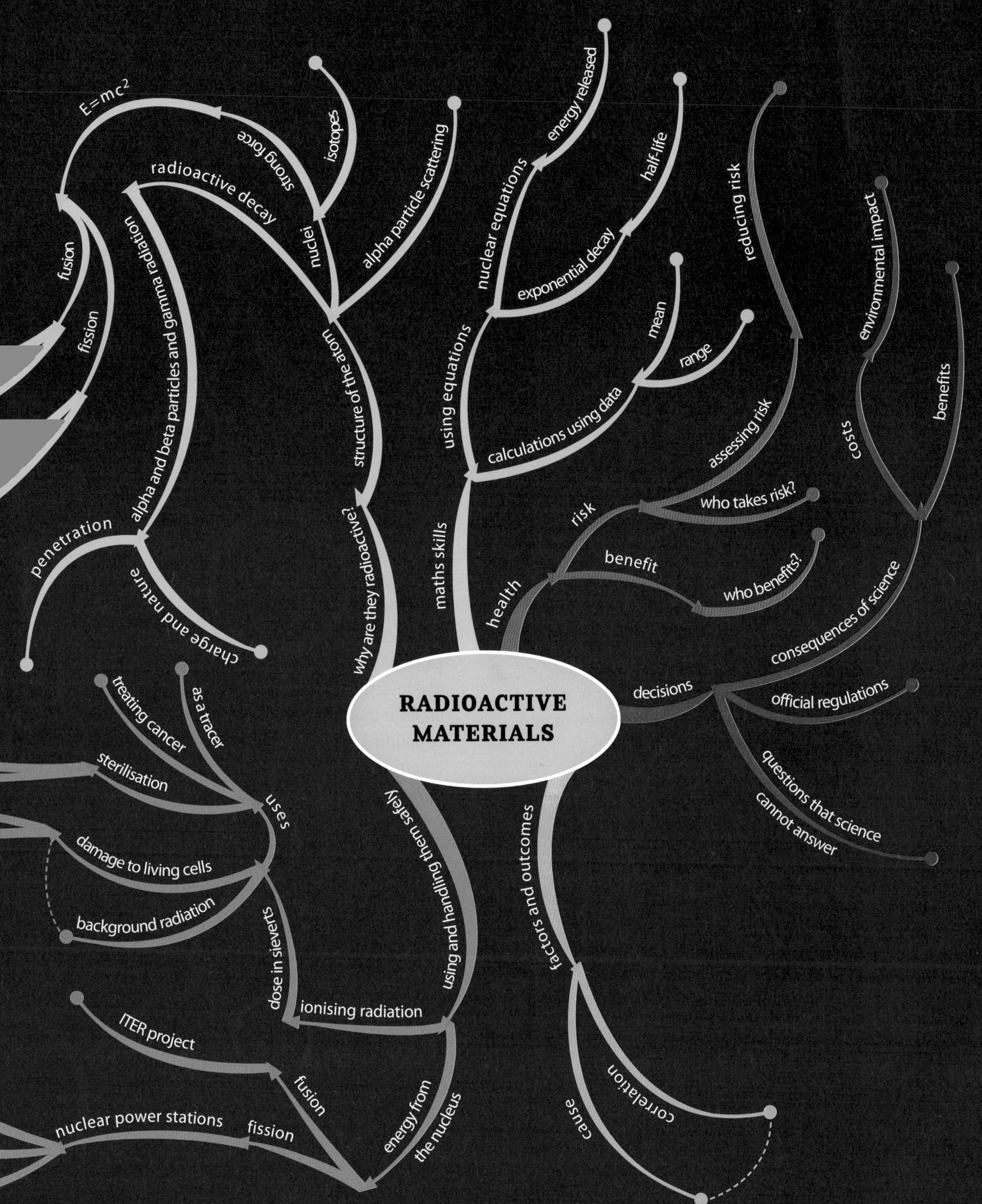
RADIOACTIVE MATERIALS
why are they radioactive?
structure of the atom
nuclei
isotopes
strong force
$E=mc^2$
radioactive decay
alpha particle scattering
fusion
fission
alpha and beta particles and gamma radiation
penetration
charge and nature
maths skills
using equations
nuclear equations
energy released
exponential decay
half-life
calculations using data
mean
range
health
risk
assessing risk
reducing risk
who takes risk?
benefit
who benefits?
decisions
consequences of science
costs
benefits
environmental impact
official regulations
questions that science cannot answer
factors and outcomes
correlation
cause
using and handling them safely
ionising radiation
dose in sieverts
uses
treating cancer
as a tracer
sterilisation
damage to living cells
background radiation
energy from the nucleus
fusion
ITER project
fission
nuclear power stations

Ideas about Science

In addition to developing an understanding of radioactive materials, it is important to appreciate the risks involved and how we make decisions about using science and technology. When considering risk, you should be able to:

- explain that nothing is completely safe. Everything we do has a certain risk. Background radiation is all around us, so there is always a risk of our cells being harmed by ionising radiation. But if the dose is low the risk is very small.
- list some of the uses of radioactive materials and the risks arising from them, both to people working with radioactive sources and to the environment.
- describe some of the ways that we reduce these risks.
- use data to compare and discuss different risks. Compare the risks of living in a high-radon area with risks of other activities, for example, smoking tobacco.
- discuss decisions involving risk. To decide whether to build a type of nuclear power station you would need to take account of the chance of contaminating the environment and how serious that would be.

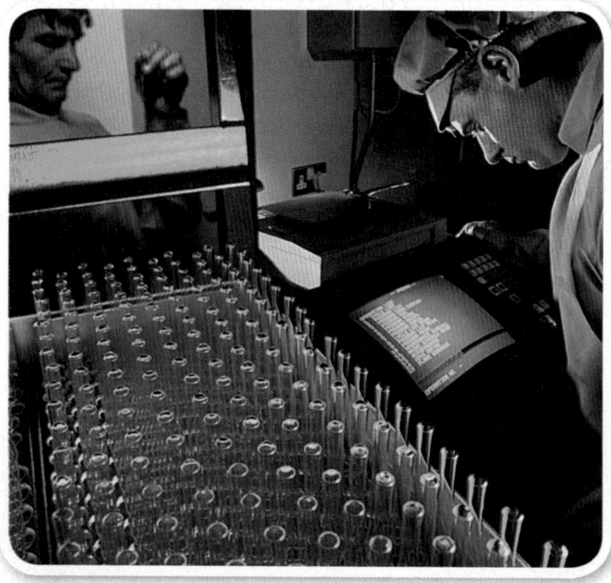

- identify risks and benefits to individuals and groups. Many medical treatments make use of radioactive isotopes. The risk and the benefit to the patient must be weighed up.
- take into account, in making decisions, who benefits and who takes the risks, for example, when deciding whether to build a nuclear power station.
- suggest benefits of activities known to have risk, for example, a scan that involves injecting a radioactive tracer into the body.
- suggest reasons why people are willing (or reluctant) to take a risk. For example, people who are ill may choose to have treatment in the hope that they will be cured. Some people may refuse treatment if doctors say they must have it, as they prefer to choose for themselves.

In making decisions about science and technology, you should be able to:

- identify the groups affected by a decision, and the main benefits and costs for each group, for example, when deciding on a location for a waste-disposal site.
- explain that different decisions may be made depending on social and economic factors. Nuclear power stations are built in remote areas where they will affect fewer people. Countries with no other resources for generating electricity may choose to build nuclear power stations.
- explain that there is official regulation of some areas of research and application of knowledge. Countries that have nuclear power stations keep account of all the radioactive waste, as this could be processed to produce nuclear weapons.
- distinguish questions that can be answered using a scientific approach from those that cannot, for example, 'Shall we build a nuclear power station at this site?' If there is no scientific reason why not, the final decision still depends on what society wants to do.

Review Questions

1 Complete the sentences about radioactivity. Use key words from the module.

____________ radiation is produced by radioactive ________________. The radiation is produced when an ______________ nucleus ________________. The three types of radiation produced are: ____________, which is made up of two protons and two neutrons, ________________, which is a high-energy electron, and ________________, which is an electromagnetic radiation.

2
- **a** What did the back-scattering alpha particles in the Rutherford scattering experiment show about atoms?
- **b** How is the back-scattering explained by your answer to part a?

3 An isotope has a half-life of 74 days. Its activity is measured at 10 000 decays per second. It emits alpha radiation.
- **a** What will its activity be after 133 days?
- **b** How long will it take for the activity to reach 625 decays per second?
- **c** For each example below, explain why the isotope would not be suitable and suggest an isotope from the module that would be suitable.
  - **i** Measuring the age of rocks.
  - **ii** A radioactive tracer in the body.

4 What is the difference between nuclear fission and nuclear fusion? Give examples of each with a nuclear equation.

5 The table shows some of the radioactive isotopes that are used in a range of applications.

Isotope	Radiation emitted	Half-life
americium-241	alpha	430 years
carbon-14	beta	5700 years
cobalt-60	gamma	5 years
iodine-123	gamma	13 hours
iodine-131	beta	8 days
strontium-90	beta	29 years
uranium-235	alpha	700 million years

For each application listed below choose the isotope you think most suitable and justify your answer, referring to both the half-life and the radiation emitted.
- **a** calculating the age of rocks
- **b** dating an ancient leather belt
- **c** monitoring the thickness of paper in a factory
- **d** monitoring uptake of iodine by the thyroid
- **e** detecting smoke
- **f** sterilising medical equipment

6 A nuclear power station uses the energy released by **nuclear fission** to generate electricity. The nuclear reactor is designed to control the **chain reaction**.
- **a** Draw diagrams to explain what is meant by **nuclear fission** and a **chain reaction**.
- **b** Describe the design features of the reactor that are used to control the chain reaction.

Glossary

abundant Abundance measures how common an element is. Silicon is abundant in the lithosphere. Nitrogen is abundant in the atmosphere.

acceleration The rate of change of an object's velocity, that is, its change of velocity per second. In situations where the direction of motion is not important, the change of speed per second tells you the acceleration.

acid A compound that dissolves in water to give a solution with a pH lower than 7. Acid solutions change the colour of indicators, form salts when they neutralize alkalis, react with carbonates to form carbon dioxide, and give off hydrogen when they react with a metal. An acid is a compound that contains hydrogen in its formula and produces hydrogen ions when it dissolves in water.

action at a distance An interaction between two objects that are not in contact, where each exerts a force on the other. Examples include two magnets, two electric charges, or two masses, for example, the Earth and the Moon.

active site The part of an enzyme that the reacting molecules fit into.

active transport Molecules are moved in or out of a cell using energy. This process is used when transport needs to be faster than diffusion, and when molecules are being moved from a region where they are at low concentration to where they are at high concentration.

activity The rate at which nuclei in a sample of radioactive material decay and give out alpha, beta, or gamma radiation.

actual yield The mass of the required chemical obtained after separating and purifying the product of a chemical reaction.

aerobic respiration Respiration that uses oxygen.

air resistance The force exerted on an object by the air, when it moves through it. Its direction is opposite to the direction in which the object is moving.

alkali A compound that dissolves in water to give a solution with a pH higher than 7. An alkali can be neutralised by an acid to form a salt. Solutions of alkalis contain hydroxide ions.

alkali metal An element in Group 1 of the periodic table. Alkali metals react with water to form alkaline solutions of the metal hydroxide.

alloy A mixture of metals. Alloys are often more useful than pure metals.

alpha radiation The least penetrating type of ionising radiation, produced by the nucleus of an atom in radioactive decay. A high-speed helium nucleus.

alternating current (a.c.) An electric current that reverses direction many times a second.

Alzheimer's disease A form of senile dementia caused by irreversible degeneration of the brain.

amino acids The small molecules that are joined in long chains to make proteins. All the proteins in living things are made from 20 different amino acids joined in different orders.

ammeter A meter that measures the size of an electric current in a circuit.

ampere (amp) The unit of electric current.

anaerobic respiration Respiration that does not use oxygen.

antibodies A group of proteins made by white blood cells to fight dangerous microorganisms. A different antibody is needed to fight each different type of microorganism. Antibodies bind to the surface of the microorganism, which triggers other white blood cells to digest them.

atmosphere The layer of gases that surrounds the Earth.

attract Pull towards.

attractive forces (between molecules) Forces that try to pull molecules together. Attractions between molecules are weak. Molecular chemicals have low melting points and boiling points because the molecules are easy to separate.

auxin A plant hormone that affects plant growth and development. For example, auxin stimulates growth of roots in cuttings.

average speed The distance moved by an object divided by the time taken for this to happen.

axon A long, thin extension of the cytoplasm of a neuron. The axon carries electrical impulses very quickly.

background radiation The low-level radiation, mostly from natural sources, that everyone is exposed to all the time, everywhere.

bacterium (plural bacteria) One type of single-celled microorganism. They do not have a nucleus. Some bacteria may cause disease.

balanced equation An equation showing the formulae of the reactants and products. The equation is balanced when there is the same number of each kind of atom on both sides of the equation.

base pairing The bases in a DNA molecule (A, C, G, T) always bond in the same way. A and T always bond together. C and G always bond together.

behaviour Everything an organism does; its response to all the stimuli around it.

beta blockers Drugs that block the receptor sites for the hormone adrenaline. They inhibit the normal effects of adrenaline on the body.

beta radiation One of several types of ionising radiation, produced by the nucleus of an atom in radioactive decay. More penetrating than alpha radiation but less penetrating than gamma radiation. A high-speed electron.

bioethanol Ethanol fuel produced by yeast fermentation of plant materials, such as cane sugar and sugar beet.

biogas Methane gas produced by the anaerobic digestion of organic material, such as farm animal manure.

bleach A chemical that can destroy unwanted colours. Bleaches also kill bacteria. A common bleach is a solution of chlorine in sodium hydroxide.

bulk chemicals Chemicals made by industry on a scale of thousands or millions of tonnes per year. Examples are sulfuric acid, nitric acid, sodium hydroxide, ethanol, and ethanoic acid.

burette A graduated tube with taps or valves used to measure the volume of liquids or solutions during quantitative investigations such as titrations.

carbohydrate A natural chemical made of carbon, hydrogen, and oxygen. An example is glucose $C_6H_{12}O_6$. Carbohydrates include sugars, starch, and cellulose.

carbonate A compound that contains carbonate ions, CO_3^{2-}. An example is calcium carbonate, $CaCO_3$.

cartilage Tough, flexible tissue found at the end of bones and in joints. It protects the end of bones from rubbing together and becoming damaged.

catalyst A chemical that starts or speeds up a chemical reaction but is not used up in the process.

cell The basic structural and functional unit of all living things.

cell membrane Thin layer surrounding the cytoplasm of a cell. It restricts the passage of substances into and out of the cell.

cell wall Rigid outer layer of plant cells and bacteria.

cellulose The chemical that makes up most of the fibre in food. The human body cannot digest cellulose.

central nervous system In mammals the brain and spinal cord.

cerebral cortex The highly folded outer region of the brain, concerned with conscious behaviour.

chain reaction A process in which the products of one nuclear reaction cause further nuclear reactions to happen, so that more and more reactions occur and more and more product is formed. Depending on how this process is controlled, it can be used in nuclear weapons or the nuclear reactors in power stations.

charged Carrying an electric charge. Some objects (such as electrons and protons) are permanently charged. A plastic object can be charged by rubbing it. This transfers electrons to or from it.

chemical change/reaction A change that forms a new chemical.

chemical equation A summary of a chemical reaction showing the reactants and products with their physical states (see balanced chemical equation).

chemical industry The industry that converts raw materials such as crude oil, natural gas, and minerals into useful products such as pharmaceuticals, fertilisers, paints, and dyes.

chemical properties A chemical property describes how an element or compound interacts with other chemicals, for example, the reactivity of a metal with water.

chemical species The different chemical forms that an element can take, for example, chlorine has three chemical species: atom, molecule, and ion. Each of these forms has distinct properties.

chlorophyll A green pigment found in chloroplasts. Chlorophyll absorbs energy from sunlight for photosynthesis.

chloroplast An organelle found in some plant cells where photosynthesis takes place.

chromosome Long, thin, threadlike structures in the nucleus of a cell made from a molecule of DNA. Chromosomes carry the genes.

clone A new cell or individual made by asexual reproduction. A clone has the same genes as its parent.

collision theory The theory that reactions happen when molecules collide. The theory helps to explain the factors that affect the rates of chemical change. Not all collisions between molecules lead to reaction.

commutator An device for changing the direction of the electric current through the coil of a motor every half turn. It consists of a ring divided into two halves (a split ring) with two contacts (brushes) touching the two halves.

concentrated solution The concentration of a solution depends on how much dissolved chemical (solute) there is compared with the solvent. A concentrated solution contains a high level of solute to solvent.

concentration The quantity of a chemical dissolved in a stated volume of solution. Concentrations can be measured in grams per litre.

conditioned reflex A reflex where the response is associated with a secondary stimulus, for example, a dog salivates when it hears a bell because it has associated the bell with food.

conditioning Reinforcement of behaviour associated with conditioned reflexes.

conscious To have awareness of surroundings and sensations.

consciousness The part of the human brain concerned with thought and decision making.

conservation of energy The fundamental idea that the total amount of energy in the universe is constant, and never increases or decreases. So if something loses energy, one or more other things must have gained the same amount of energy.

contamination (radioactive) Having a radioactive material inside the body, or having it on the skin or clothes.

control rod In a nuclear reactor, rods made of a special material that absorbs neutrons are raised and lowered to control the rate of fission reactions.

coolant In a nuclear reactor, the liquid or gas that circulates through the core and transfers heat to the boiler.

corrosive A corrosive chemical may destroy living tissue on contact.

counter-force A force in the opposite direction to something's motion.

covalent bonding Strong attractive forces that hold atoms together in molecules. Covalent bonds form between atoms of non-metallic elements.

crust (of the Earth) The outer layer of the lithosphere.

crystalline A material with molecules, atoms, or ions lined up in a regular way as in a crystal.

cutting A shoot or leaf taken from a plant, to be grown into a new plant.

cytoplasm Gel enclosed by the cell membrane that contains the cell organelles such as mitochondria.

decommissioning Taking a power station out of service at the end of its lifetime, dismantling it, and disposing of the waste safely.

denatured A change in the usual nature of something. When enzymes are denatured by heat, their structure, including the shape of the active site, is altered.

development How an organism changes as it grows and matures. As a zygote develops, it forms more and more cells. These are organised into different tissues and organs.

diamond A gemstone. A form of carbon. It has a giant covalent structure and is very hard.

diatomic A molecule with two atoms, for example, N_2, O_2, and Cl_2

diffusion Movement of molecules from a region of high concentration to a region of lower concentration.

dilute The concentration of a solution depends on how much dissolved chemical (solute) there is compared with the solvent. A dilute solution contains a low level of solute to solvent.

direct current (d.c.) An electric current that stays in the same direction.

displacement The length and direction of the straight line from the initial position of an object to its position at a later time.

displacement reaction A more reactive halogen will displace a less reactive halogen, for example, chlorine will displace bromide ions to form bromine and chloride ions.

displacement–time graph It is only possible to draw a displacement–time graph for an object moving along a straight line. It shows the straight-line distance of the object from its starting point at each moment during its journey.

dissolve Some chemicals dissolve in liquids (solvents). Salt and sugar, for example, dissolve in water.

distance The length of the path along which an object moves.

distance–time graph A useful way of summarising the motion of an object by showing how far it has moved from its starting point at every instant during its journey.

double helix The shape of the DNA molecule, with two strands twisted together in a spiral.

driving force The force pushing something forward, for example, a bicycle.

Ecstasy A recreational drug that increases the concentration of serotonin at the synapses in the brain, giving pleasurable feelings. Long-term effects may include destruction of the synapses.

effector The part of a control system that brings about a change to the system.

electric charge A fundamental property of matter. Electrons and protons are charged particles. Objects become charged when electrons are transferred to or from them, for example, by rubbing.

electric circuit A closed loop of conductors connected between the positive and negative terminals of a battery or power supply.

electric current A flow of charges around an electric circuit.

electric field A region where an electric charge experiences a force. There is an electric field around any electric charge.

electrode A conductor made of a metal or graphite through which a current enters or leaves a chemical during electrolysis. Electrons flow into the negative electrode (cathode) and out of the positive electrode (anode).

electrolysis Splitting up a chemical into its elements by passing an electric current through it.

electrolyte A chemical that can be split up by an electric current when molten or in solution is the electrolyte. Ionic compounds are electrolytes.

electromagnetic induction The name of the process in which a potential difference (and hence often an electric current) is generated in a wire, when it is in a changing magnetic field.

electron A tiny, negatively charged particle, which is part of an atom. Electrons are found outside the nucleus. Electrons have negligible mass and one unit of charge.

electron arrangement The number and arrangement of electrons in an atom of an element.

electrostatic attraction The force of attraction between objects with opposite electric charges.

embryonic stem cell Unspecialised cell in the very early embryo that can divide to form any type of cell, or even a whole new individual. In human embryos the cells are identical and unspecialised up to the eight-cell stage.

end point The point during a titration at which the reaction is just complete. For example, in an acid–alkali titration, the end point is reached when the indicator changes colour. This happens when exactly the right amount of acid has been added to react with all the alkali present at the start.

endothermic An endothermic process takes in energy from its surroundings.

energy level The electrons in an atom have different energies and are arranged at distinct energy levels.

energy-level diagram A diagram to show the difference in energy between the reactants and the products of a reaction.

enzyme A protein that catalyses (speeds up) chemical reactions in living things.

ethanol Waste product from anaerobic respiration in plants and yeast.

exothermic An exothermic process gives out energy to its surroundings.

extraction (of metals) The process of obtaining a metal from a mineral by chemical reduction or electrolysis. It is often necessary to concentrate the ore before extracting the metal.

fatty sheath Fat wrapped around the outside of an axon to insulate neurons from each other.

feral Untamed, wild.

fermentation Chemical reactions in living organisms that release energy from organic chemicals, such as yeast producing alcohol from the sugar in grapes.

fermenter A large vessel in which microorganisms are grown to make a useful product.

fetus A developing human embryo is referred to as a fetus once it reaches eight weeks after fertilization. A fetus already has all the main organs that it will have at birth.

fine chemicals Chemicals made by industry in smaller quantities than bulk chemicals. Fine chemicals are used in products such as food additives, medicines, and pesticides.

flame colour A colour produced when a chemical is held in a flame. Some elements and their compounds give characteristic colours. Sodium and sodium compounds, for example, give bright yellow flames.

food chain In the food industry this covers all the stages from where food grows, through harvesting, processing, preservation, and cooking to being eaten.

force A push or a pull experienced by an object when it interacts with another. A force is needed to change the motion of an object.

formulae (chemical) A way of describing a chemical that uses symbols for atoms. A formula gives information about the numbers of different types of atom in the chemical. The formula of sulfuric acid, for example, is H_2SO_4.

friction The force exerted on an object due to the interaction between it and another object that it is sliding over. It is caused by the roughness of both surfaces at a microscopic level.

fuel rod A container for nuclear fuel, which enables fuel to be inserted into, and removed from, a nuclear reactor while it is operating.

gametes The sex cells that fuse to form a zygote. In humans, the male gamete is the sperm and the female gamete is the egg.

gamma radiation (gamma rays) The most penetrating type of ionising radiation, produced by the nucleus of an atom in radioactive decay. The most energetic part of the electromagnetic spectrum.

gas exchange The exchange of oxygen and carbon dioxide that takes place in the lungs.

gene A section of DNA giving the instructions for a cell about how to make one kind of protein.

gene switching Genes in the nucleus of a cell switch off and are inactive when a cell becomes specialised. Only genes that the cell needs to carry out its particular job stay active.

generator A device that uses motion to generate electricity. It consists of a coil that rotates in a magnetic field. This produces a potential difference across the ends of the coil, which can then be used to provide an electric current.

genetic Factors that are affected by an organism's genes.

genetic variation Differences between individuals caused by differences in their genes. Gametes show genetic variation – they all have different genes.

giant covalent structure A giant, three-dimensional arrangement of atoms that are held together by covalent bonds. Silicon dioxide and diamond have giant covalent structures.

giant ionic lattice The structure of solid ionic compounds. There are no individual molecules, but millions of oppositely charged ions packed closely together in a regular, three-dimensional arrangement.

glands Parts of the body that make enzymes, hormones, and other secretions in the body, for example sweat glands.

glucose Sugar produced during photosynthesis.

graphite A form of carbon. It has a giant covalent structure. It is unusual for a non-metal in that it conducts electricity.

gravitational potential energy The energy stored when an object is raised to a higher point in the Earth's gravitational field.

group Each column in the periodic table is a group of similar elements.

habitat The place where an organism lives.

haemoglobin The protein molecule in red blood cells. Haemoglobin binds to oxygen and carries it around the body. It also gives blood its red colour.

half-life The time taken for the amount of a radioactive element in a sample to fall to half its original value.

halogens The family name of the Group 7 elements.

harmful A harmful chemical is one that may cause damage to health if swallowed, breathed in, or absorbed through the skin.

high-level waste A category of nuclear waste that is highly radioactive and hot. Produced in nuclear reactors and nuclear-weapons processing.

hormone A chemical messenger secreted by specialised cells in animals and plants. Hormones bring about changes in cells or tissues in different parts of the animal or plant.

hydrogen ion A hydrogen atom that has lost one electron. The symbol for a hydrogen ion is H^+. Acids produce aqueous hydrogen ions, $H^+(aq)$, when dissolved in water.

hydrosphere All the water on Earth. This includes oceans, lakes, rivers, underground reservoirs, and rainwater.

hydroxide ion A negative ion, OH^-. Alkalis give aqueous hydroxide ions when they dissolve in water.

in parallel A way of connecting electric components that makes a branch (or branches) in the circuit so that charges can flow around more than one loop.

in series A way of connecting electric components so that they are all in a single loop. The charges pass through them all in turn.

indicator A chemical that shows whether a solution is acidic or alkaline. For example, litmus turns blue in alkalis and red in acids. Universal indicator has a range of colours that show the pH of a solution.

innate Inborn, inherited from parents via genes.

insoluble Does not form a solution (dissolve) in water or other solutes.

instantaneous speed The speed of an object at a particular instant. In practice, its average speed over a very short time interval.

interaction What happens when two objects collide, or influence each other at a distance. When two objects interact, each experiences a force.

interaction pair Two forces that arise from the same interaction. They are equal in size and opposite in direction, and each acts on a different object.

intermediate-level waste A category of nuclear waste that is generally short-lived but requires some shielding to protect living organisms, for example, contaminated materials that result from decommissioning a nuclear reactor.

involuntary An automatic response made by the body without conscious thought.

ion An electrically charged atom or group of atoms.

ionic bonding Very strong attractive forces that hold the ions together in an ionic compound. The forces come from the attraction between positively and negatively charged ions.

ionic compounds Compounds formed by the combination of a metal and a non-metal. They contain positively charged metal ions and negatively charged non-metal ions.

ionic equation An ionic equation describes a chemical change by showing only the reacting ions in solution.

ionising Able to remove electrons from atoms, producing ions.

ionising radiation Radiation with photons of sufficient energy to remove electrons from atoms in its path. Ionising radiation, such as ultraviolet, X-rays, and gamma rays, can damage living cells.

irradiation Being exposed to radiation from an external source.

isotope Atoms of the same element that have different mass numbers because they have difference numbers of neutrons in the nucleus.

kinetic energy The energy that something has owing to its motion.

lactic acid Waste product from anaerobic respiration in animals.

learn To gain new knowledge or skills.

life cycle The stages an organism goes through as it matures, develops, and reproduces.

light intensity The amount of light reaching a given area.

light meter Device for measuring light intensity.

light-dependent resistor (LDR) An electric circuit component whose resistance varies depending on the brightness of light falling on it.

limiting factor The factor that prevents the rate of photosynthesis from increasing at a particular time. This may be light intensity, temperature, carbon dioxide concentration, or water availability.

line spectrum A spectrum made up of a series of lines. Each element has its own characteristic line spectrum.

lithosphere The rigid outer layer of the Earth, made up of the crust and the part of mantle just below it.

lock-and-key model In chemical reactions catalysed by enzymes, molecules taking part in the reaction fit exactly into the enzyme's active site. The active site will not fit other molecules – it is specific. This is like a key fitting into a lock.

long-term memory The part of the memory that stores information for a long period, or permanently.

low-level waste A category of nuclear waste that contains small amounts of short-lived radioactivity, for example, paper, rags, tools, clothing, and filters from hospitals and industry.

magnetic field The region around a magnet, or a wire carrying an electric current, in which magnetic effects can be detected. For example, another small magnet in this region will experience a force and may tend to move.

mantle The layer of rock between the crust and the outer core of the Earth. It is approximately 2900 km thick.

meiosis Cell division that halves the number of chromosomes to produce gametes. The four new cells are genetically different from each other and from the parent cell.

memory The storage and retrieval of information by the brain.

meristem cells Unspecialised cells in plants that can develop into any kind of specialised cell.

metal Elements on the left side of the periodic table. Metals have characteristic properties: they are shiny when polished and they conduct electricity. Some metals react with acids to give salts and hydrogen. Metals are present as positive ions in salts.

metal hydroxide A compound consisting of metal positive ions and hydroxide ions. Examples are sodium hydroxide, NaOH, and magnesium hydroxide, $Mg(OH)_2$.

metal oxide A compound of a metal with oxygen.

metallic bonding Very strong attractive forces that hold metal atoms together in a solid metal. The metal atoms lose their outer electrons and form positive ions. The electrons drift freely around the lattice of positive metal ions and hold the ions together.

mineral A naturally occurring element or compound in the Earth's lithosphere.

mitochondrion (plural mitochondria) An organelle in animal and plant cells where respiration takes place.

mitosis Cell division that makes two new cells identical to each other and to the parent cell.

models of memory Explanations for how memory is structured in the brain.

molecular models Models to show the arrangement of atoms in molecules, and the bonds between the atoms.

molecule A group of atoms joined together. Most non-metals consist of molecules. Most compounds of non-metals with other non-metals are also molecular.

molten A chemical in the liquid state. A chemical is molten when the temperature is above is melting point but below its boiling point.

momentum (plural momenta) A property of any moving object. Equal to mass multiplied by velocity.

motor A device that uses an electric current to produce continuous motion.

motor neuron A neuron that carries nerve impulses from the brain or spinal cord to an effector.

mRNA Messenger RNA, a chemical involved in making proteins in cells. The mRNA molecule is similar to DNA but single stranded. It carries the genetic code from the DNA molecule out of the nucleus into the cytoplasm.

multistore model One explanation for how the human memory works.

muscles Muscles move parts of the skeleton for movement. There is also muscle tissue in other parts of the body, for example, in the walls of arteries.

negative A label used to name one type of charge or one terminal of a battery. It is the opposite of positive.

negative ion An ion that has a negative charge (an anion).

nerve cell A cell in the nervous system that transmits electrical signals to allow communication within the body.

nerve impulses Electrical signals carried by neurons (nerve cells).

nervous system Tissues and organs that control the body's responses to stimuli. In a mammal it is made up of the central nervous system and peripheral nervous system.

neuron Nerve cell.

neuroscientist A scientist who studies how the brain and nerves function.

neutralisation reaction A reaction in which an acid reacts with an alkali to form a salt. During neutralisation reactions, the hydrogen ions in the acid solution react with hydroxide ions in the alkaline solution to make water molecules.

neutron An uncharged particle found in the nucleus of atoms. The relative mass of a neutron is 1.

newborn reflexes Reflexes to particular stimuli that usually occur only for a short time in newborn babies.

nitrate ions An ion is an electrically charged atom or group of atoms. The nitrate ion has a negative charge, NO3–.

non-ionising radiation Radiation with photons that do not have enough energy to ionise molecules.

nuclear fission The process in which a nucleus of uranium-235 breaks apart, releasing energy, when it absorbs a neutron.

nuclear fuel In a nuclear reactor, each uranium atom in a fuel rod undergoes fission and releases energy when hit by a neutron.

nuclear fusion The process in which two small nuclei combine to form a larger one, releasing energy. An example is hydrogen combining to form helium. This happens in stars, including the Sun.

nucleus (atom) The tiny central part of an atom (made up of protons and neutrons). Most of the mass of an atom is concentrated in its nucleus.

nucleus (cell) Organelle that contains the chromosomes cells of plants, animals, fungi, and some microorganisms.

ohm The unit of electrical resistance. Symbol Ω.

Ohm's law The result that the current, *I*, through a resistor, *R*, is proportional to the voltage, *V*, across the resistor, provided its temperature remains the same. Ohm's law does not apply to all conductors.

optimum temperature The temperature at which enzymes work fastest.

ore A natural mineral that contains enough valuable minerals to make it profitable to mine.

organelles The specialised parts of a cell, such as the nucleus and mitochondria. Chloroplasts are organelles that occur only in plant cells.

organs Parts of a plant or animal made up of different tissues.

osmosis The diffusion of water across a partially permeable membrane.

oxidation A reaction that adds oxygen to a chemical.

oxide A compound of an element with oxygen.

pancreas An organ in the body that produces some hormones and digestive enzymes. The hormone insulin is made here.

partially permeable membrane A membrane that acts as a barrier to some molecules but allows others to diffuse through freely.

pathway A series of connected neurones that allow nerve impulses to travel along a particular route very quickly.

percentage yield A measure of the efficiency of a chemical synthesis.

period In the context of chemistry, a row in the periodic table.

periodic In chemistry, a repeating pattern in the properties of elements. In the periodic table one pattern is that each period starts with metals on the left and ends with non-metals on the right.

peripheral nervous system The network of nerves connecting the central nervous system to the rest of the body.

pH scale A number scale that shows the acidity or alkalinity of a solution in water.

phloem A plant tissue that transports sugar throughout a plant.

photons Tiny 'packets' of electromagnetic radiation. All electromagnetic waves are emitted and absorbed as photons. The energy of a photon is proportional to the frequency of the radiation.

photosynthesis The process in green plants that uses energy from sunlight to convert carbon dioxide and water into the sugar glucose.

phototropism The bending of growing plant shoots towards the light.

physical properties Properties of elements and compounds such as melting point, density, and electrical conductivity. These are properties that do not involve one chemical turning into another.

pilot plant A small-scale chemical processing facility. A pilot plant is used to test processes before scaling up to full-scale production.

plant A chemical plant is an industrial facility used to manufacture chemicals.

plasma A collection of electrons and nuclei that can be formed when a gas has so much energy that its atoms are fully ionised.

polymer A material made up of very long molecules. The molecules are long chains of smaller molecules.

positive A label used to name one type of charge, or one terminal of a battery. It is the opposite of negative.

positive ion Ions that have a positive charge (cations).

potential difference (p.d.) The difference in potential energy (for each unit of charge flowing) between any two points in an electric circuit.

power In an electric circuit, the rate at which work is done by the battery or power supply on the components in a circuit. Power is equal to current × voltage.

precipitate An insoluble solid formed on mixing two solutions. Silver bromide forms as a precipitate on mixing solutions of silver nitrate and potassium bromide.

proportional Two variables are proportional if there is a constant ratio between them.

protein Chemicals in living things that are polymers made by joining together amino acids.

proton Tiny particle present in the nuclei of atoms. Protons are positively charged (+1).

proton number The number of protons in the nucleus of an atom (also called the atomic number). In an uncharged atom this also gives the number of electrons.

Prozac A brand name for an antidepressant drug. It increases the concentration of serotonin at the synapses in the brain.

pupil reflex The reaction of the muscles in the pupil to light. The pupil contracts in bright light and relaxes in dim light.

quadrat A square grid of a known area that is used to survey plants in a location. Quadrats come in different sizes up to 1 m^2. The size of quadrat that is chosen depends on the size of the plants and also the area that needs to be surveyed.

radiation A flow of information and energy from a source. Light and infrared are examples. Radiation spreads out from its source, and may be absorbed or reflected by objects in its path. It may also go (be transmitted) through them.

radiation dose A measure, in millisieverts, of the possible harm done to your body, which takes into account both the amount and type of radiation you are exposed to.

radioactive Used to describe a material, atom, or element that produces alpha, beta, or gamma radiation.

radioactive dating Estimating the age of an object such as a rock by measuring its radioactivity. Activity falls with time, in a way that is well understood.

radioactive decay The spontaneous change in an unstable element, giving out alpha, beta, or gamma radiation. Alpha and beta emission result in a new element.

radiotherapy Using radiation to treat a patient.

random Of no predictable pattern.

rate of photosynthesis Rate at which green plants convert carbon dioxide and water to glucose in the presence of light.

rate of reaction A measure of how quickly a reaction happens. Rates can be measured by following the disappearance of a reactant or the formation of a product.

reactants The chemicals on the left-hand side of an equation. These chemicals react to form the products.

reacting mass The masses of chemicals that react together, and the masses of products that are formed. Reacting masses are calculated from the balanced symbol equation using relative atomic masses and relative formula masses.

reaction (of a surface) The force exerted by a hard surface on an object that presses on it.

reactive metal A metal with a strong tendency to react with chemicals such as oxygen, water, and acids. The more reactive a metal, the more strongly it joins with other elements such as oxygen. So reactive metals are hard to extract from their ores.

receptor The part of a control system that detects changes in the system and passes this information to the processing centre.

receptor molecule A protein (often embedded in a cell membrane) that exactly fits with a specific molecule, bringing about a reaction in the cell.

recycling A range of methods for making new materials from materials that have already been used.

red blood cells Blood cells containing haemoglobin, which binds to oxygen so that it can be carried around the body by the bloodstream.

reducing agent A chemical that removes oxygen from another chemical. For example, carbon acts as a reducing agent when it removes oxygen from a metal oxide. The carbon is oxidised to carbon monoxide during this process.

reduction A reaction that removes oxygen from a chemical.

reflex arc A neuron pathway that brings about a reflex response. A reflex arc involves a sensory neuron, connecting neurons in the brain or spinal cord, and a motor neuron.

relative atomic mass The mass of an atom of an element compared to the mass of an atom of carbon. The relative atomic mass of carbon is defined as 12.

relative formula mass The combined relative atomic masses of all the atoms in a formula. To find the relative formula mass of a chemical, you just add up the relative atomic masses of the atoms in the formula.

relay neuron A neuron that carries the impulses from the sensory neuron to the motor neuron.

repel Push apart.

repetition Act of repeating.

repetition of information Saying or writing the same thing several times.

resistance The resistance of a component in an electric circuit indicates how easy or difficult it is to move charges through it.

respiration A series of chemical reactions in cells that release energy for the cell to use.

response Action or behaviour that is caused by a stimulus.

resultant force The sum, taking their directions into account, of all the forces acting on an object.

retina Light-sensitive layer at the back of the eye. The retina detects light by converting light into nerve impulses.

retrieval of information Collecting information from a particular source.

ribosomes Organelles in cells. Amino acids are joined together to form proteins in the ribosomes.

risk The probability of an outcome that is seen as undesirable, associated with some behaviour or process.

risk assessment A check on the hazards involved in a scientific procedure. A full assessment includes the steps to be taken to avoid or reduce the risks from the hazards identified.

rock A naturally occurring solid, made up of one or more minerals.

root hair cell Microscopic cell that increases the surface area for absorption of minerals and water by plant roots.

rooting powder A product used in gardening containing plant hormones. Rooting powder encourages a cutting to form roots.

salt An ionic compound formed when an acid neutralizes an alkali or when a metal reacts with a non-metal.

sample Small part of something that is likely to represent the whole.

scale up To redesign a synthesis to produce a chemical in larger amounts. A process might be scaled up first from a laboratory method to a pilot plant, then from a pilot plant to a full-scale industrial process.

sensory neuron A neuron that carries nerve impulses from a receptor to the brain or spinal cord.

serotonin A chemical released at one type of synapse in the brain, resulting in feelings of pleasure.

shell A region in space (around the nucleus of an atom) where there can be electrons.

short-term memory The part of the memory that stores information for a short time.

simple reflex An automatic response made by an animal to a stimulus.

slope The slope of a graph is a measure of its steepness.

small molecules Particles of chemicals that consist of small numbers of atoms bonded together. Chemicals made up of one or more non-metallic elements and that have low boiling and melting points consist of small molecules.

social behaviour Behaviour that takes place between members of the same species, including humans.

specialised A specialised cell is adapted for a particular job.

spectroscopy The use of instruments to produce and analyse spectra. Chemists use spectroscopy to study the composition, structure, and bonding of elements and compounds.

starch A type of carbohydrate found in bread, potatoes, and rice. Plants produce starch to store the energy food they make by photosynthesis. Starch molecules are a long chain of glucose molecules.

starch grains Microscopic granules of starch forming an energy store in plant cells.

static electricity Electric charge that is not moving around a circuit but has built up on an object such as a comb or a rubbed balloon.

stem cell Unspecialised animal cell that can divide and develop into specialised cells.

sterilisation The process of making something free from live bacteria and other microorganisms.

stimulus A change in the environment that causes a response.

stomata Tiny holes in the underside of a leaf that allow carbon dioxide into the leaf and water and oxygen out of the leaf.

strong (nuclear) force A fundamental force of nature that acts inside atomic nuclei.

structural proteins Proteins that are used to build cells.

sub-atomic particles The particles that make up atoms. Protons, neutrons, and electrons are sub-atomic particles.

surface area (of a solid chemical) The area of a solid in contact with other reactants that are liquids or gases.

sustainable Able to continue over long periods of time.

synapse A tiny gap between neurons that transmits nerve impulses from one neuron to another by means of chemicals diffusing across the gap.

tarnish When the surface of a metal becomes dull or discoloured because it has reacted with the oxygen in the air.

theoretical yield The amount of product that would be obtained in a reaction if all the reactants were converted to products exactly as described by the balanced chemical equation.

therapeutic cloning Growing new tissues and organs from cloning embryonic stem cells. The new tissues and organs are used to treat people who are ill or injured.

thermistor An electric circuit component whose resistance changes markedly with its temperature. It can therefore be used to measure temperature.

tissue Group of specialised cells of the same type working together to do a particular job.

tissue fluid Plasma that is forced out of the blood as it passes through a capillary network. Tissue fluid carries dissolved chemicals from the blood to cells.

titration An analytical technique used to find the exact volumes of solutions that react with each other.

toxic A chemical that may lead to serious health risks, or even death, if breathed in, swallowed, or taken in through the skin.

transect A straight line that runs through a location. Data on plant and animal distribution is recorded at regular intervals along the line.

transformer An electrical device consisting of two coils of wire wound on an iron core. An alternating current in one coil causes an ever-changing magnetic field that induces an alternating current in the other. Used to 'step' voltage up or down to the level required.

transmitter substance Chemical that bridges the gap between two neurons.

trend A description of the way a property increases or decreases along a series of elements or compounds, which is often applied to the elements (or their compounds) in a group or period.

triplet code A sequence of three bases coding for a particular amino acid in the genetic code.

unspecialised Cells that have not yet developed into one particular type of cell.

unstable The nucleus in radioactive isotopes is not stable. It is liable to change, emitting one of several types of radiation. If it emits alpha or beta radiation, a new element is formed.

velocity The speed of an object in a given direction. Unlike speed, which only has a size, velocity also has a direction.

velocity–time graph A useful way of summarising the motion of an object by showing its velocity at every instant during its journey.

voltage The voltage marked on a battery or power supply is a measure of the 'push' it exerts on charges in an electric circuit. The 'voltage' between two points in a circuit means the 'potential difference' between these points.

voltmeter An instrument for measuring the potential difference between two points in an electric circuit.

work Work is done whenever a force makes something move. The amount of work is force multiplied by distance moved in the direction of the force. This is equal to the amount of energy transferred.

working memory The system in the brain responsible for holding and manipulating information needed to carry out tasks.

xylem Plant tissue that transports water through a plant.

yeast Single-celled fungus used in brewing and baking.

yield The crop yield is the amount of crop that can be grown per area of land.

zygote The cell made when a sperm cell fertilises an egg cell in sexual reproduction.

Index

Appendices

Useful relationships, units, and data

Relationships

You will need to be able to carry out calculations using these mathematical relationships.

P4 Explaining motion

$$\text{speed} = \frac{\text{distance}}{\text{time}}$$

$$\text{acceleration} = \frac{\text{change in velocity}}{\text{time taken}}$$

momentum = mass × velocity

change of momentum = resultant force × time for which it acts

work done by a force = force × distance moved in the direction of the force

amount of energy transferred = work done

change in gravitational potential energy = weight × vertical height difference

kinetic energy = ½ × mass × velocity^2

P5 Electric circuits

power = voltage × current

$$\text{resistance} = \frac{\text{voltage}}{\text{current}}$$

$$\frac{\text{voltage across primary coil}}{\text{voltage across secondary coil}} = \frac{\text{number of turns in primary coil}}{\text{number of turns in secondary coil}}$$

C6 Chemical synthesis

$$\text{percentage yield} = \frac{\text{actual yield}}{\text{theoretical yield}} \times 100\%$$

P6 Radioactive materials

Einstein's equation: $E = mc^2$, where E is the energy produced, m is the mass lost, and c is the speed of light in a vacuum.

Units that might be used in the Additional Science course

length: metres (m), kilometres (km), centimetres (cm), millimetres (mm), micrometres (μm), nanometres (nm)

mass: kilograms (kg), grams (g), milligrams (mg)

time: seconds (s), milliseconds (ms)

temperature: degrees Celsius (°C)

area: cm^2, m^2

volume: cm^3, dm^3, m^3, litres (l), millilitres (ml)

speed and velocity: m/s, km/s, km/h

energy: joules (J), kilojoules (kJ), megajoules (MJ), kilowatt-hours (kWh), megawatt-hours (MWh)

electric current: amperes (A), milliamperes (mA)

potential difference/voltage: volts (V)

resistance: ohms (Ω)

power: watts (W), kilowatts (kW), megawatts (MW)

radiation dose: sieverts (Sv)

Prefixes for units

nano	micro	milli	kilo	mega	giga	tera
one thousand millionth	one millionth	one thousandth	× thousand	× million	× thousand million	× million million
0.000000001	0.000001	0.001	1000	1000 000	1000 000 000	1000 000 000 000
10^{-9}	10^{-6}	10^{-3}	$\times 10^3$	$\times 10^6$	$\times 10^9$	$\times 10^{12}$

Useful information and data

P4 Explaining motion

A mass of 1 kg has a weight of 10 N on the surface of the Earth.

C5 Chemicals of the natural environment

Approximate proportions of the main gases in the atmosphere: 78% nitrogen, 21% oxygen, 1% argon, and 0.04 % carbon dioxide.

P5 Electric circuits

mains supply voltage: 230 V

P6 Radioactive materials

speed of light (c) = 300 000 000 m/s

Chemical formulae

C4 Chemical patterns

water H_2O, hydrogen H_2, chlorine Cl_2, bromine Br_2, iodine I_2

lithium chloride LiCl, lithium bromide LiBr, lithium iodide LiI

sodium chloride NaCl, sodium bromide NaCl, sodium iodide NaI

potassium chloride KCl, potassium bromide KBr, potassium iodide KI

lithium hydroxide LiOH, sodium hydroxide NaOH, potassium hydroxide KOH

C5 Chemicals of the natural environment

nitrogen N_2, oxygen O_2, argon A, carbon dioxide CO_2

sodium chloride NaCl, magnesium chloride $MgCl_2$

sodium sulfate Na_2SO_4, magnesium sulfate $MgSO_4$

potassium chloride KCl, potassium bromide KBr

C6 Chemical synthesis

chlorine Cl_2, hydrogen H_2, nitrogen N_2, oxygen O_2

hydrochloric acid HCl, nitric acid HNO_3, sulfuric acid H_2SO_4

sodium hydroxide NaOH, sodium chloride NaCl, sodium carbonate Na_2CO_3, sodium nitrate $NaNO_3$, sodium sulfate Na_2SO_4, potassium chloride KCl

magnesium oxide MgO, magnesium hydroxide $Mg(OH)_2$, magnesium carbonate $MgCO_3$, magnesium chloride $MgCl_2$, magnesium sulfate $MgSO_4$

calcium carbonate$CaCO_3$, calcium chloride $CaCl_2$, calcium sulfate $CaSO_4$

Qualitative analysis data

Tests for negatively charged ions

Ion	Test	Observation
carbonate CO_3^{2-}	add dilute acid	effervesces, and carbon dioxide gas is produced (the gas turns lime water milky)
chloride (in solution) Cl^-	acidify with dilute nitric acid, then add silver nitrate solution	white precipitate
bromide (in solution) Br^-	acidify with dilute nitric acid, then add silver nitrate solution	cream precipitate
iodide (in solution) I^-	acidify with dilute nitric acid, then add silver nitrate solution	yellow precipitate
sulfate (in solution) SO_4^{2-}	acidify, then add barium chloride solution or barium nitrate solution	white precipitate

Tests for positively charged ions

Ion	Test	Observation
calcium Ca^{2+}	add sodium hydroxide solution	white precipitate (insoluble in excess)
copper Cu^{2+}	add sodium hydroxide solution	light-blue precipitate (insoluble in excess)
iron(II) Fe^{2+}	add sodium hydroxide solution	green precipitate (insoluble in excess)
iron(III) Fe^{3+}	add sodium hydroxide solution	red–brown precipitate (insoluble in excess)
zinc Zn^{2+}	add sodium hydroxide solution	white precipitate (soluble in excess, giving a colourless solution)

OXFORD
UNIVERSITY PRESS

Great Clarendon Street, Oxford OX2 6DP

Oxford University Press is a department of the University of Oxford. It furthers the University's objective of excellence in research, scholarship, and education by publishing worldwide in

Oxford New York

Auckland Cape Town Dar es Salaam Hong Kong Karachi Kuala Lumpur Madrid Melbourne Mexico City Nairobi New Delhi Shanghai Taipei Toronto

With offices in
Argentina Austria Brazil Chile Czech Republic France Greece Guatemala Hungary Italy Japan Poland Portugal Singapore South Korea Switzerland Thailand Turkey Ukraine Vietnam

First published 2011.

British Library Cataloguing in Publication Data.

Data available.

ISBN 978-0-19-913821-0

10 9 8 7 6 5 4 3 2

Printed in Great Britain by Bell and Bain, Glasgow.

Paper used in the production of this book is a natural, recyclable product made from wood grown in sustainable forests. The manufacturing process conforms to the environmental regulations of the country of origin.

Acknowledgements
The publisher and authors would like to thank the following for their permission to reproduce photographs and other copyright material:
P13: Andrew Lambert Photography/Science Photo Library; **P14:** Andrew Dorey/Istockphoto; **P17t:** Tfoxfoto/Shutterstock; **P17b:** James King Holmes/Science Photo Library; **P18:** J.C. Revy, Ism/Science Photo Library; **P19:** Eric and David Hosking/Corbis; **P20t:** Rick Price/Corbis; **P20b:** Simon Fraser/Science Photo Library; **P22:** J.C. Revy, Ism/Science Photo Library; **P23:** Dr Jeremy Burgess/Science Photo Library; **P24t:** Zooid Pictures; **P24b:** Manuel Blondeau/Photo & Co./Corbis; **P27t:** Biophoto Associates/Science Photo Library; **P27b:** Biophoto Associates/Science Photo Library; **P30:** Sinclair Stammers/Science Photo Library; **P31t:** Dr Jeremy Burgess/Science Photo Library; **P32:** P Phillips/Shutterstock; **P33:** Martyn F. Chillmaid/Science Photo Library; **P34:** K.R. Porter/Science Photo Library; **P35:** Kimimasa Mayama/Reuters; **P36l:** Aflo Foto Agency/Alamy; **P36r:** Blickwinkel/Alamy; **P37:** Power And Syred/Science Photo Library; **P38t:** Viktor1/Shutterstock; **P38b:** Azpworldwide/Shutterstock; **P39l:** Ashley Cooper/Specialist Stock/Splashdown Direct/Rex Features; **P39r:** Eye Of Science/Science Photo Library; **P44:** Dirk Wiersma/Science Photo Library; **P46t:** The Print Collector/Photolibrary; **P46b:** Andrew Lambert Photography/Science Photo Library; **P48:** Andrew Lambert Photography/Science Photo Library: **P49:** Charles D. Winters/Science Photo Library; **P52:** Herve Berthoule/Jacana/Science Photo Library; **P53:** Andrew Lambert Photography/Science Photo Library; **P54t:** William B. Jensen/Oesper Collections: University of Cincinnati; **P54bl:** Martyn F. Chillmaid; **P54br:** Dept. Of Physics, Imperial College/Science Photo Library; **P55:** Roger Ressmeyer/Corbis; **P57:** David Parker/Science Photo Library; **P58:** Dept. Of Physics, Imperial College/Science Photo Library; **P62tl:** Arnold Fisher/Science Photo Library; **P62ml:** José Manuel Sanchis Calvete/Corbis; **P62bl:** Dirk Wiersma/Science Photo Library; **P62tm:** Andrew Lambert Photography/Science Photo Library, Charles D. Winters/Science Photo Library; **P62tr:** Andrew Lambert Photography/Science Photo Library; **P63t:** Andrew Lambert Photography/Science Photo Library; **P63b:** Charles D. Winters/Science Photo Library; **P64:** Science Photo Library; **P68t:** Charles D. Winters/Science Photo Library; **P68b:** NASA/Science Photo Library; **P69t:** Andrew Lambert Photography/Science Photo Library; **P69m:** Arnold Fisher/Science Photo Library; **P69b:** Martyn F. Chillmaid/Science Photo Library; **P72:** Roger Ressmeyer/Corbis; **P74:** Paul Mckeown/Istockphoto; **P76:** Vakhrushev Pavel/Shutterstock; **P77:** Photodisc/Photolibrary; **P78l:** NASA; **P78r:** Greg McCracken/Istockphoto; **P79:** P B images/Alamy; **P80l:** Fine Shine/Shutterstock; **P80r:** Power And Syred/Science Photo Library; **P83:** Kenneth Eward/Biografx/Science Photo Library; **P84:** Santiago Cornejo/Shutterstock; **P91:** Offside/Rex Features; **P92:** EuroNCAP Partnership; **P98:** Adrian Houston/Photolibrary; **P102:** P B images/Alamy; **P104:** Neil Bromhall/Science Photo Library; **P106tl:** Oxford Scientific Films/Photolibrary; **P106tm:** Oxford Scientific Films/Photolibrary; **P106tr:** Corel/Oxford University Press; **P106bl:** Michael & Patricia Fogden/Corbis; **P106bm:** Corel/Oxford University Press; **P106br:** Corel/Oxford University Press; **P108t:** Edelmann/Science Photo Library; **P108m:** Alexander Tsiaras/Science Photo Library; **P108b:** Science Photo Library; **P109t:** Bob Rowan; Progressive Image/Corbis; **P109b:** Leo Batten/Frank Lane Picture Agency/Corbis; **P110t:** Joe McDonald/Corbis; **P110b:** Astrid & Hanns-Frieder Michler/Science Photo Library; **P111t:** M J Higginson/Science Photo Library; **P111b:** Martyn F. Chillmaid/Science Photo Library; **P112l:** Oxford Scientific Films/photolibrary; **P112r:** John Kaprielian/Science Photo Library; **P115:** Rob Lewine/Corbis; **P117t:** Francis Leroy/Biocosmos/Science Photo Library; **P117b:** Corel/Oxford University Press; **P118tl:** Carl & Ann Purcell/Corbis; **P118tm:** Holt Studios International; **P118tr:** Martin Harvey/Corbis; **P118bl:** CNRI/Science Photo Library; **P118br:** Dr Jeremy Burgess/Science Photo Library; **P120t:** Science Photo Library; **P120b:** A. Barrington Brown/Science Photo Library; **P124l:** Bob Gibbons/Holt Studios International; **P124r:** Helen Mcardle/Science Photo Library; **P125tl:** Anthony Bannister/Gallo Images/Corbis; **P125tr:** Marko Modic/Corbis; **P125bl:** Russ Munn/Agstock/Science Photo Library; **P125br:** Foodfolio/Alamy; **P126t:** Biophoto Associates/Science Photo Library; **P126m:** Dr Yorgos Nikas/Science Photo Library; **P126b:** Dr. Y. Nikas/Phototake Inc/Alamy; **P127:** Edelmann/Science Photo Library; **P128:** Mauro Fermariello/Science Photo Library; **P129t:** Massimo Brega, The Lighthouse/Science Photo Library; **P129b:** Rex Features; **P132:** CNRI/Science Photo Library; **P134:** Dane Steffes/Istockphoto; **P139:** Nina Towndrow/Nuffield Curriculum Centre; **P140:** Alen Rowell/Corbis; **P142l:** Joe Gough/Shutterstock; **P142m:** Martin Garnham/Dreamstime; **P142r:** Dogi/Shutterstock; **P143:** Dirk Wiersma/Science Photo Library; **P144:** Andrew Lambert Photography/Science Photo Library; **P145t:** Andrew Lambert Photography/Science Photo Library; **P145b:** Kesu/Shutterstock; **P146l:** George Bernard/Science Photo Library; **P146m:** Roberto de Gugliemo/Science Photo Library; **P146r:** Jos E Manuel Sanchis Calvete/Corbis; **P147l:** Pascal Goetgheluck/Science Photo Library; **P147r:** Peter Falkner/Science Photo Library; **P148:** Charles D. Winters/Science Photo Library; **P149t:** Mikeuk/Istockphoto; **P149bl:** Sinclair Stammers/Science Photo Library; **P149br:** Richard Megna/Fundamental/Science Photo Library; **P150t:** Layne Kennedy/Corbis; **P150b:** James L. Amos/Corbis; **P151:** Kevin Fleming/Corbis; **P156tl:** H. David Seawell/Corbis; **P156tm:** Alexis Rosenfeld/Science Photo Library; **P156tr:** John Van Hasselt/Sygma/Corbis; **P156b:** Charles E. Rotkin/Corbis; **P159:** Nik Wheeler/Corbis; **P162t:** Dirk Wiersma/Science Photo Library; **P162b:** H. David Seawell/Corbis; **P164:** Blackred/Istockphoto; **P166:** NOAA Central Library; OAR/ERL/National Severe Storms Laboratory (NSSL)/NOAA Photo Library; **P172:** Anthony Redpath/Corbis; **P173:** Zooid Pictures; **P176:** Masterfile; **P177:** Ilya Zlatyev/Shutterstock; **P183:** Phillppe Hays/Rex Features; **P186:** Sheila Terry/Science Photo Library; **P188t:** Anthony Vizard/Eye Ubiquitous/Corbis; **P188b:** T H Longannet/Scottish Power; **P189l:** Hink General 01 HR/British Energy; **P189m:** Nicholas Bailey/Rex Features; **P189r:** Sub_station3/Scottish Power; **P192:** Ilya Zlatyev/Shutterstock; **P194:** Sovereign, Ism/Science Photo Library; **P196tl:** Dennis Kunke/Phototake Inc./Alamy; **P196bl:** Astrid & Hanns-Frieder Michler/Science Photo Library; **P196br:** Eye Of Science/Science Photo Library; **P197tl:** Jeff Rotman/Nature Picture Library; **P197tr:** Tobias Bernhard/Oxford Scientific Films/Photolibrary; **P197bl:** Manfred Danegger/NHPA; **P197br:** Clive Druett/Papilio/Corbis; **P198l:** Sheila Terry/Science Photo Library; **P198m:** Owen Franken/Corbis; **P198r:** Laura Dwight/Corbis; **199l:** BSIP Astier/Science photo Library; **P199m:** Jennie Woodcock/Reflections Photolibrary/Corbis; **P199r:** Morgan McCauley/Corbis; **P203t:** Adam Hart-Davis/Science Photo Library; **P204t:** Greg Fiume/Corbis; **P204b:** Sipa Press/Rex Features; **P207t:** S. Kramer/Custom Medical Stock Photo/Science Photo Library; **P207b:** Mark Lythgoe and Steve Smith/Wellcome Trust; **P208:** Steve Bloom Images/Alamy; **P209:** Anthony Bannister/Gallo Images/Corbis; **P210t:** Lawrence Manning/Corbis; **P210m:** Kim Kulish/Corbis; **P210b:** Sally and Richard Greenhill/Alamy; **P213:** Jerry Wachter/Science Photo Library; **P214:** Karen Kasmauski/Corbis; **P215:** Roger Ressmeyer/Corbis; **P218:** C. Warner Br/Everett/Rex Features; **P222:** Anthony Bannister/Gallo Images/Corbis; **P224:** Michael Rosenfeld/Science Faction/Corbis; **P226t:** Maximilian Stock Ltd/Science Photo Library; **P226b:** Alexander Raths/Shutterstock; **P227l:** Geoff Tompkinson/Science Photo Library; **P227r:** William Taufic/Corbis; **P228l:** Dave Bartruff/Corbis, Andrew Lambert Photography/Science Photo Library; **P228r:** Georgy Markov/Shutterstock, Martyn F. Chillmaid/Science Photo Library; **P229t:** Lurgi Metallurgie/Outokumpu; **P229bl:** Martyn F. Chillmaid/Science Photo Library; **P229br:** Fotex/Rex Features; **P230l:** Charles D. Winters/Science Photo Library; **P230m:** Marketa Mark/Shutterstock; **P230r:** Kojiro/Shutterstock; **P230b:** Zooid Pictures; **P231:** Andrew Lambert Photography/Science Photo Library; **P235:** Martyn F. Chillmaid/Science Photo Library;

P236t: John Casey/Fotolia; **P236b:** AJ Photo/Science Photo Library; **P237:** Martyn F. Chillmaid/Science Photo Library; **P238:** Peter Bowater/Alamy; **P242:** Gary Banks/BP Saltend; **P244:** Bsip/Beranger/Science Photo Library; **P245:** Holt Studios International; **P247:** Sidney Moulds/Science Photo Library; **P252:** Peter Bowater/Alamy; **P254:** Efda-Jet/Science Photo Library; **P256t:** Kletr/Shutterstock; **P256bl:** Health Protection Agency/Science Photo Library; **P256br:** Radiation Protection Division/Health Protection Agency/Science Photo Library; **P257t:** Davies and Starr/Stone/Getty Images; **P257bl:** Matthias Kulka/Corbis; **P257br:** Peter Thorne, Johnson Matthey/Science Photo Library; **P260:** Amrit G/Fotolia; **P263t:** Geoff Tompkinson/Science Photo Library; **P263b:** Krzysztof Slusarczyk/Shutterstock; **P265:** Julia Hedgecoe; **P266t:** Josh Sher/Science Photo Library; **P266b:** Health Protection Agency/Science Photo Library; **P267:** Health Protection Agency/Science Photo Library; **P272:** Prof. J. Leveille/Science Photo Library; **P275t:** Jerry Mason/Science Photo Library; **P275b:** Ria Novosti/Science Photo Library; **P276t:** Steve Allen/Science Photo Library; **P276b:** Keith Beardmore/The Point/British Nuclear Fuels Limited; **P279:** ITER/Science Photo Library; **P282:** Geoff Tompkinson/Science Photo Library.

Illustrations by IFA Design, Plymouth, UK, Clive Goodyer, and Q2A Media.

The publisher and authors are grateful for permission to reprint the following copyright material:
Although we have made every effort to trace and contact all copyright holders before publication this has not been possible in all cases. If notified, the publisher will rectify any errors or omissions at the earliest opportunity.

Project Team acknowledgements
These resources have been developed to support teachers and students undertaking the OCR GCSE Science Twenty First Century Science suite of specifications. They have been developed from the 2006 edition of the resources.
We would like to thank David Curnow and Alistair Moore and the examining team at OCR, who produced the specifications for the Twenty First Century Science course.

Authors and editors of the first edition
We thank the authors and editors of the first edition, Jenifer Burden, Simon Carson, John Holman, Anna Grayson, Angela Hall, John Holman, Andrew Hunt, Bill Indge, John Lazonby, Allan Mann, Jean Martin, Robin Millar, Nick Owens, Stephen Pople, Carol Tear, and Linn Winspear.
Many people from schools, colleges, universities, industry and the professions contributed to the production of the first edition of these resources. We also acknowledge the invaluable contribution of the teachers and students in the Pilot Centres.
The first edition of Twenty First Century Science was developed with support from the Nuffield Foundation, The Salters Institute, and The Wellcome Trust. A full list of contributors can be found in the Teacher and Technician resources.

The continued development of *Twenty First Century Science* is made possible by generous support from:
- The Nuffield Foundation
- The Salters' Institute